Ecohydrology in the Context of Climate Change: Strategies for Management, Monitoring, and Modeling

Ecohydrology in the Context of Climate Change: Strategies for Management, Monitoring, and Modeling

Guest Editors

Carmen Maftei
Ashok Vaseashta

MDPI

Basel • Beijing • Wuhan • Barcelona • Belgrade • Novi Sad • Cluj • Manchester

Guest Editors

Carmen Maftei
Civil Engineering Faculty
Transilvania University of
Brasov
Brasov
Romania

Ashok Vaseashta
Applied Research
International Clean Water
Institute
Manassas, VA
USA

Editorial Office
MDPI AG
Grosspeteranlage 5
4052 Basel, Switzerland

This is a reprint of the Special Issue, published open access by the journal *Water* (ISSN 2073-4441), freely accessible at: https://www.mdpi.com/journal/water/special_issues/OFGAW44163.

For citation purposes, cite each article independently as indicated on the article page online and as indicated below:

Lastname, A.A.; Lastname, B.B. Article Title. *Journal Name* **Year**, *Volume Number*, Page Range.

ISBN 978-3-7258-7783-6 (Hbk)
ISBN 978-3-7258-7784-3 (PDF)
https://doi.org/10.3390/books978-3-7258-7784-3

Contents

About the Editors

Carmen Maftei

Carmen Maftei (born Petrescu) attended the Polytechnic Institute of Iasi, where she majored in Land Reclamation in 1988, and from 1996 to 1998 she attended a specialization in water resources and protection of water resources, equivalent to a Master of Science. In 2002, she finished her PhD studies in Water Science in the Continental Environment. Since 2016 she has been coordinating doctoral theses in the field of civil engineering. Her research activity is focused on (i) hydrology and hydrological modelling; (ii) Geographic Information Systems applications in environmental sciences; and (iii) remote sensing applications in environmental sciences. As a result of her research, she has 20 books/book chapters, more than 100 scientific papers published in various journals and conference proceedings, one patent, two international projects, 13 national grants, and 10 research contracts with economic partners. Her 30 years of teaching experience include hydrology, hydraulic, irrigation and drainage, Geographic Information Systems and Applications at "Ovidius" University of Constanta and Transilvania University of Brasov.

Ashok Vaseashta

Ashok Vaseashta received a Ph.D. from Virginia Tech, Blacksburg, VA, in 1990, followed by a postdoctoral fellowship at Kobe. He was awarded an Honoris Causa doctorate from Riga Technical University, honorary membership of the Academy of Sciences of Moldova, membership of the Academy of Sciences of Georgia, membership of the Euro Mediterranean Academy of Arts and Sciences, and a Gold Medal for his leadership in Nanotechnology from the National Polytechnic University of Armenia. Currently, he serves as the Executive Director of Strategic Research at the International Clean Water Institute in Virginia, USA. He served as Professor at Virginia Tech and Marshall University, Director of Research at the Institute for Advanced Sciences Convergence and International Clean Water Institute for Norwich University Applied Research Institutes, Vice-Provost (Rector) for Research in South Carolina, and Executive Director and Chair of the Institutional Review Board at a State University in New Jersey. He held several visiting positions in Latvia, Romania, and Türkiye and was a scientist at the Weizmann Institute of Science, Israel. He is a member of CIRET, France. He served the U.S. Department of State in two rotations as a strategic S&T advisor. He served as NATO project director for eight activities funded under NATO's Science for Peace and Security program. Inspired by nature and guided by societal needs, he strives to advance technology to address the global challenges of the 21st century. He is a Fulbright Specialist through 2026 and serves as a research fellow at the Center for Disruptive Technologies and Future Warfare in the Institute for National Strategic Studies at the National Defense University. He is the author/editor of 30 books and has published over 400 articles in scientific journals, book chapters, and conferences. He serves on the editorial boards of several international journals and is an active member of various professional organizations.

Preface

Global anthropogenic activities and associated climate extremes are disrupting closely coupled hydroecological systems, altering catchment-scale water balances, biogeochemical fluxes, and the resilience of ecosystem services. This reprint aims to address current ecohydrological research that integrates data-driven hydrological modeling, remote-sensing assimilation, and process-based carbon–nitrogen coupling frameworks to resolve scale-dependent dynamics under climate-change scenarios. The reprint covers several key areas, including climate change-related hydroecological effects; assessment and characterization of ecohydrological dynamics, such as the effects of hydrology on soil erosion; adaptive, nature-based ecohydrological solutions to droughts and floods; and strategies for management, monitoring, and modeling, with the hope that this reprint will initiate new research topics and directions, to address this topic from global perspectives.

Carmen Maftei and Ashok Vaseashta
Guest Editors

water

MDPI

Editorial

Ecohydrology in the Context of Climate Change: Strategies for Management, Monitoring, and Modeling

Carmen Maftei [1,*] and Ashok Vaseashta [2,3,*]

[1] Civil Engineering Faculty, Transilvania University of Brasov, 500152 Brasov, Romania
[2] Strategic Research, International Clean Water Institute, Manassas, VA 20108, USA
[3] Research Institute, University of Bucharest, 050663 Bucharest, Romania
* Correspondence: carmen.maftei@unitbv.ro (C.M.); prof.vaseashta@ieee.org (A.V.)

1. Introduction to the Special Issue

Global anthropogenic activities and conjoint climate extremes are fundamentally disrupting closely coupled hydroecological systems, altering catchment-scale water balances, biogeochemical fluxes, and the resilience of ecosystem services [1–4]. Progressive degradation of system integrity through mechanisms such as land-use conversion, groundwater overexploitation, nutrient loading, and hydroclimatic extremes has induced nonlinear feedback between hydrological processes and ecosystem functions, manifesting as diminished stability, accelerated soil erosion, impaired carbon sequestration, and disrupted nutrient spiraling [5–8]. Contemporary ecohydrological research must therefore transcend disciplinary silos, integrating data-driven hydrological modeling, isotope tracing, remote sensing assimilation, and process-based carbon–nitrogen coupling frameworks to resolve scale-dependent dynamics under climate change scenarios. In particular, attention must be paid to quantifying hydroecological services, optimizing soil–water resource governance, and developing adaptive, nature-based ecohydrological solutions that enhance system buffering capacity against droughts and floods [9]. Advances in urban ecohydrology, water-quality diagnostics, and data-driven modeling architectures—including machine learning (ML) augmented watershed simulations—are essential for robust monitoring, predictive analytics, and sustainable management strategies [10–12]. A rigorous, systems-oriented approach that links mathematical modeling to empirical validation will be indispensable for safeguarding water security and ecosystem functionality in this modern age of accelerating environmental change.

To bridge the knowledge gap, this Special Issue titled "Ecohydrology in the Context of Climate Change: Strategies for Management, Monitoring, and Modeling" focuses on the latest research on ecohydrological processes and water balance, climate change and ecosystem responses, carbon storage and nutrient transfer, and sustainable solutions. Research topics include, but are not limited to, the following: understanding the role of hydroecological services; water and soil resource management; ecohydrology solutions; climate change impacts on the hydrological cycle; effects of hydrology on soil erosion; ecosystem sensitivity to extreme events such as drought and flood; mathematical modeling aspects; urban ecohydrology; water quality assessment; and strategies for management, monitoring and modeling.

Since the call for papers was announced in October 2024, 12 original papers have been accepted for publication after a rigorous peer-review process. While most presented unique perspectives, the papers can be classified into several key areas: climate change-related hydroecological effects; assessment and characterization of ecohydrological dynamics, such as the effects of hydrology on soil erosion; adaptive, nature-based ecohydrological solutions against droughts and floods; and strategies for management, monitoring, and modeling. The highlights of this Special Issue are summarized in the following.

2. Overview of the Contributions

The first paper was submitted by Pronin et al.; they present an interesting study of tracking the changes in saltwater lakes with specific, rare, and protected aquatic plant vegetation, primarily using sediment values along the alkalization gradient. The authors studied changes that might be reflected in the stable carbon and nitrogen isotopes of the organic matter of those plants and sediments. They found a positive correlation between the $\delta^{13}C$ values of plants and sediments, confirming that the macrophytes were a significant source of organic matter in sediments. Regarding $\delta^{15}N$ values, they found no significant differentiation between plants and sediments across ecological groups, however, the sediment did consistently exhibit ^{15}N-enrichment compared to plant material. This enrichment is likely attributed to the accumulation of ^{15}N during the decomposition of the deposited material. In a study by Attou et al., stable isotope monitoring was reported in a semi-arid region of Morocco. The majority of water withdrawals in Morocco are for irrigation, and optimizing irrigation practices in agriculture is a national priority amid recurring droughts and growing pressure on groundwater resources. Since the hydrological impacts of different drip-irrigation systems in the soil–plant–atmosphere continuum is insufficiently understood, the authors monitored the stable isotope composition (δ^2H, $\delta^{18}O$) across two agricultural plots in Marrakech, Morocco, with surface and subsurface drip-irrigation treatments for a complete hydrologic year (June 2022 to June 2023). Their results indicate that subsurface drip irrigation can significantly improve drought resilience and water-use efficiency in the expanding olive sector of the Maghreb. At the same time, continuous isotope monitoring is a practical approach to enhancing sustainable and adaptive water management in water-limited regions. Another study by Makhmudova et al. presents a scenario-based assessment of the future sensitivity of minimal low-water runoff to climate change in Western Kazakhstan. The authors applied an ensemble of global climate models from the Coupled Model Intercomparison Project Phase 6 (CMIP6), combined with dynamically down-scaled projections for Central Asia, to estimate the minimal monthly runoff during the summer–autumn and winter low-water periods for rivers in the Zhaiyk–Caspian water management basin. Covering three future time horizons spanning 2031–2090 and under two greenhouse gas concentration scenarios, SSP3-7.0 (moderately high emissions) and SSP5-8.5 (high emissions), their results reveal a pronounced seasonal contrast in the projected hydrological response. Their findings emphasize the importance of developing seasonally differentiated, climate-resilient water management strategies to mitigate low-flow risks and ensure water security under future climate conditions in arid and semi-arid regions.

A study by Jia et al. simulates vegetation productivity in the Biome-BGC Model in the Yellow River Basin (YRB), China. The authors studied the quantification of carbon absorption by gross primary productivity (GPP), which underpins the carbon cycle in terrestrial ecosystems. The Biome-BGC model is a widely used biogeochemical process model for simulating the stocks and fluxes of water, carbon, and nitrogen between ecosystems and the atmosphere. The parameter optimization method that couples the differential evolution algorithm (DE) with the Biome-BGC model was used to calibrate and validate the eco-physiological parameters. The authors accurately estimated GPP, which provided a foundation for analyzing spatiotemporal trends in GPP across the YRB. This study provides a reference for optimizing Biome-BGC model parameters and for simulating diverse vegetation types at large scales. An investigation by Rivas et al. evaluated the

Snowmelt-Runoff Model (SRM) and the Génie Rural à X Paramètres Journalier (GRxJ) model family, analyzing both independently and in combination with the CemaNeige snow module. Accurate streamflow estimation in snow- and rain-dominated basins is crucial for water resource management, especially in the Andes where climate variability and glacier retreat threaten long-term water availability. The analysis was conducted in two Chilean watershed basins with contrasting regimes: the snow-dominated Aconcagua and the mixed rain–snow Duqueco basins. Using daily data (2012–2020) on precipitation, temperature, evapotranspiration, snow cover (MODIS), and streamflow, the authors demonstrated that SRM performed best in both analyzed catchments. However, it reproduced extreme streamflow events less successfully. In contrast, the GRxJ models did not adequately capture the hydrological dynamics of the snow-dominated Aconcagua catchment; however, their performance improved considerably when applied to the mixed regime of the Duqueco River. These findings highlight the importance of adapting modeling strategies to local hydrological conditions with limited data availability, offering practical guidance for water management and climate change adaptation in Andean catchments. In the context of efficient water resource management, Zhen et al. applied variational quantum regression (VQR), a novel machine learning (ML) approach inspired by quantum computing principles, to a series of river discharges in Romania. The models were evaluated against the quantum neural network (QNN) and other classic artificial intelligence (AI) outputs on the same dataset. Performance was assessed based on the coefficient of determination (R2), mean absolute error (MAE), and mean squared error (MSE). VQR outperformed classical neural networks and hybrid models in terms of MSE and MAE, demonstrating superior accuracy and generalization. The models exhibited exceptional skill in capturing monthly maxima, underscoring the potential of VQR as a powerful and reliable tool for hydrological forecasting, particularly for nonlinear and high-variability data series. Another research study by Lin et al. reports a method for watershed eco-functional zoning based on distributed hydrological modeling. Using the Jinjiang Basin in Southeast China as a case study, they applied the Soil and Water Assessment Tool (SWAT) model to delineate the basic zoning units and simulate their hydrological processes. By integrating natural environmental indicators, like topography, vegetation, meteorology, and hydrology, their approach enabled them to conduct both first-level and second-level eco-functional zoning of the watershed. The results indicated that the Jinjiang Basin can be categorized into three main groups consisting of six first-level aquatic ecological zones, which reflect the spatial variability of terrestrial natural environmental factors and their influence on aquatic ecosystems. The first-level aquatic ecological regions were further divided into five categories comprising 18 second-level aquatic ecological functional zones, emphasizing the impact of human activities on aquatic ecosystems and their associated first-level ecological service functions. The application of hydrological simulation techniques enables a comprehensive assessment of the spatial variability of hydrological processes, thereby enhancing the validity of ecological function zoning results and providing robust technical support for watershed ecological function zoning.

An article by Serban et al. examines a study that assessed the severity of droughts and their changes in the Dobrogea region of southeastern Romania from 2001 to 2021 and develops a high-resolution (1 km) Composite Drought Index (CDI) dataset. The effectiveness of the index was evaluated using correlation analyses between the CDI, the Standardized Precipitation Index (SPI), and the Standardized Precipitation Evapotranspiration Index (SPEI), which showed a strong positive relationship among these indices. These results show the potential of the CDI as a significant index for assessing drought, providing useful insights for future drought monitoring. Additionally, the authors considered that the

https://doi.org/10.3390/w18050643

GPM IMERG satellite product can be used to implement Drought Management Plans and calculate drought indices derived from remote sensing data.

A study by Khachatryan et al. aims to assess the influence of climate change on the basin's thermal dynamics by analyzing both historical and projected temperature variations. For their study, they examined Lake Sevan, one of the most important lakes in Armenia, that has experienced a marked rise in air temperatures for the past three decades. Seasonal variability revealed distinct contrasts between winter and summer, with winter exhibiting greater fluctuations, ranging from 1.67 to 2.41 °C, compared to the more stable summer range of 0.81 to 1.41 °C. Based on their ongoing project CMIP6 (Coupled Model Intercomparison Project phase 6) and the ensemble outputs from four Shared Socioeconomic Pathways (SSP1–2.6, SSP2–4.5, SSP3–7.0, and SSP5–8.5), the authors suggest a persistent warming trend throughout the 21st century. The authors project that the most significant increases are expected during summer months, with an anticipated mean annual temperature rise of up to 6 °C by the end of the century under the high-emission scenario (SSP5–8.5). A somewhat similar study by Xi et al. evaluates the spatiotemporal evolution of compound dry–hot events and their impacts on vegetation NPP (net primary productivity) in the YRB (Yangtze River Basin), China. Using ERA5-Land and MODIS NPP (2002–2024), the authors developed climate indices and a Copula-based standardized compound dry–hot index (SCDHI) to detect events and examine spatiotemporal patterns. Trend and correlation analyses quantified NPP sensitivity and lag, and an NPP–SCDHI coupling framework assessed resistance and resilience across major vegetation types. The mean NPP response time was approximately 2 months, with forests and croplands typically lagging by 2–3 months. Under extreme stress, forests showed high resistance but limited recovery, whereas shrublands showed moderate resistance and low resilience. Cultivated vegetation exhibited the lowest resistance and weak resilience; grasslands had low resistance but relatively rapid recovery; and alpine vegetation showed moderate resistance and the highest resilience, indicating that cultivated vegetation and grasslands may pose a high risk to ecological management. Extending climate change investigations, Bărbulescu et al. report shifts in precipitation variability near the DDBR (Danube Delta Biosphere Reserve). Their study examines the long-term precipitation record (1965–2019) from the Tulcea station, located just 4 km from the DDBR. The study aims to evaluate the impact of climate change on precipitation variability, which can significantly affect biodiversity in this protected area. They integrated change point detection (CPD), stationarity tests, trend analysis, and series decomposition to characterize shifts and patterns in the time series. The STL (Seasonal-Trend decomposition using Loess) found the highest seasonality indices in June and July. The Ensemble Empirical Mode Decomposition (EEMD) emphasizes a substantial difference in the seasonal cycle. Their results indicate high variability in precipitation patterns, with periods of high precipitation followed by dry periods.

An article by Gevorgyan et al. on hydrochemical characterization assesses surface-water quality for irrigation in the Voghji River catchment, Armenia. Nine key parameters were measured at seven sampling points. The chemical analysis of surface waters in the Voghji River catchment basin revealed dominant Ca^{2+}-HCO_3^- and mixed Ca^{2+}-K^+-SO_4^{2-}-Cl^- facies, with key geochemical processes including carbonate and gypsum dissolution, silicate weathering, and cation exchange. Ionic correlations indicated that Na+ and Cl^-, both their natural and anthropogenic inputs, influenced sources, while Ca^{2+} and Mg^{2+} primarily originated from carbonate dissolution. The Gibbs diagram suggested that rock–water interactions were the primary natural mechanism controlling the water chemistry, with evaporation also playing a significant role. Various indices, including the Kelly index, magnesium adsorption ratio, sodium percentage, sodium adsorption ratio, permeability index, potential salinity, residual sodium carbonate, soluble sodium percentage, and irri-

https://doi.org/10.3390/w18050643

gation water quality index, were applied, along with the US Salinity Laboratory diagram and the Wilcox diagram, to assess the suitability of irrigation. Most indices confirmed the suitability of the waters for irrigation; however, the Achanan River near the mouth and the Voghji River downstream of Kapan Town exhibited moderate salinity, underscoring the need for water management to prevent soil degradation.

3. Conclusions

The Guest Editors anticipate that the contributions to this Special Issue will initiate a rigorous, integrative, and forward-looking reference for researchers, doctoral scholars, practitioners, and policy architects operating at the nexus of climate dynamics and water-ecosystems science. Collectively, the published works advance contemporary understanding of hydroecological coupling, climate perturbation diagnostics, and adaptive resource governance by delivering both conceptual refinement and methodological innovation. A defining feature of this Special Issue is its emphasis on technological convergence: satellite-based Earth observation systems, advanced remote sensing analytics, hybrid process-based–numerical simulations, machine learning-enabled data assimilation, and parametric and multi-objective optimization frameworks tailored to ecohydrological applications. The integration of high-resolution Earth observation platforms with physically based and data-driven models enhances the quantification of water-balance components, evapotranspiration flux partitioning, vegetation–soil–atmosphere feedback mechanisms, carbon–nutrient transport pathways, and groundwater–surface-water interactions under non-stationary climatic regimes. Such integrative approaches materially strengthen predictive skill, model transferability, and uncertainty characterization across watershed, regional, and transboundary domains. Several contributions introduce advanced algorithmic architectures—including stochastic optimization, ensemble forecasting, Bayesian calibration, and data-centric scenario modeling—to improve the robustness of hydroclimatic risk assessment and decision-support systems. These analytical advancements are particularly consequential for diagnosing compound extremes, reducing vulnerability to drought and floods, safeguarding groundwater sustainability, and optimizing ecosystem-service resilience under intensifying anthropogenic and climatic stressors.

The translational value of these technological and analytical developments is substantial. By coupling monitoring infrastructures with scalable modeling frameworks and governance-relevant metrics, this Special Issue equips scientists, water-resource managers, and development planners with quantitative tools to evaluate climate-change impacts and design adaptive management strategies anchored in empirical evidence and systems-level insight. This collection of research-intensive contributions underscores the imperative of integrating monitoring technologies, modeling architectures, and policy-responsive governance strategies within ecohydrological systems research. In light of sustained scholarly engagement and the accelerating pressures of climate variability and socio-environmental change, a follow-up issue is envisioned to further catalyze interdisciplinary synthesis, methodological advancement, and actionable science in this strategically vital domain.

Author Contributions: Conceptualization, C.M., A.V.; methodology, C.M., A.V.; validation, C.M., A.V., formal analysis, C.M., A.V., resources, C.M., A.V., data curation, C.M., A.V., writing—original draft preparation, C.M., A.V.; writing—review and editing, C.M., A.V.; visualization, C.M., A.V. All authors have read and agreed to the published version of the manuscript.

Funding: This research received no external funding.

Acknowledgments: All reviewers are acknowledged for their professional reviews and comments.

Conflicts of Interest: The authors declare no conflicts of interest.

https://doi.org/10.3390/w18050643

List of Contributions:

1. Khachatryan, G.; Gevorgyan, A.; Vaseashta, A.; Misakyan, A.; Rinke, K.; Gevorgyan, A.; Ghukasyan, L.; Gevorgyan, G. Climate Change and Thermal Dynamics of the Lake Sevan Basin (Armenia): Observational Insights and Future Projections. *Water* **2026**, *18*, 352. https://doi.org/10.3390/w18030352.
2. Xi, H.; Zhang, G.; Wang, H. Spatiotemporal Evolution of Compound Dry–Hot Events and Their Impacts on Vegetation Net Primary Productivity in the Yangtze River Basin. *Water* **2026**, *18*, 276. https://doi.org/10.3390/w18020276.
3. Rivas, B.; Osores, V.; González, D.; Gualtieri, C.; Yépez, S. Performance Evaluation of the SRM and GRxJ—CemaNeige Models for Daily Streamflow Simulation in Two Catchments with Snow and Rain Dominated Hydrological Regimes. *Water* **2025**, *17*, 3413. https://doi.org/10.3390/w17233413.
4. Altou, T.; Kharrou, M.H.; Kuppel, S.; Ait Brahim, Y.; Bouchaou, L.; Demarez, V.; Lehmann, M.M.; Raibi, F.; Elghali, T.; Elazhari, A.; et al. Stable Isotope Monitoring in a Semi-Arid Olive Orchard Suggests Changes in Ecohydrological Dynamics from Contrasting Drip Irrigation Regimes. *Water* **2025**, *17*, 3029. https://doi.org/10.3390/w17213029.
5. Bărbulescu, A.; Dumitriu, C.Ș. Shifts in Precipitation Variability near the Danube Delta Biosphere Reserve (1965–2019). *Water* **2025**, *17*, 2692. https://doi.org/10.3390/w17182692.
6. Lin, B.; Chen, Y.; Lin, M.; Ye, L.; Cai, M. Research on the Zoning of Watershed Aquatic Ecological Functions Based on a Distributed Hydrological Model. *Water* **2025**, *17*, 2464. https://doi.org/10.3390/w17162464.
7. Makhmudova, L.; Alimkulov, S.; Tursunova, A.; Birimbayeva, L.; Talipova, E.; Alzhanov, O.; Rodrigo-Clavero, M.E.; Rodrigo-Ilarri, J. Seasonally Contrasting Sensitivity of Minimal River Runoff to Future Climate Change in Western Kazakhstan: A CMIP6 Scenario Analysis. *Water* **2025**, *17*, 2417. https://doi.org/10.3390/w17162417.
8. Zhen, L.; Bărbulescu, A. Variational Quantum Regression Application in Modeling Monthly River Discharge. *Water* **2025**, *17*, 1836. https://doi.org/10.3390/w17121836.
9. Gevorgyan, G.; Khachatryan, G.; Varagyan, A.; Varagyan, V.; Vaseashta, A. Hydrochemical Characterization, Source Identification, and Irrigation Water Quality Assessment in the Voghji River Catchment Area, Southern Armenia. *Water* **2025**, *17*, 854. https://doi.org/10.3390/w17060854.
10. Serban, C.; Maftei, C. Spatiotemporal Drought Analysis Using the Composite Drought Index (CDI) over Dobrogea, Romania. *Water* **2025**, *17*, 481. https://doi.org/10.3390/w17040481.
11. Jia, L.; Zhang, B. Simulating the Vegetation Gross Primary Productivity by the Biome-BGC Model in the Yellow River Basin of China. *Water* **2024**, *16*, 3468. https://doi.org/10.3390/w16233468.
12. Pronin, E.; Banaś, K.; Chmara, R.; Ronowski, R.; Merdalski, M.; Santoni, A.-L.; Mathieu, O. Characteristics of Stable Carbon and Nitrogen Isotopes in Different Ecological Plant Groups and Sediments Collected from 14 Softwater Lakes in Poland. *Water* **2024**, *16*, 3403. https://doi.org/10.3390/w16233403.

References

1. Vaseashta, A.; Gevorgyan, G.; Kavaz, D.; Ivanov, O.; Jawaid, M.; Vasović, D. Exposome, Biomonitoring, Assessment, and Data Analytics to Quantify Universal Water Quality. In *Water Safety, Security and Sustainability*; Vaseashta, A., Maftei, C., Eds.; Advanced Sciences and Technologies for Security Applications; Springer: Cham, Switzerland, 2021. [CrossRef]
2. Maftei, C.; Muntean, R.; Vaseashta, A. (Eds.) *Modeling and Monitoring Extreme Hydrometeorological Events*; IGI Global Scientific Publishing: Hershey, PA, USA, 2024. [CrossRef]
3. Vaseashta, A.; Duca, G.; Covaliova, O.; Romanciuc, L. Water Safety, Security, and Sustainability: Emerging Trends and Future Pathways. In *Handbook of Research on Water Sciences and Society*; Vaseashta, A., Duca, G., Travin, S., Eds.; IGI Global Scientific Publishing: Hershey, PA, USA, 2022; pp. 1–36. [CrossRef]
4. Duca, G.; Vaseashta, A. (Eds.) *Handbook of Research on Emerging Developments and Environmental Impacts of Ecological Chemistry*; IGI Global Scientific Publishing: Hershey, PA, USA, 2020. [CrossRef]
5. Kaushal, S.S.; Gold, A.J.; Mayer, P.M. Land Use, Climate, and Water Resources—Global Stages of Interaction. *Water* **2017**, *9*, 815. [CrossRef] [PubMed]

https://doi.org/10.3390/w18050643

6. Morrison, R.R.; Simonson, K.; McManamay, R.A.; Carver, D. Degradation of floodplain integrity within the contiguous United States. *Commun. Earth Environ.* **2023**, *4*, 215. [CrossRef]
7. Khan, S.; Masoodi, T.H.; Islam, M.A.; Arjumand, T.; Raja, A.; Parrey, A.A.; Pallavi, A. Ecosystem Degradation to Restoration: A Challenge. In *Climate Crisis: Adaptive Approaches and Sustainability*; Chatterjee, U., Shaw, R., Kumar, S., Raj, A.D., Das, S., Eds.; Sustainable Development Goals Series; Springer: Cham, Switzerland, 2023. [CrossRef]
8. Schroeter, S.A.; Orme, A.M.; Lehmann, K.; Lehmann, R.; Chaudhari, N.M.; Küsel, K.; Wang, H.; Hildebrandt, A.; Totsche, K.U.; Trumbore, S.; et al. Hydroclimatic extremes threaten groundwater quality and stability. *Nat. Commun.* **2025**, *16*, 720. [CrossRef] [PubMed]
9. Maftei, C.; Dobrica, G.; Barbulescu, A.; Serban, C. Analysis Methods for Hydrological Drought. In *Modeling and Monitoring Extreme Hydrometeorological Events*; Maftei, C., Muntean, R., Vaseashta, A., Eds.; IGI Global Scientific Publishing: Hershey, PA, USA, 2024; pp. 90–113. [CrossRef]
10. Msigwa, A.; Makinde, A.S. Application of Machine Learning Techniques in Hydrometeorological Event Prediction. In *Modeling and Monitoring Extreme Hydrometeorological Events*; Maftei, C., Muntean, R., Vaseashta, A., Eds.; IGI Global Scientific Publishing: Hershey, PA, USA, 2024; pp. 135–161. [CrossRef]
11. Brčeski, I.; Vaseashta, A. Environmental Forensic Tools for Water Resources. In *Water Safety, Security and Sustainability*; Vaseashta, A., Maftei, C., Eds.; Advanced Sciences and Technologies for Security Applications; Springer: Cham, Switzerland, 2021. [CrossRef]
12. Maftei, C.; Muntean, R.; Vaseashta, A. Monitoring and Modelling of Extreme Hydrometeorological Events. In *Modeling and Monitoring Extreme Hydrometeorological Events*; Maftei, C., Muntean, R., Vaseashta, A., Eds.; IGI Global Scientific Publishing: Hershey, PA, USA, 2024; pp. 1–21. [CrossRef]

https://doi.org/10.3390/w18050643

water

MDPI

Article

Characteristics of Stable Carbon and Nitrogen Isotopes in Different Ecological Plant Groups and Sediments Collected from 14 Softwater Lakes in Poland

Eugeniusz Pronin [1,*], Krzysztof Banaś [1], Rafał Chmara [1], Rafał Ronowski [1], Marek Merdalski [1], Anne-Lise Santoni [2] and Olivier Mathieu [2]

1 Department of Plant Ecology, Faculty of Biology, University of Gdansk, 80-309 Gdańsk, Poland; krzysztof.banas@ug.edu.pl (K.B.); rafal.chmara@ug.edu.pl (R.C.); rafal.ronowski@ug.edu.pl (R.R.); marek.merdalski@ug.edu.pl (M.M.)

2 Biogéosciences, UMR 6282 CNRS, Université Bourgogne Franche-Comté, F-21000 Dijon, France; anne-lise.santoni@u-bourgogne.fr (A.-L.S.); olivier.mathieu@u-bourgogne.fr (O.M.)

* Correspondence: eugeniusz.pronin@ug.edu.pl; Tel.: +48-58-523-61-27

Abstract: Softwater lakes with specific, rare, and protected aquatic plant vegetation are very sensitive to increased trophic and water chemical changes, especially alkalization. These changes might be reflected in the stable carbon and nitrogen isotopes of the organic matter (OM) of those plants ($\delta^{13}C_{ORG}$ and $\delta^{15}N_{ORG}$) and sediments ($\delta^{13}C_{ORG}$ and $\delta^{15}N_{TN}$) which they cocreated. To recognize the relationship between OM in the aquatic plants of softwater lakes and the cocreated sediments, we analyzed the plants $\delta^{13}C_{ORG}$ and $\delta^{15}N_{ORG}$ in light of four ecological groups (as well as the sediments $\delta^{13}C_{ORG}$ and $\delta^{15}N_{TN}$ in which these plants thrived). Studies were performed in July 2020 on 14 softwater lakes in northern Poland with varying pH gradients from 4.86 to 9.20. For each lake, a single stand was examined for each species detected. The goal was to investigate each species at 10 sites, but this target was not reached for several species. Among the investigated ecological groups, isoetids and mosses showed the least variation in obtained values. In contrast, elodeids exhibited the highest variation in $\delta^{13}C$ results due to their diverse carbon uptake strategies, involving both CO_2 and HCO_3^- forms. Moreover, the $\delta^{13}C$ values of charophytes and elodeids were highly related to the increasing pH of water. The slight differences between the $\delta^{13}C$ results of plants and sediments in isoetid stands further support this finding. Furthermore, we noticed an increase in sediment $\delta^{13}C_{ORG}$ values along the alkalization gradient, suggesting that macrophytes were the primary source of OM for the sediments. The positive correlation between the $\delta^{13}C$ values of plants and sediments ($r = 0.69$, $p < 0.05$) might also confirm that the macrophytes were a significant source of OM in sediments. Regarding $\delta^{15}N$ values, we did not find significant differentiation between plants and sediments across ecological groups. However, sediment consistently exhibited ^{15}N-enrichment compared to plant material. This enrichment is likely attributed to the accumulation of ^{15}N during the decomposition of the deposited material. This study confirms the possibility of tracking changes in the aquatic vegetation of softwater lakes based primarily on the sediment $\delta^{13}C_{ORG}$ values along the alkalization gradient.

Keywords: lobelia lakes; isoetids; ecological groups of macrophytes; stable isotope; pH gradient; lake alkalization

1. Introduction

Softwater lake vegetation is highly sensitive to changes in water quality, especially concerning pH levels and nutrient concentrations [1,2]. Numerous studies have focused on the physiology of specific plants growing in these ecosystems and the relations among plant communities (e.g., [3–6]). However, our understanding of the biogeochemistry of these water bodies, especially in light of the stable isotope compositions of carbon and nitrogen

($\delta^{13}C$ and $\delta^{15}N$), remains limited. This is especially visible in the lack of a study comparing the isotopic signal from plants and sediments, which is crucial to reliable paleolimnological interpretation based on the isotopic data from these specific ecosystems. Although some reports have investigated the geochemistry of water and sediment (e.g., [7–9]) and plant organic matter (OM) (e.g., [9,10]), only a few studies have focused on $\delta^{13}C$ and $\delta^{15}N$ [10–15]. Furthermore, these studies have yielded only limited results on the $\delta^{13}C$ and $\delta^{15}N$ values of aquatic plants in softwater lakes [10–12,16]. Additionally, to the best of the authors' knowledge, there is a lack of studies on the variation in the $\delta^{13}C$ and $\delta^{15}N$ values of both plant and sediment OM along environmental gradients in softwater lake ecosystems, especially including such an amount of investigated aquatic species.

In softwater lakes, primary production is usually limited by nutrient availability [17–19]. While it is commonly stated that phosphorus is the main limitation in freshwater ecosystems, in the case of oligotrophic softwater lakes, nitrogen and carbon also play significant roles as limiting factors [17,20]. Softwater lakes possess specific conditions, such as low calcium concentrations (<1 to ~15 mg/L^{-1}) and low water mineralization (conductivity < 150 µScm^{-1}; [21]), which support the existence of particular groups of plants. These groups include evergreen plants from the species *Isoëtes lacustris* L., collectively known as isoetids. Another characteristic isoetid species found in these lakes, often alongside elodeids (submerged plants from the *Elodea* genus attached to the lake bottom), is *Lobelia dortmanna* L. Charophytes (here, species of *Chara* and *Nitella* genera—macroscopic green algae) and mosses are also frequently found in these diverse lakes, although they less commonly create monospecific large-area communities in softwater lakes compared to isoetids [2].

Macrophytes play a crucial role in carbon uptake for photosynthesis processes, obtaining carbon from both water and sediment pore water, as well as from sediments themselves [18]. The pH of water is particularly significant for macrophytes, as it determines the available carbon forms (CO_2 in acidic water, and HCO_3^- in neutral and alkaline water to a pH of about 9) used by plants in the process of photosynthesis [22,23]. The $\delta^{13}C$ values of CO_2 and HCO_3^- dissolved in the water differ by approximately 8–12‰ depending on water temperature [24–26]. In addition, intersediment pore water is typically enriched in ^{12}C, available due to the decomposition of OM [27]. However, this may not always be the case in anoxic environments where methane is formed. The ^{12}C, which favors methane production, could be removed from the reservoir along with methane bubbles [28].

Macrophytes also significantly modify the environment, including water and sediment parameters [6,29,30]. By doing this, they create and thrive in specific conditions, which is characteristic of oligotrophic-oriented species from the isoetid group. These species depend on particular water parameter values, such as Ca^{2+} ion concentration and water pH [18]. Especially in lobelia lakes (softwater lakes with characteristic species such as *I. lacustris*, *L. dortmanna*, *Luronium natans* (L.) Raf., and *Littorella uniflora* (L.) Asch.), the presence of submerged macrophytes is pH-dependent [2]. Moreover, the vegetation structure of lobelia lakes is specific and highly sensitive to changes in habitat, especially alkalization and increased eutrophication [2,31–33].

Vegetation in water primarily uses two forms of nitrogen to construct its structures—NO_3^- and NH_4^+ [34,35]—along with organic nitrogen, such as amino acids (Chappuis et al. [10] and reference therein). The $\delta^{15}N$ values of these nitrogen forms exhibit an extensive range from approximately −11‰ for NO_3^- to up to +25‰, and their values depend on whether the nitrogen circulating in the lake is of mineral or organic origin [36,37]. Recent studies revealed that the nitrogen input from domestic sewage and pollutants or organic fertilizer runoff increases the $\delta^{15}N$ values of aquatic plants [14,38,39]. This ^{15}N-enriched isotopic signal of the aquatic plant seems to be a good indicator of the anthropogenic N input, but it probably is more reliable for river ecosystems than lakes [14]. In softwater lakes, the proportion of OM in mineral sediment varies from a minor contribution in the typically mineral sandy shallow littoral zone to a very high contribution in the deeper, more pelagic zone (Table S1 in Supplementary Materials). Limited studies [40,41] indicate that OM can dominate the sediment, especially when macrophytes, particularly isoetid plants (e.g., *I. lacustris* L., *L. dortmanna* L., *L. natans* (L.) Raf., and *L. uniflora* (L.) Asch.), overgrow these types of lakes. Isoetids have adapted well to oligotrophic conditions, developing extensive and strong roots that efficiently collect the available carbon and nitrogen from nutrient-poor water and sediments, because these plants readapted to living in water conditions [18].

Furthermore, in degraded oligotrophic lobelia lakes, when submerged vegetation decreases, phytoplankton may become dominant, and this shift can be detected in the sediment and identified through a comparison of the C/N ratios of OM buried in deposits or other paleoecological analyses, such as diatoms, pollen, and Cladocera remains [40,41]. The C/N ratio of aquatic plants and phytoplankton differs [42], with the phytoplankton C/N ratio being lower than that of macroscopic aquatic plants known as macrophytes [22,42]. As a result, the C/N ratio of sediments deposited by these biotic components also varies. Meyers and Ishiwatari [42] proposed using these differences, along with support from stable isotopic data, to identify the origin of sediments more accurately. This approach is commonly employed to recognize the main components of OM in lacustrine sediments [42–47].

We hypothesized that along an alkalization gradient, the $\delta^{13}C_{ORG}$ and $\delta^{15}N_{ORG}$ values of different ecological groups of aquatic vegetation would vary systematically due to differences in plant carbon acquisition strategies and the availability of carbon from diverse sources. Additionally, we hypothesized that the $\delta^{13}C_{ORG}$ and $\delta^{15}N_{ORG}$ values of the organic matter (OM) of sediments beneath the plants would closely correspond to the $\delta^{13}C_{ORG}$ and $\delta^{15}N_{ORG}$ values of the plants, indicating a strong connection between plant-derived OM and sediment OM. Thus, our study aimed to test these hypotheses by examining whether environmental variables influence the relationship between plant-derived OM and sediment OM and whether the $\delta^{13}C_{ORG}$ and $\delta^{15}N_{ORG}$ values, along with the C/N ratio in sediments, align with those of the plants. Confirming these hypotheses would support the idea that submerged aquatic vegetation is the primary contributor to sediment OM in low-productivity lake ecosystems.

Notably, our research presents the first $\delta^{13}C_{ORG}$ and $\delta^{15}N_{ORG}$ records of vegetation and sediments concerning water parameters provided for the softwater ecosystem in Poland. The obtained results might be necessary for other researchers who are interested in tracing the food web with a focus on specific groups of organisms such as benthic macroinvertebrates, fishes, ducks, swans, and other animals for which aquatic plants and their decaying parts might be the source of food in their diet [48]. Moreover, in the last decade, interest in the application of mixing models based on the OM isotopic data of varied sediment components to distinguish which are the primary sources of OM in the aquatic ecosystem has increased [43–45]. Thus, more specific isotopic data must be available for the broader scientific community and for paleolimnologists who support their interpretations by applying stable isotopic analyses to increase validity [14,49,50].

2. Materials and Methods

2.1. Study Sites

This study was performed in the middle of the vegetation season (July 2020) at 14 softwater lakes located in the Pomeranian Lakeland in north-western Poland (Figure 1). These lakes represent a broad spectrum of softwater habitats, exhibiting varying water pH levels ranging from 4.86 to 9.20, along with other distinct parameters (Supplementary Materials Table S1).

In each lake, we investigated stands of aquatic plants among ten previously selected macrophytes, including species characteristic of softwater lakes such as *I. lacustris*, *L. dortmanna*, *L. natans*, and *L. uniflora*, as well as frequently occurring species in these types of lakes, namely *Sphagnum denticulatum* Brid., *Myriophyllum alterniflorum* DC., *Nitella flexilis* (L.) AG, *Fontinalis antipyretica* Hedw., *Elodea canadensis* Michx., and *Chara globularis* Thuiller, which display differentiated carbon acquisition strategies and belong to different ecological groups (Figure 2). Several of these listed plants are protected in Poland and are found in natural reserves. Therefore, before commencing the study and collecting the plants and sediments, we obtained the necessary permits from the Regional Director for Environmental Protection in Gdańsk, Poland (for further details, see Table S1 in the Supplementary Materials).

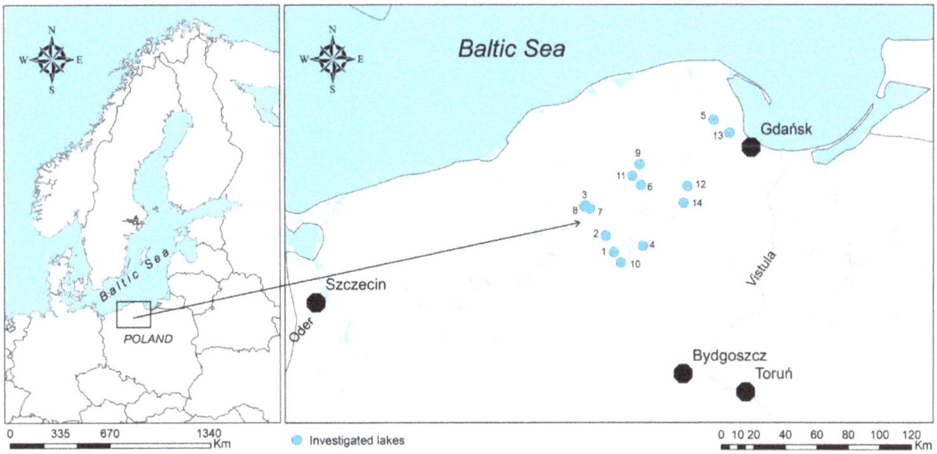

Figure 1. Localization of the investigated lakes: 1. Linowskie; 2. Krasne; 3. Smołowe; 4. Moczadło; 5. Piasek; 6. Kamień; 7. Łąkie; 8. Zawiad; 9. Okoń Duży; 10. Obrowo Małe; 11. Jeleń; 12. Dobrogoszcz; 13. Osowskie; 14. Zakrzewie.

Figure 2. Investigated plant species. Ecological group: (**a**,**b**)—mosses; (**c**–**f**)—isoetids; (**g**,**h**)—elodeids; (**i**,**j**)—charophytes. The photographs of the plants listed above were taken by the authors.

2.2. Description of Investigated Aquatic Plant Sites

Aquatic plants tend to form plant communities mainly dominated by one or two species, often creating dense patches and overgrowing the bottom sediments. This phenomenon was observed in 73 out of the 85 investigated sites. Environmental samples (individuals of aquatic plants, sediment from the sites where the plants were collected, water from the plant community) were usually collected from areas dominated by the investigated plant or sites dominated by different macrophytes where the chosen plant was a co-occurring species. At 43 locations out of the 85 sites, the chosen plant species represented a dominant species, constituting from 30% to 100% of the total bottom area (~1.0 m^2). Still, in a few cases, it co-occurred with other species, reaching more than 30% coverage (Table S1 in Supplementary Materials). For this reason, water and sediment samples were collected from 80 sites, as at 5 sites, both species of interest occurred together. Due to the site area being 100 cm × 100 cm, collecting two separate sediment samples (using a Ø15 cm plexiglass core sampler) was purposeless.

At each study site (*N* = 80), we collected plants (*N* = 85) and sediments (*N* = 80) for further $\delta^{13}C_{ORG}$ and $\delta^{15}N_{ORG}$ analyses. A team of experienced SCUBA divers collected plants (usually 10 individuals) and sediments (top 5 cm layer using the plexiglass core sampler). Before collecting the plants and sediments, we conducted field measurements of pH using the YSI 650 MDS with Multiparameter Water Quality Sonde 6600 V2 (Yellow Springs, OH, USA)., as well as measurements of the depth of the stands. Additionally, we collected one plastic 0.5 l^{-1} bottle of water from the plants' surroundings. These water samples were intended for further chemical laboratory analyses, including of nutrient levels (nitrite NO_3^-; total nitrogen—TN; and total phosphorus—TP), as well as concentrations of Ca^{2+}, CO_2, HCO_3^-, and CO_3^{2-}, which corresponded to dissolved inorganic carbon—DIC.

2.3. Laboratory Analysis

In the collected water samples (*N* = 80), the concentrations of dissolved forms of inorganic carbon (DIC—CO_2, HCO_3^-, and CO_3^{2-}) were assessed through titration. Calcium concentration (Ca^{2+}) was measured using a complexometric method with disodium edetate in the presence of calconcarboxylic acid sodium salt as an indicator, as described in [51]. Nitrate (NO_3^-), TN, and TP concentrations were determined using photometric methods with the MERCK Spectroquant cuvette test on a UV–VIS spectrophotometer (Aquamate, Thermo Electron Corporation, Waltham, MA, USA). TP was analyzed after mineralizing water samples in a mixture of nitric and sulfuric acid (1:2) in the microwave digestion system Mars 5 CEM (Matthews, NC, USA).

The OM content of the sediments was calculated by measuring the difference between the dry sediment weight before and after combustion at 550 °C (Thermolyne 62700 muffle furnace Waltham, MA, USA). The C/N ratios were calculated based on the percentage of C and N in the sediment and plant samples (EA VarioMicro Cube, Elementar, Langenselbold, Germany and Flash Smart EA, Thermo Scientific, Waltham, MA, USA, respectively).

The collected plant samples were washed in the field to remove epiphytes and other contaminants (e.g., sand, sediments). Both plant and sediment samples were dried at 60 °C for 48 h and stored in dry conditions. In the laboratory, the dry plant samples were crushed to a powder using a mixer mill (MM 400 Retsch, Haan, Germany). The presence of carbonates was checked by HCl acid, and due to the deficient concentration of Ca^{2+} and other ions in softwater lakes, we did not observe any reactions in the plant and sediment samples. Thus, we did not apply the desiccator HCl treatment to all subsamples in the analysis as we did in hardwater lake samples [13]. Furthermore, Fourier-Transform Infrared Spectroscopy with Attenuated Total Reflection (FTIR-ATR) was applied to all samples following the Pronin [14] procedure and confirmed the absence of $CaCO_3$ in the samples. Bulk $\delta^{13}C_{ORG}$ and $\delta^{15}N_{ORG}$ analyses of the plants were then performed on 85 powdered samples (triplicated as a standard laboratory procedure, resulting in a total of 255 samples analyzed). The analysis was conducted in the GISMO platform, Biogéosciences laboratory of the University of Burgundy (Dijon, France), using an Elementar analyzer

Flash Smart EA (Thermo Scientific, Waltham, MA, USA) coupled to a Delta V stable isotope ratio mass spectrometer (Thermo Scientific, Waltham, MA, USA). Calibration and control were achieved using the standard USG40 (glutamic acid, $\delta^{13}C = -26.39‰$, $\delta^{15}N = -4.5‰$) and the standard Wheat Flour B2157 (Elemental Microanalysis, Okehampton, UK) certified reference materials. The $\delta^{13}C$ and $\delta^{15}N$ values were expressed in ‰ relative to V-PDB for carbon and atmospheric N_2 for nitrogen. The precision of the analysis was validated by the external reproducibility of replicate analyses of USG40 and B2157 standards, which were better than ±0.15‰ for $\delta^{13}C$ and ±0.20‰ for $\delta^{15}N$ (2σ).

We removed larger particles from the dry sediment samples to lower the probability of allochthonous materials not decomposing, such as terrestrial plants, leaves, and wood. After this, the sediment samples were crushed and homogenized using a mixer mill (MM 400 Retsch, Haan, Germany). Moreover, the aforementioned bigger particles of terrestrial plants were also checked for their isotopic composition in different laboratories due to the same logistic problems. These analyses were combined in the isotopic equipment (Thermo Scientific Delta V plus isotope ratio mass spectrometer Waltham, MA, USA coupled with a Flash 2000 Elemental Analyzer Waltham, MA, USA via continuous flow) in the Laboratory of the Biogeochemistry and Environmental Protection, Biological and Chemical Research Centre, University of Warsaw. Due to the small size of the site samples, we combined samples from all sites into one sample for each lake ($N = 14$).

For the $\delta^{13}C_{ORG}$ and $\delta^{15}N_{TN}$ analyses of OM in the sediment samples ($N = 80$, each analyzed in three repetitions, resulting in a total of 240 samples analyzed), a dry combustion method was employed. This analysis was performed on an elemental analyzer linked to an isotope ratio mass spectrometer (EA-IRMS, VarioMicro Cube linked to an Isoprime, Elementar, Langenselbold, Germany). The precision of the analysis was validated using the external reproducibility of replicate standard analyses of USG40, and the in-house standards were better than ±0.15‰ for $\delta^{13}C$ and ±0.2‰ for $\delta^{15}N$ (2σ).

2.4. Statistical Analyses

The normality of the distributions of the analyzed stable isotope values and physicochemical variables was evaluated using the Shapiro–Wilk test, conducted with R 4.2.1 software [52]. The results showed that most of the analyzed parameters, especially $\delta^{13}C$ and $\delta^{15}N$, did not follow a normal distribution. Therefore, nonparametric analyses were applied. The Wilcoxon Signed-Rank test was utilized to compare the values of $\delta^{13}C$ and $\delta^{15}N$ of plants and the sediment OM for each investigated group of plants. The dependent paired group test was chosen, as the isotopic signal of sediment might be related to the isotopic signal of plants. The Kruskal–Wallis test was used when comparing the values of the $\delta^{13}C$ and the $\delta^{15}N$ of ecological plant groups, and a Dunn's post-hoc test was performed using the Dunn.test R package to identify differences between them [53]. For all statistical tests, $p < 0.05$ was accepted as the significance level. Additionally, Spearman rank correlations were used to check the relationships between the $\delta^{13}C_{ORG}$ and $\delta^{15}N_{ORG}$ values of plants and sediment based on ecological groups of plants. Moreover, Spearman rank correlation was applied to check the relationship between the $\delta^{13}C_{ORG}$ and $\delta^{15}N_{ORG}$ values of plants and sediments with the pH of ambient water. The results were visualized using R 4.2.1 software [52] with the ggplot2 package [54]. Furthermore, a principal component analysis (PCA) was performed for isotopic data and other investigated parameters for ambient water. The data were assigned to the ecological groups of the investigated plants. PCA was carried out using the FactoMineR package [55], and the obtained results were visualized using the factoextra package [56]. Moreover, PERMANOVA analysis of the data sets of ambient waters and sediment variables was performed to check if there were differences between the investigated plant groups. This analysis was done using the R.4.2.1 software and vegan package [57].

3. Results

3.1. $\delta^{13}C_{ORG}$ and $\delta^{15}N_{ORG}$ Characteristics of Plants and Sediments of Softwater Lakes

The obtained results for the $\delta^{13}C_{ORG}$ values of plants ranged from −33.80 to −9.90‰, while the sediment OM exhibited a similar but smaller range from −30.10 to −13.60‰. For the $\delta^{15}N_{ORG}$ values of plants and the $\delta^{15}N_{TN}$ values of sediment, the ranges were from −5.50 to 6.00‰ and −2.90 to 4.80‰, respectively (Table S1 in the Supplementary Materials; Figure 3). Notably, the plants showed higher variations in $\delta^{13}C_{ORG}$ and $\delta^{15}N_{ORG}$ compared to the sediments (Figure 3a,b). Additionally, the data from the sediment samples with bigger particles of terrestrial plants showed a smaller range of values from −21.34 to −29.97‰ and from −2.70 to 3.66‰ for $\delta^{13}C_{ORG}$ and $\delta^{15}N_{ORG}$, respectively (Table S1 in the Supplementary Materials; Figure 3).

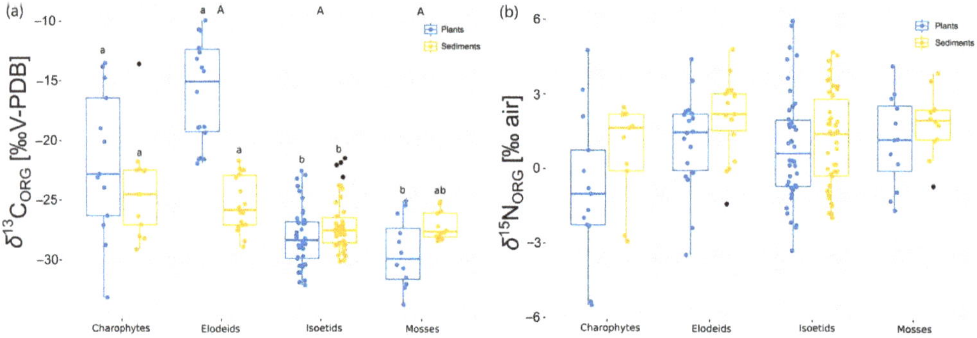

Figure 3. Comparison of isotopic results of OM from plants and sediments of the investigated group of species: mosses (N = 12), isoetids (N = 42), elodeids (N = 18), charophytes (N = 13). (**a**) $\delta^{13}C$ values of plants (blue) and sediments (orange); (**b**) $\delta^{15}N$ values of plants and sediments. "A" above boxplots means $p < 0.05$ for the Wilcoxon Signed-Rank test of plants and sediment within the group; different lowercase letters means the statistical significance $p < 0.05$ for the Dunn post-hoc test after the Kruskal–Wallis test. The box represents 25–75%, the line in the box represents the median, the whiskers represent the min and max of the non-outlier values, and the black dots represent the outlier values.

The isotopic differences between plants and sediments displayed similar trends for two ecological groups, but opposite trends were observed for the other two ecological groups. In the case of mosses and isoetids, we found lower plant $\delta^{13}C_{ORG}$ values compared to the $\delta^{13}C_{ORG}$ values of the sediments they grew on, and this difference was statistically significant ($p < 0.05$, Wilcoxon Signed-Rank test). Conversely, elodeids and charophytes exhibited the opposite pattern, with values of $\delta^{13}C_{ORG}$ in the sediments being lower than those in the investigated aquatic plants (Figure 3a). The observed differences in $\delta^{13}C_{ORG}$ values between plants and sediments were statistically significant only in the elodeid group ($p < 0.05$, U Wilcoxon Signed-Rank test).

Additionally, there were statistically significant differences in $\delta^{13}C$ values between mosses and isoetids compared to elodeids and charophytes (Figure 3a). Similarly, when comparing the sediments associated with different ecological groups of plants, statistically significant differences were found between isoetids and elodeids, as well as charophytes (Figure 3a). These findings suggest that the $\delta^{13}C_{ORG}$ values of mosses and isoetids were better preserved in the sediment compared to those of charophytes and elodeids (Figure 3a).

The $\delta^{15}N_{ORG}$ values of plants were found to be enriched in ^{14}N compared to the values recorded for the sediment deposits, and this pattern was consistent across all investigated ecological groups of plants (Figure 3b). The highest plant $\delta^{15}N_{ORG}$ values were found for isoetids and the lowest for charophytes (Figure 3b). Moreover, when considering the ecological group of plants, we did not find any statistically significant differences in the recorded $\delta^{15}N_{ORG}$ values (Figure 3b).

It is noteworthy that the isoetid plants in our study (i.e., *L. dortmana*, *I. lasuctris*, *L. uniflora*, *L. natans*) had significantly lower $\delta^{13}C_{ORG}$ values compared to other groups of plants, such as charophytes (*N. flexilis*, *C. globularis*) and elodeids (*E. canadensis*, *M. alterniflorum*), as shown in Figure 3a. However, we observed no statistical differences between mosses and isoetids. The $\delta^{13}C_{ORG}$ values of isoetid plants were slightly lower than those obtained for sphagnum moss (*S. denticulatum*) and higher than those of bryophyte (*F. anthypyretica*), as shown in Figure S1a in the Supplementary Materials.

The correlation analysis performed between the $\delta^{13}C_{ORG}$ values of plants and sediment samples revealed a strong relationship, indicating that increases in the values of $\delta^{13}C_{ORG}$ in plants were also reflected in the $\delta^{13}C_{ORG}$ values of the sediments (Spearman rank correlation $r = 0.69$; $p < 0.05$; Table 1). The strongest relationship between the $\delta^{13}C_{ORG}$ of plants and sediments was found for isoetids, followed by for charophytes and elodeids. In the case of mosses, the relationship between the $\delta^{13}C_{ORG}$ of plants and sediments was weak and statistically insignificant (Table 1). On the other hand, a moderate relationship was found between the $\delta^{15}N_{ORG}$ values of plants and the $\delta^{15}N_{TN}$ values of sediment (Spearman rank correlation $r = 0.31$; $p < 0.05$; Table 1). This relationship was only moderate and significant in the isoetid group (Table 1).

Table 1. Spearman correlation coefficient (r) and *p*-value for $\delta^{13}C$ and $\delta^{15}N$ values of plants vs. $\delta^{13}C$ and $\delta^{15}N$ values of sediments in each group. The statistically significant relationships were bolded.

	Group	r	*p*-Value
$\delta^{13}C$	All groups (N = 85)	**0.688467**	**0.0001**
	Mosses (N = 12)	0.195804	0.5400
	Isoetids (N = 42)	**0.745786**	**0.0001**
	Elodeids (N = 18)	**0.656686**	**0.0030**
	Charophytes (N = 13)	**0.662999**	**0.0200**
$\delta^{15}N$	All groups (N = 85)	**0.311075**	**0.0037**
	Mosses (N = 12)	0.251748	0.4301
	Isoetids (N = 42)	**0.400097**	**0.0086**
	Elodeids (N = 18)	−0.006198	0.9805
	Charophytes (N = 13)	−0.129299	0.6738

Furthermore, we observed differences in the C/N ratio among the investigated groups of macrophytes. The highest C/N ratio was found in the isoetid group, and this ratio was significantly higher than in elodeids and charophytes (Figure 4). Additionally, the C/N ratio of charophytes was significantly different compared to that of mosses (Dunn's post-hoc test after Kruskal–Wallis test, $p < 0.05$; Figure 4).

The analysis of the C/N ratio of OM revealed that the isoetid group had a C/N ratio that differed significantly from the C/N ratio recorded in the OM of the sediments collected below these plants ($p < 0.05$, Wilcoxon Signed-Rank test, Figure 4). Similarly, for elodeids, the C/N ratio of plants was significantly higher than in the sediments (Figure 4). Interestingly, for charophytes, the sediments had a slightly higher C/N ratio compared to that recorded in plants, although this difference was relatively small (Figure 4).

We plotted the C/N ratios of plants and sediments against their respective isotopic values to explore the relationships between $\delta^{13}C_{ORG}$ and $\delta^{15}N_{ORG}$ in plants and sediments (Figure 5). A clear relationship was observed between the $\delta^{13}C_{ORG}$ values and C/N ratios of plant samples, which were reflected in the corresponding sediment samples (Figure 5a,c). However, for $\delta^{15}N_{ORG}$ values combined with C/N ratios, no consistent pattern was detected between plants and sediments (Figure 5b,d). These findings suggest that carbon dynamics between plants and sediments are more closely linked than nitrogen dynamics in the investigated ecosystems.

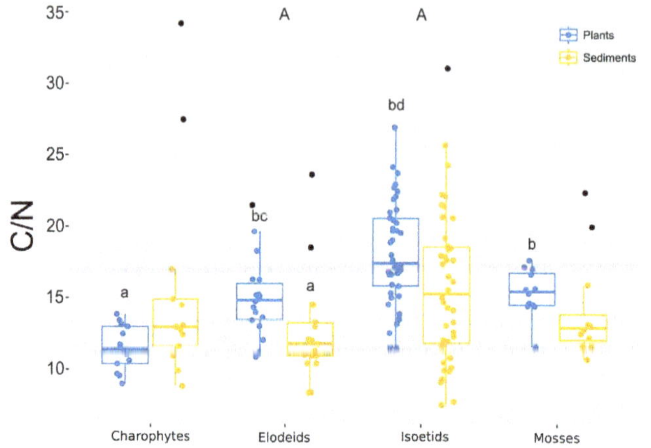

Figure 4. C/N ratios of investigated plant and sediment samples. "A" above boxplots means $p < 0.05$ for U of the Wilcoxon Signed-Rank test between plants and sediments within the group; different lowercase letters mean the statistical significance $p < 0.05$ for Dunn's post-hoc test after Kruskal–Wallis. The box represents 25–75%, the line in the box represents the median, the whiskers represent the min and max of the non-outlier values, and the black dots represent the outlier values.

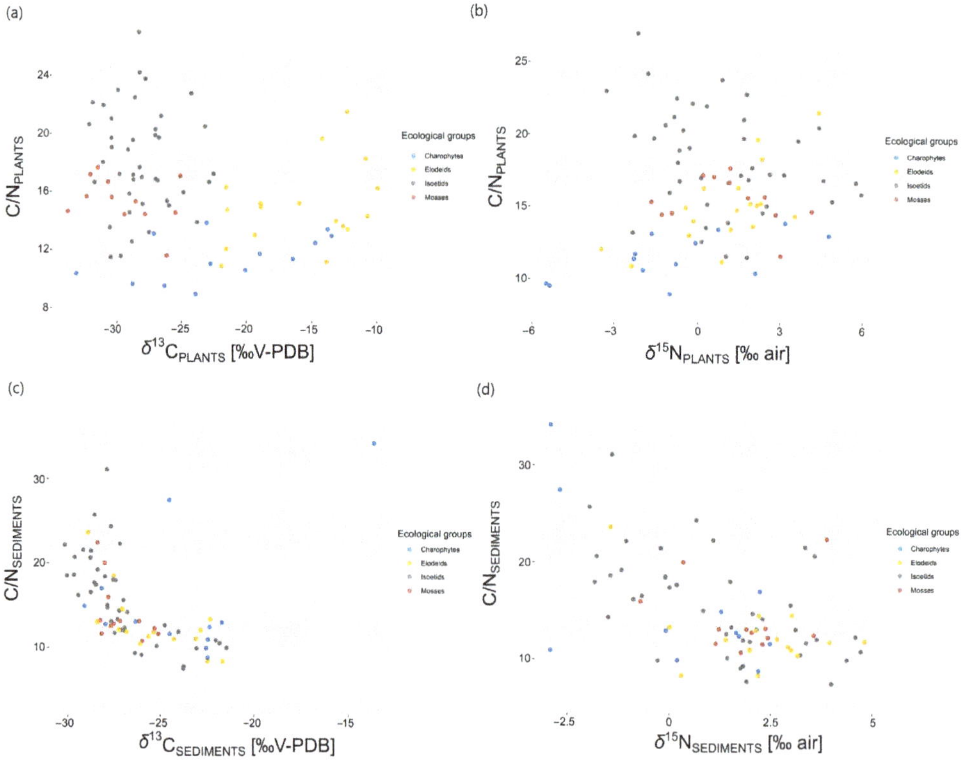

Figure 5. Scatterplot of the stable isotope signal of plants and sediments and their C/N ratios: (**a**) $\delta^{13}C_{ORG}$ of plants and C/N_{PLANTS}; (**b**) $\delta^{15}N_{ORG}$ of plants and C/N_{PLANTS}; (**c**) $\delta^{13}C_{ORG}$ of sediment and $C/N_{SEDIMENTS}$; (**d**) $\delta^{15}N_{TN}$ of sediment and $C/N_{SEDIMENTS}$.

3.2. Relationships Between $\delta^{13}C_{ORG}$ and $\delta^{15}N_{ORG}$ of Plant and Sediment OM and Other Investigated Variables

PCA showed several significant relationships between isotopic variables and water and sediment parameters (Figure 6a,b). The most strongly related variables were the $\delta^{13}C_{ORG}$ of plants and sediments. Furthermore, negative relationships were observed between the $\delta^{13}C_{ORG}$ and $\delta^{15}N_{ORG}$ values of plants and sediments and the C/N ratio of sediments. Additionally, moderate positive relationships were found for plants and sediments' pH, Ca^{2+}, and $\delta^{13}C_{ORG}$ values (Figure 6a,b). A moderate negative relationship was also observed between $\delta^{13}C_{ORG}$ and the OM content in sediments (Figure 6a,b). Correlations with the third dimension were found, as the parameters related to the light condition were negatively correlated with PAR and DOC concentration (Figure 6b). Moreover, DOC was related to CO_2 concentration, suggesting that when humic substances increase, the pH is lower; thus, the CO_2 concentration in these stands was higher.

However, the percentage of explained variance in the PCA results, including the three dimensions of the ordination plot for water, was not very high and did not exceed 50% (Figure 6a,b). PCA did not distinctly segregate the samples by groups of plants. Nevertheless, subtle patterns were observed in which isoetids occupied the left side of the ordination space, being more associated with a higher C/N ratio of sediments, lower pH of water, and higher CO_2 concentration. On the right side of the ordination plot, there were more elodeid and charophyte sites, which were more related to higher pH, HCO_3^-, and Ca^{2+} concentrations. Furthermore, some isoetids and mosses were also present in this region (Figure 6a,b). Thus, this segregation was not that evident. The PERMANOVA test, conducted to determine if the ecological groups of plants differ based on all investigated variables, revealed a significant difference ($F = 8.07$, $p < 0.001$).

To further investigate our hypothesis regarding the high variability of $\delta^{13}C_{ORG}$ values along the pH gradient, we checked the Spearman correlations among these variables while considering the ecological groups of plants (Table 2). The correlation coefficient of the relationship between plants' $\delta^{13}C_{ORG}$ values and pH was found to be higher for elodeids and charophytes, which confirms our expectations. The isoetid group exhibited relatively low variation, and its $\delta^{13}C_{ORG}$ values showed less dependence on the pH of water, which was statistically insignificant (Table 2). Interestingly, for mosses, we even observed some negative relationships, in which the lowest $\delta^{13}C_{ORG}$ values were found for the highest pH values of water (Table 2). This may be attributed to the replacement of *S. denticulatum* with *F. antipyretica* along the alkalization gradient of lakes.

Table 2. Spearman correlation coefficient (r) and *p*-value for $\delta^{13}C$ values of plants and sediments vs. ambient water pH $\delta^{15}N$ in each group. The statistically significant relationships were bolded.

	Group	r	*p*-Value
$\delta^{13}C_{PLANTS}$ vs. pH	All groups (*N* = 85)	**0.373492**	**0.0004**
	Mosses (*N* = 12)	**−0.748252**	**0.0073**
	Isoetids (*N* = 42)	0.246069	0.1162
	Elodeids (*N* = 18)	**0.708678**	**0.0009**
	Charophytes (*N* = 13)	**0.577717**	**0.0386**
$\delta^{13}C_{SEDIMENT}$ vs. pH	All groups (*N* = 85)	**0.279709**	**0.0095**
	Mosses (*N* = 12)	−0.300699	0.3425
	Isoetids (*N* = 42)	0.126641	0.4242
	Elodeids (*N* = 18)	0.391731	0.1079
	Charophytes (*N* = 13)	0.349862	0.2413

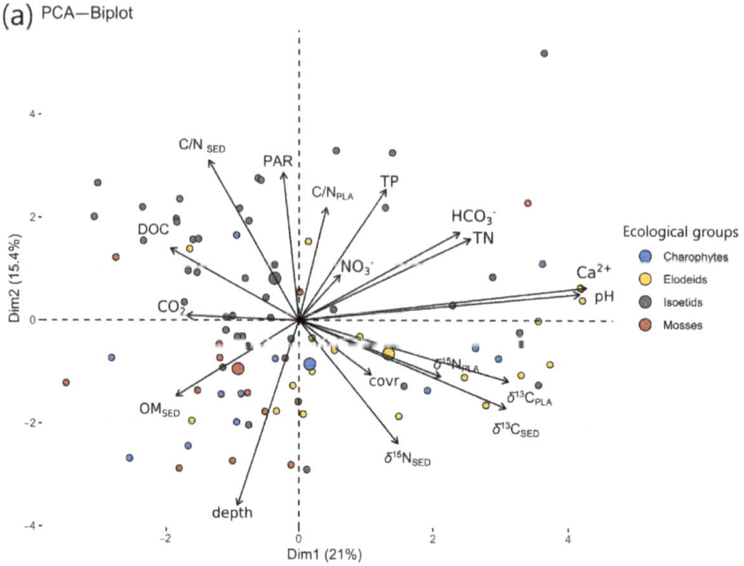

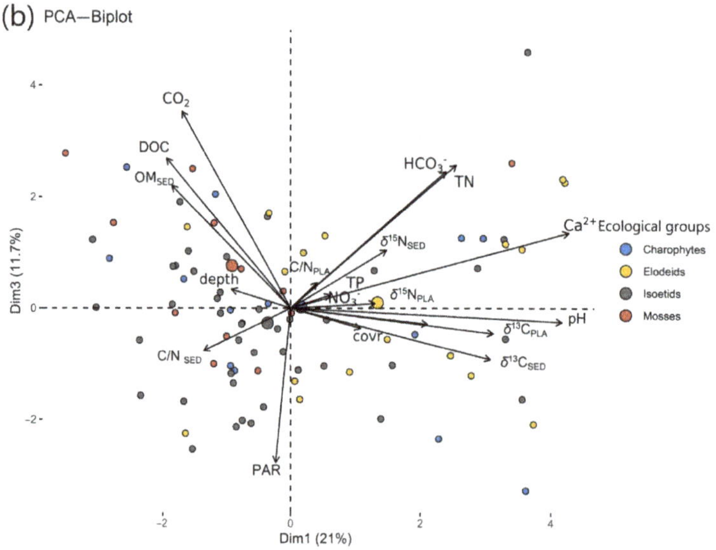

Figure 6. PCA of (**a**) ambient water variables and the $\delta^{13}C$ and $\delta^{15}N$ of plants and sediments and other measured parameters (N = 85) of dimensions 1 and 2; and (**b**) ambient water variables and the $\delta^{13}C$ and $\delta^{15}N$ of plants and sediments and other measured parameters (N = 85) of dimensions 1 and 3. DOC—dissolved organic carbon; PAR—photosynthetically active radiation; TP—total phosphorus; TN—total nitrogen; $\delta^{13}C_{PLA}$—$\delta^{13}C$ values of plants; $\delta^{13}C_{SED}$—$\delta^{13}C$ values of sediments; $\delta^{15}N_{PLA}$—$\delta^{15}N$ values of plants; $\delta^{15}N_{SED}$—$\delta^{15}N$ values of sediments; OM_{SED}—organic matter content (%) in sediments; depth—depth of the stands; covr—plant coverage; C/N_{PLA}—C/N ratio of plant OM; C/N_{SED}—C/N ratio of sediment OM; pH—pH of the water.

4. Discussion

4.1. The $\delta^{13}C_{ORG}$ of Aquatic Macrophytes and Sediment OM in Softwater Lakes

The range of $\delta^{13}C_{ORG}$ values identified in our macrophytes was broad (Table S1 in Supplementary Materials; i.e., −33.8 to −9.9‰) and showed differentiation between the investigated species and ecological groups of plants (Figure 2a and Figure S1a in Supplementary Materials). These findings are consistent with those of other authors who have focused on the isotopic study of submerged macrophytes and macroscopic algae from the charophyte group [10–12,16,58–65].

Furthermore, the recorded $\delta^{13}C_{ORG}$ values of isoetid plants do not vary as much as those of elodeids and charophytes (Figure 2a). This difference is probably due to the physiological variations in these plants, which enable isoetids to utilize only CO_2 from the sediments [18]. In addition, isoetids are commonly found in well-oxygenated littoral zones, where the decomposition of OM in sediments favors the release of ^{13}C-depleted CO_2 [66]. Consequently, more ^{12}C is likely utilized during the photosynthesis process by this type of vegetation [16]. Moreover, the utilization of only CO_2 by isoetids for photosynthesis was observed, resulting in a weak and statistically insignificant correlation between $\delta^{13}C_{ORG}$ values and the pH of ambient water.

In the case of elodeids and charophytes, the matter is more complex. These species primarily prefer carbon from water for photosynthesis and tissue development, but they can also utilize carbon from sediment water [17]. Furthermore, elodeids and charophytes may use two sources of carbon, CO_2 and HCO_3^-, which have different $\delta^{13}C$ values (CO_2 in the water is 8–12‰ ^{13}C-depleted compared to HCO_3^-, which dominates at pH levels above 7.0 [24]). The availability of these different forms of carbon in water depends on its pH. Consequently, the presence of plants in an environment with specific proportions of carbon forms in the water influences their $\delta^{13}C$ values. Knowing the $\delta^{13}C$ values of DIC and OM in sediments and OM in plants makes it possible to infer which source and form of carbon plants are utilizing for photosynthesis. However, we encountered issues in obtaining the $\delta^{13}C$ of DIC results due to preservation problems with the water samples for $\delta^{13}C$ analyses.

Despite this limitation, we found significant positive correlations between the $\delta^{13}C_{ORG}$ of elodeids and charophytes and the pH of the surrounding water. These results indirectly confirm that the $\delta^{13}C_{ORG}$ values of charophytes and elodeids depend on water pH, affecting the proportions of CO_2 and HCO_3^- in the water. Conversely, we observed an opposite relationship for the moss group, likely due to the replacement of *S. denticulatum* by *F. antipyretica* along the alkalization gradient. *S. denticulatum* was predominantly found in shallower sites compared to *F. antipyretica*, which usually occurred in the deeper sites (Table S1 in Supplementary Materials) and on sandy sediments. Mineralization might be faster in those sediments compared to the more organic compounds on which *S. denticulatum* was often found (Table S1 in Supplementary Materials). Moreover, in the literature on *F. antipyretica*, authors have reported extremely low $\delta^{13}C$ values (−50.7 and −45.0‰, [11,67]), but it is worth noting that these low values were found in rivers, not observed in lakes. However, investigations in the Polish hardwater lake Lednica, conducted by Apolinarska et al. [63], also found *F. antipyretica* to have almost the lowest values (−32.5‰) among other investigated species, but this was found in shallow water (2 m depth).

Additionally, elodeid species such as *M. alterniflorum* and *E. canadensis* and charophytes such as *C. globularis* and *N. flexilis* exhibited the highest $\delta^{13}C$ values in OM. These high values were recorded in more alkaline lakes where the proportion of CO_2 to HCO_3^- favors ^{13}C-enriched HCO_3^-. Conversely, lower values in more acidic lakes might be related to the occurrence of these species in deeper sites where they form dense patches, which might decrease the water mixing probability and also increase the dependency of photosynthesis in these plants on CO_2 released during sediment OM decomposition, as suggested by Pronin et al. [60]. Additionally, in deeper sites, anoxic conditions might have been present, leading to the release of methane bubbles, resulting in ^{13}C-enrichment of CO_2 [68]. Consequently, under specific conditions, this process might increase the $\delta^{13}C$ values of plants growing in deep sites, particularly in locations with high sediment OM content.

As mentioned previously, various factors influence the determination of $\delta^{13}C$ values in submerged aquatic plants, and it is important to consider differences in plant morphology and community structure created by these plants [13,60,61]. In addition, the strategies of carbon uptake and the photosynthesis pathway (mainly C3 and C4, but also CAM) play a significant role [6,16,59,69]. Other environmental variables, such as Ca^{2+} concentration, as found in this study, are also relevant factors to consider. All these differences collectively contribute to what can be referred to as biosynthesis fractionation, which has been suggested to explain the significant differences in $\delta^{13}C$ values between two different plants investigated by Liu et al. [64], in addition to the inherent variations between the $\delta^{13}C_{ORG}$ of plants and the $\delta^{13}C$ of DIC.

Regarding sediments collected from beneath the investigated macrophytes, the range of $\delta^{13}C_{ORG}$ values was slightly narrower compared to the results obtained for plants (Table S1 in Supplementary Materials; i.e., sediment values ranged from –30.1 to −13.6‰, while plant values ranged from −33.8 to −9.9‰), and these values differentiated between the investigated species and ecological groups of plants (Figure 2a).

For isoetids, the $\delta^{13}C_{ORG}$ values of sediments were higher compared to the $\delta^{13}C_{ORG}$ values of plants, although the variability in sediment values was not significant (Figure 2a). This pattern may be attributed to the co-occurrence of elodeids in isoetid sites; as mentioned in the Section 1, elodeids become more frequent as the alkalization gradient increases [2]. Consequently, the sediment, which is a mixture of the decomposed remains of all species occurring at the investigated site, might have higher $\delta^{13}C_{ORG}$ values due to the presence of elodeids. Another reason for the low variability in sediment $\delta^{13}C_{ORG}$ values in isoetid sites is related to their specific carbon acquisition strategy, which is primarily oriented towards collecting CO_2 solely from sediment.

Moreover, the observed modest increase (averaging 1.05‰) in the sediment $\delta^{13}C_{ORG}$ values of isoetids compared to plants aligns with the findings of [70]. The study investigated diagenesis processes influencing the $\delta^{13}C_{ORG}$ values of sediment OM in a softwater lake in Sweden. The authors reported a 0.4–1.5‰ increase in OM $\delta^{13}C_{ORG}$ values, especially during the first 5 years after deposition. These results further support the notion of subtle changes in sediment $\delta^{13}C_{ORG}$ values over time.

A different pattern emerged when comparing the $\delta^{13}C_{ORG}$ values of plants and sediments in the elodeid and charophyte groups. The sediments had lower $\delta^{13}C_{ORG}$ values on average, by 9.29‰ for elodeids and 2.34‰ for charophytes, compared to the plants (Figure 2). Additionally, the difference between plant and sediment $\delta^{13}C_{ORG}$ values was more substantial, and the variation in values was higher than what was found for isoetids. This difference could be related to the fact that these species co-occurred with plants that have lower $\delta^{13}C_{ORG}$ values or potentially replaced sites previously occupied by isoetids due to increasing alkalization [2].

Supporting this, it was observed that for elodeids, as water pH increased, the $\delta^{13}C_{ORG}$ values of the plants also significantly increased, which had a corresponding effect on sediments, leading to a slight increase in the $\delta^{13}C_{ORG}$ values of sediments. It is worth mentioning that one of the species in the elodeid group, *E. canadensis*, is not native to Poland, and the first recorded occurrence in the actual Polish territory dates back to 1867 [71]. Despite this relatively short period compared to the thousands of years of lake history, *E. canadensis* may have contributed to the increase in the $\delta^{13}C_{ORG}$ values of the sediments, considering that the study only considered the top 5 cm layer of sediments. Furthermore, the $\delta^{13}C_{ORG}$ values of *E. canadensis*, belonging to the elodeids group, were much higher than those of isoetids (Figure 2).

For the group of mosses, the differences between the $\delta^{13}C_{ORG}$ values of plants and sediments were significantly higher, with an average of 2.27‰, compared to isoetids (Figure 3a). This discrepancy can be attributed to the different ecology of the two investigated moss species. *S. denticulatum* is found at a wide range of water acidity levels, from acidic to neutral, while *F. antipyretica* prefers more alkaline lakes, usually close to neutral, with sandy sediments, as mentioned earlier. This might prompt the mineralization process, increasing the $\delta^{13}C_{ORG}$ of OM buried in sediment [70]. Furthermore, the presence of *F. antipyretica* often alongside elodeids and charophytes (Table S1 in Supplementary Materials) might also contribute to a slight increase in the $\delta^{13}C_{ORG}$ values of the sediment OM.

It should be noted that other autochthonous and allochthonous sources of OM could also play a role in influencing the $\delta^{13}C_{ORG}$ values of the sediment OM presented in this study. While these interactions are relevant, they extend beyond the primary focus of this manuscript and will be explored in more detail in future research. In brief, based on a comparison between the C/N ratios of plants and sediments, it was confirmed that the OM of the deposits was mostly derived from aquatic plants, as sediment C/N ratios in most investigated sites were not lower than 10, a threshold indicating that phytoplankton primarily contributed to the OM of sediments [42]. It is worth noting that charophytes also have C/N ratios close to this value [13] because they are macroscopic algae. Therefore, the presence and distribution of charophytes might also lower the C/N ratio of the OM of sediments in some sites of the softwater lakes.

4.2. $\delta^{15}N_{ORG}$ *of Aquatic Macrophytes and* $\delta^{15}N_{TN}$ *in Softwater Lakes*

The general trend of sediments' ^{15}N enrichment, except for the stands of two investigated species (*C. globularis* and *L. natans*, Figure S1b in Supplementary Materials), compared to the $\delta^{15}N_{ORG}$ of the investigated aquatic plants, aligns with the findings of [20]. Similar to their study, the $\delta^{15}N_{ORG}$ of the macrophytes exhibited greater variability than the $\delta^{15}N_{TN}$ of the sediment OM. Furthermore, the range of recorded $\delta^{15}N_{TN}$ values for sediments was wider (−2.91 to 4.81‰ vs. 0.22 to 6.09‰) than those recorded for upland lakes where aquatic vegetation, including several lakes with isoetids, was investigated [17,20]. However, we recorded a similar $\delta^{15}N_{ORG}$ range for the macrophytes (−5.49 to 5.96‰ vs. −3.06 to 7.31‰ as provided by Jones et al. 2004).

Additionally, Chappuis et al. [10] recorded values for several of the same investigated species that fall within the ranges presented in our study, with some exceptions, such as charophyte species, for which they also recorded values exceeding our maximum range (upper limit of about 10.00‰ vs. our upper limit of 5.96‰) for investigated macrophytes. This comparison shows that the recorded $\delta^{15}N_{TN}$ values of the OM of sediments and macrophytes were within the range of values presented in the literature. However, the results of $\delta^{15}N_{TN}$ for sediments were ^{15}N-depleted compared to those recorded by Jones et al. [20] and correspond more to the values recorded here for macrophytes (Figure 3).

Several variables influence the plants and sediments' $\delta^{15}N_{ORG}$ values. For macrophytes, the most important factors include the source of the N used to build their tissues (NO_3^-, NH_4^+, NH_3, N_{ORG}) and temperature [10,34,35,72]. Additionally, the fractionation of $\delta^{15}N_{ORG}$ tends to favor ^{14}N [73], although this becomes less important and is often not recorded when the demand for N is higher than its availability in plants [20,74].

In this study, no significant relationships were found between NO_3^- concentration and $\delta^{15}N_{ORG}$ in both macrophytes and sediments (Figure 6a,b). Moreover, in many stands, the concentrations of NO_3^-, and in almost all stands, the concentrations of NH_4^+, were below the quantification level of the applied spectroscopic methods (Table S1 in Supplementary Materials), which might influence the high variability in obtained $\delta^{15}N_{ORG}$ results. However, the PCA for sediment water does reveal some relationship between TN concentration and $\delta^{15}N_{ORG}$ (Figure 6b). Nevertheless, without more precise information about the particular concentrations of TN components, the correct interpretation of this relationship is impossible.

On the other hand, the values of $\delta^{15}N_{TN}$ for sediment OM might be ^{15}N-enriched due to N limitation, which was previously mentioned as a fractionation limitation [17,20]. Additionally, C limitation might also play a role in decreasing plant growth rates during vegetation seasons [18,19]. However, we recorded the availability of DIC forms in all investigated sites (Table S1 in Supplementary Materials), indicating that C limitation may not be a major factor in our study.

Other processes and variables also influence the $\delta^{15}N_{TN}$ of sediments. For instance, decomposition and denitrification processes might lead to the ^{15}N-enrichment of sediments [75]. However, in one softwater lake, the mineralization processes of OM in the sediments showed almost no fractionation effects in the case of the $\delta^{15}N_{ORG}$ of the sediments (from 0.3 to 0.7‰), especially during the first 5 years of diagenesis [70].

Our results indicated a negative relationship between the $\delta^{15}N_{ORG}$ of sediment and their C/N ratios (Figure 5a,b). Higher values of $\delta^{15}N_{TN}$ for sediment OM were noted for the samples with lower C/N ratios, suggesting that there was more N in these samples relative to C. This situation is likely related to the presence of different plant types and the decomposition stage rather than ^{15}N-enrichment from denitrification processes. Furthermore, our study was primarily conducted in shallow sites (Table S1 in Supplementary Materials) with well-mixed water and without oxygen deficiencies. Additionally, the oxidative capability of isoetid plants, which dominated the investigated lakes [18], might limit anoxic denitrification processes in the sediments.

The $\delta^{15}N_{ORG}$ values of plants, as demonstrated above, are consistent with values recorded in the literature. When combined with $\delta^{13}C_{ORG}$ values and C/N ratios, these findings strongly support the statement that macrophytes were likely the dominant source of OM in the sediment of softwater lakes (Figure 5a,c).

Softwater lakes that harbor protected species from the isoetid group are particularly sensitive to alkalization and eutrophication. As a result, many of these lakes in Poland are already degraded [32]. The presented results of $\delta^{13}C_{ORG}$ values show significant differences between the isoetid and elodeid groups, which are more dependent on the pH of the water. These findings can be valuable in understanding the direction of vegetation changes recorded in the $\delta^{13}C_{ORG}$ values for sediment OM. Moreover, our results improve the precision of data interpretation for potentially investigated sediment profiles in the littoral zones of lakes where $\delta^{13}C_{ORG}$ will be applied. However, it is worth noting that in the softwater littoral zones of lakes, the preservation of macro remains is often poor compared to in pelagic water, where remains of plants and animals are better preserved [76].

5. Conclusions

This study highlights the potential of using sediment $\delta^{13}C_{ORG}$ values to track changes in aquatic vegetation along alkalization gradients in softwater lakes. The findings suggest that isotopic signals in sediments are influenced by the carbon acquisition strategies of specific plant groups, reflecting their contribution to sediment OM. However, the role of nitrogen isotopes in these systems remains less clear due to the low differentiation in $\delta^{15}N_{ORG}$ values. Thus, this study underscores the need for further research. Future work could focus on applying and verifying isotopic mixing models to better quantify the contributions of different plant groups to sediment OM. Additionally, integrating multiple isotopic tracers and coupling these with biogeochemical modeling could enhance the understanding of isotopic variability. Expanding the investigation to a broader range of lake types and integrating advanced isotopic analyses could provide deeper insights into the interactions between aquatic vegetation and sediment OM across diverse ecosystems.

Supplementary Materials: The following supporting information can be downloaded at https://www.mdpi.com/article/10.3390/w16233403/s1, Figure S1: Comparison of isotopic results of OM from plants and sediments of investigated species; Table S1: The characteristics of investigated lakes, measured in variables.

Author Contributions: E.P.: Conceptualization, Supervision, Methodology, Project administration, Funding acquisition, Formal analysis, Visualization, Writing—original draft, Writing—review and editing. K.B.: Methodology, Writing—review and editing. R.C.: Methodology, Writing—review and editing. M.M.: Methodology, Writing—review and editing. R.R.: Writing—review and editing. A.-L.S.: Methodology, Writing—review and editing. O.M.: Methodology, Writing—review and editing. All authors have read and agreed to the published version of the manuscript.

Funding: The Polish National Science Centre financed this study under project No. 2019/32/C/NZ8/00147. The funder had no role in the study design, data collection and analysis, publication decisions, or manuscript preparation.

Data Availability Statement: This article and its Supplementary Materials include most of the data generated or analyzed during this study. The authors will make the rest of the included data available upon reasonable request.

Acknowledgments: Małgorzata Pronin and Józef Szmeja are highly acknowledged for their help in manuscript improvement and their valuable comments. The authors thank the GISMO platform and its staff (Biogeosciences, University Bourgogne Franche-Comté, UMR CNRS 682) for their collaboration. We thank the Translmed Publishing Group (TPG), a proofreading and copyediting company, for helping to proofread and copyedit this manuscript. The authors also thank the Regional Director of Environmental Protection in Gdańsk for allowing them to collect protected species from the investigated lakes and natural reserves (decision no. RDOŚ-Gd-WZG.6400.92.2020.AB.2 and no. RDOŚ-Gd-WOC.6205.24.2020. MaK.2). The Polish National Science Centre financed the studies under project No. 2019/32/C/NZ8/00147. Two anonymous peer reviewers and an Editor are kindly acknowledged for their comments and suggestions, which helped improve the manuscript.

Conflicts of Interest: The authors declare no conflicts of interest.

References

1. Brouwer, E.; Roelofs, J.G.M. Degraded Softwater Lakes: Possibilities for Restoration. *Restor. Ecol.* **2001**, *9*, 155–166. [CrossRef]
2. Ronowski, R.P.; Banaś, K.; Merdalski, M.; Szmeja, J. Plant Replacement Trend in Soft-Water Lakes with Isoetids. *Oceanol. Hydrobiol. Stud.* **2020**, *49*, 157–167. [CrossRef]
3. Szmeja, J. The Structure of a Population of *Lobelia dortmanna* L. along a Gradient of Increasing Depth in an Oligotrophic Lake. *Aquat. Bot.* **1987**, *28*, 1–13. [CrossRef]
4. Sand-Jensen, K.; Møller, C.L.; Borum, J. High Resistance of Oligotrophic Isoetid Plants to Oxic and Anoxic Dark Exposure. *Freshw. Biol.* **2015**, *60*, 1044–1051. [CrossRef]
5. Chmara, R.; Szmeja, J.; Banaś, K. The Relationships between Structural and Functional Diversity within and among Macrophyte Communities in Lakes. *J. Limnol.* **2018**, *77*, 100–108. [CrossRef]
6. Chmara, R.; Pronin, E.; Szmeja, J. Functional Macrophyte Trait Variation as a Response to the Source of Inorganic Carbon Acquisition. *PeerJ* **2021**, *9*, e12584. [CrossRef]
7. Jackson, T.A.; Kipphut, G.; Hesslein, R.H.; Schindler, D.W. Experimental Study of Trace Metal Chemistry in Soft-Water Lakes at Different PH Levels. *Can. J. Fish. Aquat. Sci.* **1980**, *37*, 387–402. [CrossRef]
8. Roelofs, J.G.M.; Schuurkes, J.A.A.R.; Smits, A.J.M. Impact of Acidification and Eutrophication on Macrophyte Communities in Soft Waters. II. Experimental Studies. *Aquat. Bot.* **1984**, *18*, 389–411. [CrossRef]
9. Raun, A.L.; Borum, J.; Sand-Jensen, K. Influence of Sediment Organic Enrichment and Water Alkalinity on Growth of Aquatic Isoetid and Elodeid Plants. *Freshw. Biol.* **2010**, *55*, 1891–1904. [CrossRef]
10. Chappuis, E.; Seriñá, V.; Martí, E.; Ballesteros, E.; Gacia, E. Decrypting Stable-Isotope (δ^{13}C and δ^{15}N) Variability in Aquatic Plants. *Freshw. Biol.* **2017**, *62*, 1807–1818. [CrossRef]
11. Osmond, C.B.; Valaane, N.; Haslam, S.M.; Uotila, P.; Roksandic, Z. Comparisons of Δ13C Values in Leaves of Aquatic Macrophytes from Different Habitats in Britain and Finland; Some Implications for Photosynthetic Processes in Aquatic Plants. *Oecologia* **1981**, *50*, 117–124. [CrossRef] [PubMed]
12. Richardson, K.; Griffiths, H.; Reed, M.L.; Raven, J.A.; Griffiths, N.M. Inorganic Carbon Assimilation in the Isoetids, *Isoetes lacustris* L. and *Lobelia dortmanna* L. *Oecologia* **1984**, *61*, 115–121. [CrossRef] [PubMed]
13. Pronin, E.; Banaś, K.; Chmara, R.; Ronowski, R.; Merdalski, M. Do Stable Carbon and Nitrogen Isotope Values of Nitella Flexilis Differ between Softwater and Hardwater Lakes? *Aquat. Sci.* **2023**, *85*, 79. [CrossRef]
14. Pronin, E. The Possible Use of Stable Carbon and Nitrogen Isotope Signal and Spectral Analysis to Identify Habitat Condition of Aquatic Plants. *Limnol. Rev.* **2024**, *24*, 17–29. [CrossRef]
15. Pronin, E.; Merdalski, M.; Ronowski, R.; Banaś, K. Variation of Carbon and Nitrogen Stable Isotope Composition in Leaves and Roots of *Littorella uniflora* (L.) Asch. in Relation to Water PH and Nutrient Availability. *Aquat. Bot.* **2025**, *196*, 103832. [CrossRef]
16. Pronin, E.; Banaś, K.; Chmara, R.; Ronowski, R.; Merdalski, M.; Santoni, A.-L.; Mathieu, O. Lobelia Lakes' Vegetation and Its Photosynthesis Pathways Concerning Water Parameters and the Stable Carbon Isotopic Composition of Plants' Organic Matter. *Plants* **2024**, *13*, 2529. [CrossRef]
17. King, L.; Maberly, S.C.; De Ville, M.M.; Kitschke, M.; Gibson, C.E.; Jones, R.I. Nitrogen Stable Isotope Ratios of Lake Macrophytes in Relation to Growth Form and Nutrient-Limitation. *Fundam. Appl. Limnol.* **2009**, *175*, 307–315. [CrossRef]
18. Smolders, A.J.P.; Lucassen, E.C.H.E.T.; Roelofs, J.G.M. The Isoetid Environment: Biogeochemistry and Threats. *Aquat. Bot.* **2002**, *73*, 325–350. [CrossRef]
19. Kragh, T.; Sand-Jensen, K. Carbon Limitation of Lake Productivity. *Proc. R. Soc. B Biol. Sci.* **2018**, *285*, 20181415. [CrossRef]
20. Jones, R.I.; King, L.; Dent, M.M.; Maberly, S.C.; Gibson, C.E. Nitrogen Stable Isotope Ratios in Surface Sediments, Epilithon and Macrophytes from Upland Lakes with Differing Nutrient Status. *Freshw. Biol.* **2004**, *49*, 382–391. [CrossRef]
21. Murphy, K.J. Plant Communities and Plant Diversity in Softwater Lakes of Northern Europe. *Aquat. Bot.* **2002**, *73*, 287–324. [CrossRef]

22. Cloern, J.E.; Canuel, E.A.; Harris, D. Stable Carbon and Nitrogen Isotope Composition of Aquatic and Terrestrial Plants of the San Francisco Bay Estuarine System. *Limnol. Oceanogr.* **2002**, *47*, 713–729. [CrossRef]
23. Leng, M.J.; Marshall, J.D. Palaeoclimate Interpretation of Stable Isotope Data from Lake Sediment Archives. *Quat. Sci. Rev.* **2004**, *23*, 811–831. [CrossRef]
24. Zhang, J.; Quay, P.D.; Wilbur, D.O. Carbon Isotope Fractionation during Gas-Water Exchange and Dissolution of CO_2. *Geochim. Cosmochim. Acta* **1995**, *59*, 107–114. [CrossRef]
25. Thomson, S.J.J.; Table, P.; Urey, H.C. Chapter 3 Stable Isotope Fractionation. *Elsevier Oceanogr. Ser.* **2001**, *65*, 141–250. [CrossRef]
26. Shmit, F.A.; Walker, N.A. Photosynthesis by Aquatic Plants: Effects of Unstirred Layers in Relation to Assimilation of CO_2 and HCO_3^- and to Carbon Isotopic Discrimination. *New Phytol.* **1980**, *86*, 245–259. [CrossRef]
27. Havig, J.R.; Hamilton, T.L.; McCormick, M.; McClure, B.; Sowers, T.; Wegter, B.; Kump, L.R. Water Column and Sediment Stable Carbon Isotope Biogeochemistry of Permanently Redox-Stratified Fayetteville Green Lake, New York, U.S.A. *Limnol. Oceanogr.* **2018**, *63*, 570–587. [CrossRef]
28. Schenk, J.; Sawakuchi, H.O.; Sieczko, A.K.; Pajala, G.; Rudberg, D.; Hagberg, E.; Fors, K.; Laudon, H.; Karlsson, J.; Bastviken, D. Methane in Lakes: Variability in Stable Carbon Isotopic Composition and the Potential Importance of Groundwater Input. *Front. Earth Sci.* **2021**, *9*, 722215. [CrossRef]
29. Baar, J.; Roelofs, J.G.M. Distribution of Plant Species in Relation to PH of Soil and Water. In *Handbook of Plant Growth pH as the Master Variable*; Zdenko Rengel, Ed.; Marcel Dekker, Inc.: New York, NY, USA, 2003; Volume 43, pp. 1575–1576, ISBN 9780367447076.
30. Pulido, C.; Sand-Jensen, K.; Lucassen, E.C.H.E.T.; Roelofs, J.G.M.; Brodersen, K.P.; Pedersen, O. Improved Prediction of Vegetation Composition in NW European Softwater Lakes by Combining Location, Water and Sediment Chemistry. *Aquat. Sci.* **2012**, *74*, 351–360. [CrossRef]
31. Klimaszyk, P.; Borowiak, D.; Piotrowicz, R.; Rosińska, J.; Szeląg-Wasielewska, E.; Kraska, M. The Effect of Human Impact on the Water Quality and Biocoenoses of the Soft Water Lake with Isoetids: Jeleń Lake, NW Poland. *Water* **2020**, *12*, 945. [CrossRef]
32. Kolada, A.; Piotrowicz, R.; Wilk-Woźniak, E.; Dynowski, P.; Klimaszyk, P. Conservation Status of the Natura 2000 Habitat 3110 in Poland: Monitoring, Classification and Trends. *Limnol. Rev.* **2017**, *17*, 215–222. [CrossRef]
33. Čtvrtlíková, M.; Kopáček, J.; Nedoma, J.; Znachor, P.; Hekera, P.; Vrba, J. Aquatic Quillworts, Isoëtes Echinospora and I. Lacustris under Acidic Stress—A Review from a Temperate Refuge. *Ecol. Evol.* **2023**, *13*, e9878. [CrossRef]
34. Barko, J.W.; Gunnison, D.; Carpenter, S.R. Sediment Interactions with Submersed Macrophyte Growth and Community Dynamics. *Aquat. Bot.* **1991**, *41*, 41–65. [CrossRef]
35. Peipoch, M.; Gacia, E.; Blesa, A.; Ribot, M.; Riera, J.L.; Martí, E. Contrasts among Macrophyte Riparian Species in Their Use of Stream Water Nitrate and Ammonium: Insights from 15N Natural Abundance. *Aquat. Sci.* **2014**, *76*, 203–215. [CrossRef]
36. Vrzel, J.; Vuković-Gačić, B.; Kolarević, S.; Gačić, Z.; Kračun-Kolarević, M.; Kostić, J.; Aborgiba, M.; Farnleitner, A.; Reischer, G.; Linke, R.; et al. Determination of the Sources of Nitrate and the Microbiological Sources of Pollution in the Sava River Basin. *Sci. Total Environ.* **2016**, *573*, 1460–1471. [CrossRef]
37. Bartrons, M.; Camarero, L.; Catalan, J. Isotopic Composition of Dissolved Inorganic Nitrogen in High Mountain Lakes: Variation with Altitude in the Pyrenees. *Biogeosciences* **2010**, *7*, 1469–1479. [CrossRef]
38. Heredia, C.; Guédron, S.; Point, D.; Perrot, V.; Campillo, S.; Verin, C.; Espinoza, M.E.; Fernandez, P.; Duwig, C.; Achá, D. Anthropogenic Eutrophication of Lake Titicaca (Bolivia) Revealed by Carbon and Nitrogen Stable Isotopes Fingerprinting. *Sci. Total Environ.* **2022**, *845*, 157286. [CrossRef] [PubMed]
39. Guo, H.R.; Wu, Y.; Hu, C.C.; Liu, X.Y. Elevated Nitrate Preference Over Ammonium in Aquatic Plants by Nitrogen Loadings in a City River. *J. Geophys. Res. Biogeosci.* **2022**, *127*, e2021JG006614. [CrossRef]
40. Jensen, T.C.; Zawiska, I.; Oksman, M.; Słowiński, M.; Woszczyk, M.; Luoto, T.P.; Tylmann, W.; Nevalainen, L.; Obremska, M.; Schartau, A.K.; et al. Historical Human Impact on Productivity and Biodiversity in a Subalpine Oligotrophic Lake in Scandinavia. *J. Paleolimnol.* **2020**, *63*, 1–20. [CrossRef]
41. Zawiska, I.; Dimante-Deimantovica, I.; Luoto, T.P.; Rzodkiewicz, M.; Saarni, S.; Stivrins, N.; Tylmann, W.; Lanka, A.; Robeznieks, M.; Jilbert, T. Long-Term Consequences of Water Pumping on the Ecosystem Functioning of Lake Sekšu, Latvia. *Water* **2020**, *12*, 1459. [CrossRef]
42. Meyers, P.A.; Ishiwatari, R. Lacustrine Organic Geochemistry—An Overview of Indicators of Organic Matter Sources and Diagenesis in Lake Sediments. *Org. Geochem.* **1993**, *20*, 867–900. [CrossRef]
43. Douglas, P.M.J.; Stratigopoulos, E.; Park, S.; Keenan, B. Spatial Differentiation of Sediment Organic Matter Isotopic Composition and Inferred Sources in a Temperate Forest Lake Catchment. *Chem. Geol.* **2022**, *603*, 120887. [CrossRef]
44. Duan, L.; Zhang, H.; Chang, F.; Li, D.; Liu, Q.; Zhang, X.; Liu, F.; Zhang, Y. Isotopic Constraints on Sources of Organic Matter in Surface Sediments from Two North–South Oriented Lakes of the Yunnan Plateau, Southwest China. *J. Soils Sediments* **2022**, *22*, 1597–1608. [CrossRef]
45. Guo, Q.; Wang, C.; Wei, R.; Zhu, G.; Cui, M.; Okolic, C.P. Qualitative and Quantitative Analysis of Source for Organic Carbon and Nitrogen in Sediments of Rivers and Lakes Based on Stable Isotopes. *Ecotoxicol. Environ. Saf.* **2020**, *195*, 110436. [CrossRef] [PubMed]
46. Thompson, H.A.; White, J.R.; Pratt, L.M. Spatial Variation in Stable Isotopic Composition of Organic Matter of Macrophytes and Sediments from a Small Arctic Lake in West Greenland. *Arctic. Antarct. Alp. Res.* **2018**, *50*, S100017. [CrossRef]

47. Gu, B.; Chapman, A.D.; Schelske, C.L. Factors Controlling Seasonal Variations in Stable Isotope Composition of Particulate Organic Matter in a Soft Water Eutrophic Lake. *Limnol. Oceanogr.* **2006**, *51*, 2837–2848. [CrossRef]
48. Veselý, L.; Ercoli, F.; Ruokonen, T.J.; Bláha, M.; Duras, J.; Haubrock, P.J.; Kainz, M.; Hämäläinen, H.; Buřič, M.; Kouba, A. Strong Temporal Variation of Consumer Δ13C Value in an Oligotrophic Reservoir Is Related to Water Level Fluctuation. *Sci. Rep.* **2023**, *13*, 3642. [CrossRef]
49. Wang, G.; Feng, X.; Han, J.; Zhou, L.; Tan, W.; Su, F. Paleovegetation Reconstruction Using Δ13C of Soil Organic Matter. *Biogeosciences* **2008**, *5*, 1325–1337. [CrossRef]
50. Herzschuh, U.; Mischke, S.; Meyer, H.; Plessen, B.; Zhang, C. Lake Nutrient Variability Inferred from Elemental (C, N, S) and Isotopic (Δ13C, Δ15N) Analyses of Aquatic Plant Macrofossils. *Quat. Sci. Rev.* **2010**, *29*, 2161–2172. [CrossRef]
51. Merdalski, M.; Banaś, K.; Ronowski, R. Environmental Factors Affecting Pondweeds in Water Bodies of Northwest Poland. *Biodivers. Res. Conserv.* **2019**, *56*, 13–28. [CrossRef]
52. R Core Team. *R: A Language and Environment for Statistical Computing*; R Foundation for Statistical Computing: Vienna, Austria, 2022. Available online: http://www.R-project.org/ (accessed on 10 May 2022).
53. Dinno, A. Package 'Dunn.Test'. *CRAN Repos.* **2017**, *10*, 1–7.
54. Wickham, H. *Ggplot2*; Springer: New York, NY, USA, 2009; ISBN 978-0-387-98140-6.
55. Lê, S.; Josse, J.; Husson, F. FactoMineR: An R Package for Multivariate Analysis. *J. Stat. Softw.* **2008**, *25*, 253–258. [CrossRef]
56. Kassambara, A.; Mundt, F. Factoextra: Extract and Visualize the Results of Multivariate Data Analyses. R Package Version 1.0.7. 2020. Available online: https://cran.r-project.org/web/packages/factoextra/index.html (accessed on 10 May 2022).
57. Oksanen, J.; Blanchet, F.G.; Friendly, M.; Kindt, R.; Legendre, P.; Mcglinn, D.; Minchin, P.R.; O'hara, R.B.; Simpson, G.L.; Solymos, P.; et al. Package "vegan" Title Community Ecology Package. *Community Ecol. Packag.* **2019**, *2*, 1–297.
58. Liu, J.; Xue, B.; Yao, S.; Cheng, L. Stable Carbon Isotope Signatures of Cellulose in Macrophytes Collected from Three Eutrophic Lakes in China. *Limnologica* **2020**, *81*, 125758. [CrossRef]
59. Jia, J.; Jiang, H.; Zhu, X.; Wang, S.; Wang, L.; Liu, C.; Li, W.; Huang, W. Inorganic Carbon Utilization Strategies of Plateau Aquatic Plants in Response to Native Habitats. *Photosynth. Res.* **2024**, *162*, 47–62. [CrossRef]
60. Pronin, E.; Pełechaty, M.; Apolinarska, K.; Pukacz, A.; Frankowski, M. Sharp Differences in the δ^{13}C Values of Organic Matter and Carbonate Encrustations but Not in Ambient Water DIC between Two Morphologically Distinct Charophytes. *Hydrobiologia* **2016**, *773*, 177–191. [CrossRef]
61. Pronin, E.; Panettieri, M.; Torn, K.; Rumpel, C. Stable Carbon Isotopic Composition of Dissolved Inorganic Carbon (DIC) as a Driving Factor of Aquatic Plants Organic Matter Build-up Related to Salinity. *Ecol. Indic.* **2019**, *99*, 230–239. [CrossRef]
62. Mendonça, R.; Kosten, S.; Lacerot, G.; Mazzeo, N.; Roland, F.; Ometto, J.P.; Paz, E.A.; Bove, C.P.; Bueno, N.C.; Gomes, J.H.C.; et al. Bimodality in Stable Isotope Composition Facilitates the Tracing of Carbon Transfer from Macrophytes to Higher Trophic Levels. *Hydrobiologia* **2013**, *710*, 205–218. [CrossRef]
63. Apolinarska, K.; Pełechaty, M.; Pronin, E. Discrepancies between the Stable Isotope Compositions of Water, Macrophyte Carbonates and Organics, and Mollusc Shells in the Littoral Zone of a Charophyte-Dominated Lake (Lake Lednica, Poland). *Hydrobiologia* **2016**, *768*, 1–17. [CrossRef]
64. Liu, H.; Liu, J.; Hu, J.; Cao, Y.; Xiao, S.; Liu, W. Systematical δ 13 C Investigations of TOC in Aquatic Plants, DIC and Dissolved CO 2 in Lake Water from Three Tibetan Plateau Lakes. *Ecol. Indic.* **2022**, *140*, 109060. [CrossRef]
65. Rodrigo, M.A.; García, A.; Chivas, A.R. Carbon Stable Isotope Composition of Charophyte Organic Matter in a Small and Shallow Spanish Water Body as a Baseline for Future Trophic Studies. *J. Limnol.* **2016**, *75*, 226–235. [CrossRef]
66. Myrbo, A.; Shapley, M.D. Seasonal Water-Column Dynamics of Dissolved Inorganic Carbon Stable Isotopic Compositions (Δ13CDIC) in Small Hardwater Lakes in Minnesota and Montana. *Geochim. Cosmochim. Acta* **2006**, *70*, 2699–2714. [CrossRef]
67. Mechora, Š.; Kanduč, T. Environmental Assessment of Freshwater Ecosystems of the Sava River Watershed and Cerkniško Lake, Slovenia, Using the Bioindicator Species Fontinalis Antipyretica: Insights from Stable Isotopes and Selected Elements. *Isotopes Environ. Health Stud.* **2016**, *52*, 239–257. [CrossRef]
68. Nissenbaum, A.; Presley, B.J.; Kaplan, I.R. Early Diagenesis in a Reducing Fjord, Saanich Inlet, British Columbia-I. Chemical and Isotopic Changes in Major Components of Interstitial Water. *Geochim. Cosmochim. Acta* **1972**, *36*, 1007–1027. [CrossRef]
69. Keeley, J.E.; Sternberg, L.O.; Deniro, M.J. The Use of Stable Isotopes in the Study of Photosynthesis in Freshwater Plants. *Aquat. Bot.* **1986**, *26*, 213–223. [CrossRef]
70. Gälman, V.; Rydberg, J.; Bigler, C. Decadal Diagenetic Effects on Δ13C and Δ15N Studied in Varved Lake Sediment. *Limnol. Oceanogr.* **2009**, *54*, 917–924. [CrossRef]
71. Kolada, A.; Kutyła, S. *Elodea canadensis* (Michx.) in Polish Lakes: A Non-Aggressive Addition to Native Flora. *Biol. Invasions* **2016**, *18*, 3251–3264. [CrossRef]
72. Zhang, P.; Kong, X.; Bakker, E.S.; Xu, J.; Zhang, M. Temperature Affects Carbon and Nitrogen Stable Isotopic Signatures of Aquatic Plants. *Aquat. Sci.* **2021**, *83*, 39. [CrossRef]
73. Evans, R.D. Physiological Mechanisms Influencing Plant Nitrogen Isotope Composition. *Trends Plant Sci.* **2001**, *6*, 121–126. [CrossRef]
74. Pennock, J.R.; Velinsky, D.J.; Ludlam, J.M.; Sharp, J.H.; Fogel, M.L. Isotopic Fractionation of Ammonium and Nitrate during Uptake by Skeletonema Costatum: Implications for Δ15N Dynamics under Bloom Conditions. *Limnol. Oceanogr.* **1996**, *41*, 451–459. [CrossRef]

75. Li, L.; Yu, Z.; Moeller, R.E.; Bebout, G.E. Complex Trajectories of Aquatic and Terrestrial Ecosystem Shifts Caused by Multiple Human-Induced Environmental Stresses. *Geochim. Cosmochim. Acta* **2008**, *72*, 4338–4351. [CrossRef]
76. Kuliński, K.; Święta-Musznicka, J.; Staniszewski, A.; Pempkowiak, J.; Latałowa, M. Lignin Degradation Products as Palaeoenvironmental Proxies in the Sediments of Small Lakes. *J. Paleolimnol.* **2007**, *38*, 555–567. [CrossRef]

water

MDPI

Article

Simulating the Vegetation Gross Primary Productivity by the Biome-BGC Model in the Yellow River Basin of China

Lige Jia [1,2] and Bo Zhang [1,*]

1 College of Tourism, Inner Mongolia Normal University, Hohhot 010028, China; jialige@imnu.edu.cn
2 College of Geography and Environmental Science, Northwest Normal University, Lanzhou 730070, China
* Correspondence: zhangbo@nwnu.edu.cn

Abstract: In terrestrial ecosystems, the quantification of carbon absorption is primarily represented by the gross primary productivity (GPP), which signifies the initial substances and energy acquired by the ecosystem. The GPP also serves as the foundation for the carbon cycle within the entire terrestrial ecosystem. The Biome-BGC model is a widely used biogeochemical process model for simulating the stocks and fluxes of water, carbon, and nitrogen between ecosystems and the atmosphere. However, it is the abundance of eco-physiological parameters that lead to challenges in calibrating the model. The parameter optimization method of coupling the differential evolution algorithm (DE) with the Biome-BGC model was used to calibrate and validate the eco-physiological parameters of the seven typical vegetation types in the Yellow River Basin (YRB). And then we used the calibrated parameters to simulate the GPP by way of grid-based simulation. Finally, we conducted model adaptability testing and spatiotemporal analysis of GPP variations in the YRB. The results of the validation (R^2, RMSE) were: temperate grasses (0.94, 24.33 g C m^{-2}), alpine meadows (0.94, 18.13 g C m^{-2}), shrubs (0.94, 29.20 g C m^{-2}), evergreen needle leaf forests (0.96, 27.88 g C m^{-2}), deciduous broad leaf forests (0.94, 32.09 g C m^{-2}), one crop a year (0.96, 16.19 g C m^{-2}), and two crops a year (0.90, 38.15 g C m^{-2}). After adaptability testing, the average R^2 value between the simulated GPP values and the GPP product values in the YRB was 0.85, and the average RMSE value was as low as 50.92 g C m^{-2}. Overall, the model exhibited strong simulation accuracy. Therefore, after calibrating the model with the DE algorithm, the Biome-BGC model could effectively adapt to the ecologically complex YRB. Moreover, it was able to accurately estimate the GPP, which establishes a foundation for analyzing the spatiotemporal trends of the GPP in the YRB. This study provides a reference for optimizing Biome-BGC model parameters and simulating diverse vegetation types on a large scale.

Keywords: terrestrial ecosystems; gross primary productivity; Yellow River Basin; Biome-BGC model; differential evolution algorithm

1. Introduction

Hunam activities, including the burning of fossil fuels, deforestation, and logging, have led to an increase in carbon dioxide emissions and raised atmospheric carbon dioxide levels since the Industrial Revolution, which further enhance the global temperature [1]. The atmospheric concentration of carbon dioxide has risen from 278 ppm (1 ppm = 1 μ L L^{-1}) before industrialization to the current level of 415 ppm [2]. In order to cope with climate change, China announced its national goal to reach the peak of carbon emissions by 2030 and achieve carbon neutrality by 2060 [3].

As a general system of interactions between organisms and the environment, terrestrial ecosystems can achieve interrelated connections, interactions, and mechanisms of internal self-regulation through energy flow and matter cycling [4]. Faced with global climate change, terrestrial ecosystems serve as the largest carbon sink for atmospheric carbon dioxide, offsetting more than a quarter of fossil fuel emissions [5]. This has a significant mitigating effect on global warming [6]. Therefore, terrestrial ecosystems play an important role in the "carbon neutrality" strategy that provides a cost-effective carbon sequestration pathway in the short term to reduce pressure on emissions. [1]. The Gross Primary Productivity (GPP) is the quantity of organic carbon fixed by terrestrial plants through photosynthesis per unit area within a specific time period. It constitutes the largest carbon flux between the terrestrial biosphere and the atmosphere, having a significant influence on global carbon cycling and the climate system. The GPP drives various ecosystem functions, such as respiration and growth, and holds a pivotal role in maintaining the ecological balance [7]. The primary object for quantifying carbon flux within terrestrial ecosystems is the GPP, which serves as the initial material and energy acquisition for terrestrial ecosystems. Meanwhile, the GPP forms the foundation of the carbon cycle in terrestrial ecosystems [8]. There is no doubt that the GPP reflects the capacity of vegetation to absorb and fix atmospheric carbon dioxide. Not only is the accurate comprehension of the spatiotemporal variability of the GPP within terrestrial ecosystems crucial for the global carbon cycle and forecasting the impacts of future climate change, but the GPP serves as a vital indicator for precisely assessing the capacity of terrestrial ecosystems to support sustainable human societal development [9].

At present, the methods of estimating a terrestrial ecosystem's GPP are mainly divided into eddy covariance techniques and model simulations. Among these, model simulations include statistical models, light use efficiency models, and process-based models. The statistical models, including the Miami model, SIF correlation model, NDVI correlation model, and so on, establish a correlation equation between the vegetation GPP and meteorological factors or other elements and then utilize the statistical relationship to estimate the GPP [10,11]. The light use efficiency (LUE) models estimate the GPP by taking solar radiation as the main driving variable. Next, meteorological factors, such as temperature factors and precipitation factors, are used to adjust LUE models. For instance, models like the CASA model, PEM model, VPM model, and so forth [12,13]. The process models are constructed based on the physiological processes of vegetation, taking into account the characteristics and eco-physiological parameters of different ecosystems comprehensively. They can accurately simulate the carbon, water, and nitrogen fluxes of different vegetation types under various environmental conditions. Currently, widely applied process models include the Biome-BGC model, BEPS model, TEM model, Century model, etc. [14,15] Among them, Biome-BGC (BioGeoChemical Cycles) is a process-based biogeochemical model that can estimate the water and carbon fluxes of ecosystems [16]. It has the advantages of strong flexibility and portability and performs better in explaining mechanisms [17]. Therefore, the Biome-BGC model is widely used for simulating ecosystem GPP due to its detailed description of vegetation ecology and physiological processes.

However, there are 43 parameters in the eco-physiological constant file of the Biome-BGC model. It is the numerous parameters that result in great uncertainty during the simulation process [18]. Therefore, parameter calibration and validation are essential for accurate simulation in Biome-BGC. In the most current research, eco-physiological parameters are often partially substituted with empirical or measured values and optimized to target levels using the manual adjustment method. However, these approaches are not only inefficient but also lack reliability [19,20]. While some studies have attempted to optimize parameters by coupling PEST with Biome-BGC [21], there are problems such as complex operation, easily obtaining the local optima, and low efficiency. How to efficiently optimize the eco-physiological parameters of Biome-BGC for different vegetation types is a primary challenge when using the model. For regional-scale application of the Biome-BGC model, researchers commonly employed eddy covariance flux data to calibrate model parameters. Subsequently, they conducted flux simulations for the region [22,23]. Although the Biome-BGC model performs well at the single-point level, its simulation performance at the regional scale, where it faces diverse environmental conditions, remains uncertain when using the same parameters. In previous studies at the regional scale, the majority have only focused their subject on a single vegetation type in the study area, such as studies of forests [24], crops [25], and grasses [26]. For the study of flux simulation in a large-scale region, if there is no comparison among different vegetation types, it cannot show the difference among them.

The Yellow River Basin constitutes a crucial ecological barrier and a significant economic zone in China. However, the fragile ecosystems have caused many questions about the YRB. It is of great significance to strengthen protection of the eco-environment and promote high-quality development in the YRB. Consequently, we firstly used the DE algorithm to calibrate the parameters of different vegetation ecosystems, and then we simulated the GPP of representative vegetation ecosystems within the gridded YRB from 2000 to 2021. Lastly, this study tested the model's adaptability and analyzed the spatiotemporal trends of the GPP for various typical vegetation types in the YRB. Two main questions were solved through this article. (1) Can the Biome-BGC model be applied to the Yellow River Basin with complex vegetation types by DE parameter optimization? (2) Over the past 22 years, what are the spatiotemporal changes in the GPP for the typical vegetations in the Yellow River Basin under the constant vegetation coverage data?

2. Materials and Methods

2.1. Study Area

The Yellow River, originating on the Qinghai-Tibet Plateau, is the second largest river in China, with a total length of approximately 5464 km [27], The Yellow River Basin refers to the geographical region, which is influenced by the Yellow River from its source to its mouth, with a total area of 795,000 km^2 (Figure 1a) [28]. The topography of the YRB fluctuates greatly (Figure 1b). In the western source area, the elevation is the highest, averaging over 4000 m. It is characterized by perennial snow cover and a glacier-formed topography. The central region exhibits an average elevation ranging from 200 to 3000 m, the loess landscape of which suffers from serious soil erosion issues. The eastern region is primarily composed of the Yellow River alluvial plain [29]. The climate within the YRB has diverse types, including arid, semi-arid, and semi-humid. The upper reaches exhibit an average surface temperature range of 1–8 °C, while the ranges of the middle and lower reaches are 8–14 °C and 12–14 °C, respectively. The multi-year average precipitation is 466 mm [30].

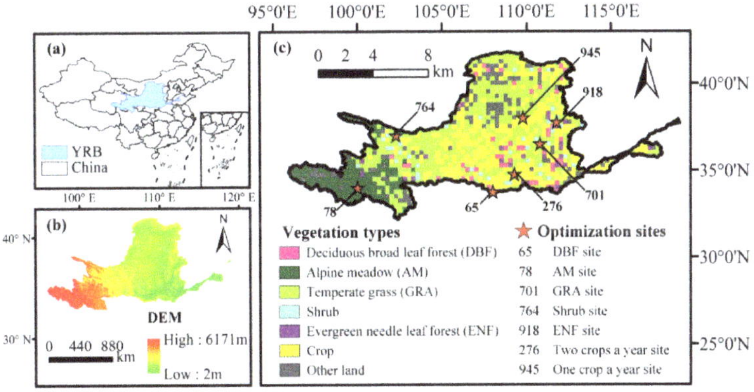

Figure 1. The Yellow River Basin's location, elevation, and parameter optimization sites. Figure (**a**) shows the Yellow River Basin and its location; Figure (**b**) shows the elevation information of the Yellow River Basin; Figure (**c**) reflects the distribution information of multiple vegetation types and the location of model parameter optimization sites in the Yellow River Basin.

The Yellow River Basin serves as an ecological corridor that connects the Qinghai-Tibet Plateau, Loess Plateau, and North China Plain, having numerous ecological functional zones [31]. In this study, the YRB was divided into 1277 grid cells, each measuring 0.25° × 0.25° in size. The vegetation grids (1138 grids) were primarily classified into seven vegetation communities: temperate grasses (GRA), alpine meadows (AM), shrubs, evergreen needle leaf forests (ENF), deciduous broad leaf forests (DBF), one crop a year (Crop1), and two crops a year (Crop2) [32]. In this study, Biome-BGC parameter optimization sites were selected for seven representative vegetation types, allowing for the calibration and validation of the model parameters (Figure 1c). The latitude and longitude of the seven points are: 65 (108°02′ E, 33°47′ N), 78 (100°15′ E, 34°02′ N), 701 (110°46′ E, 36°32′ N), 764 (102°13′ E, 37°02′ N), 918 (111°45′ E, 37°47′ N), 276 (109°17′ E, 34°45′ N), and 945 (109°46′ E, 38°03′ N).

2.2. Dataset

2.2.1. Meteorological Data

The gridded meteorological data used in this study were sourced from the CN05 gridded observational dataset. The gridded daily dataset, with the resolution of 0.25° latitude by 0.25° longitude, was based on the interpolation from over 2400 observing stations in China and included 4 variables: daily mean, minimum and maximum temperatures, and daily precipitation [33]. However, the meteorological data required for the Biome-BGC model includes the daily maximum temperature, daily minimum temperature, mean temperature, daily precipitation, saturation vapor pressure deficit, solar radiation, and day length. So, we needed to use the MTCLIM model to expand the basic meteorological data.

The MTCLIM model, a mountain microclimate simulation model, is based on the topographic conditions and climate theory and uses the basic observation data of nearby sites to estimate the local daily near-surface meteorological parameters. It is designed to handle the special problems encountered in mountainous terrain, but it is applicable in flat terrain as well [34]. Therefore, this study used the MTCLIM model to expand the weather data based on the latitude, elevation, slope, and aspect data of each grid.

2.2.2. Elevation Data

This study utilized the PIE-Engine online editing tool to acquire SRTM 250 m resolution DEM data (https://engine.piesat.cn/dataset-list, accessed on 3 January 2023). The SRTM (Shuttle Radar Topography Mission) data were from the processed radar images that were measured by NASA and NIMA. It provided elevation data for the Biome-BGC model and the MTCLIM model.

2.2.3. Soil Data

The soil data required for this study included the volumetric percentage content of sand, silt, and clay particles, as well as the effective soil depth. All of these data were sourced from the Harmonized World Soil Database version (HWSD (Version 1.2)) [35].

2.2.4. Vegetation Type Data

The vegetation type data used in this study were from a 1 km plant functional type map of China from the National Cryosphere Desert Data Center (https://www.ncdc.ac.cn/, accessed on 3 January 2023). Chinese vegetation coverage data were derived from the land cover map, MICLCover. On the basis of MICLCover, Ran and Ma [32] converted the data to obtain the vegetation type cover data according to the climate rule of land cover and plant function type conversion.

2.2.5. GPP Product Data

Due to the lack of enough eddy covariance data, this study regarded the VPM GPP dataset as the observed GPP data for the YRB. The VPM GPP dataset contains global GPP data with a spatial resolution of 500 m and a temporal resolution of month from 2000 to 2016. In view of an improved light use efficiency model, the GPP dataset was simulated with the input data, including satellite data from MODIS and climate data from NCEP Reanalysis II. In addition, the state-of-the-art vegetation index (VI) gap-filling and smoothing algorithm as well as separate treatment for C3/C4 photosynthesis pathways were used for this product. Finally, through the verification of GPP estimations, it provides a product with excellent performance for regional and global carbon cycle research [36].

Our study selected 2009–2011 GPP product data for model calibration and 2012–2014 GPP product data for model validation.

2.3. *Methods*

2.3.1. Biome-BGC Model

The Biome-BGC model is a typical terrestrial ecosystem process model that comprehensively accounts for differences in ecological characteristics and parameters. It accurately describes the cycles of carbon, nitrogen, and water, as well as energy flow for the different vegetation types [16]. The Biome-BGC model was jointly developed and maintained by the National Center for Atmospheric Research (NCAR) and the Numerical Terradynamic Simulation Group (NTSG) at the University of Montana. It originated from the Forest-BGC model initially [37]. Currently, the model is stable at version 4.2. It can simulate the state of ecosystem carbon and water fluxes on a daily scale from 1 m^2 to a global scale, with given geographical parameters, meteorological conditions. and specific vegetation eco-physiological parameters [18]. The Biome-BGC model has been widely used to simulate and analyze the fluxes of different ecosystems, such as forests, grasses, and crops [18].

The Biome-BGC model's structure is mainly composed of physical processes and eco-physiological processes. The physical processes primarily include solar radiation transmission and evapotranspiration, while the eco-physiological processes contain photosynthesis, respiration, and flux calculations. These processes are interconnected and mutually influence each other. The Biome-BGC model runs in two steps: (1) The first step is to carry out a spin-up model, starting from low levels of carbon and nitrogen until the system reaches equilibrium, which provides the initial state of the simulation; (2) Then, a normal model is conducted to simulate the GPP flux at each grid during the study period.

2.3.2. Differential Evolution Algorithm

The differential evolution (DE) algorithm comes from a genetic annealing algorithm, a simple and powerful global optimization algorithm, which was introduced by Storn and Price [38]. In comparison with the traditional genetic algorithm, the differential evolution algorithm possesses several primary advantages: (1) it is user-friendly and exhibits efficient memory utilization; (2) it has lower computational complexity, particularly in the context of scaling to handle large-scale problems, making it suitable for optimizing the abundant parameters of the Biome-BGC model; and (3) it requires a lower computational workload, resulting in faster convergence [39].

In this paper, the differential evolution optimization function in the Deoptim package of R language was used to implement the algorithm. The detailed steps are as follows: (1) the main function can replace the Biome-BGC model parameters and run the model in batch; and (2) an optimization objective function is established based on the method of least squares.

$$ObjeValue = \sum_{i=1}^{36} (Obs_i - Sim_i)^2 \tag{1}$$

where ObjeValue is an optimal target value between simulated and observed values for the period 2009–2011; Obsi is the monthly GPP observation from 2009 to 2011; and Simi is the monthly GPP simulation from 2009 to 2011. We calculated the value of objective function through Equation (1).

(3) Finally, we set the parameters' initial values, upper limits, and lower limits. The upper and lower limits of the parameters in this study were determined through a literature search [17,18,40–43]. For example, population number (NP) $\geq 10 \times$ number of parameters to optimize, setting the maximum number of iterations to 100, etc. The DE-optim function was employed to globally search for optimal parameters. Meanwhile, the determination

coefficient (R^2) and root mean square error (RMSE) were used as evaluation indicators for the model validation process.

2.3.3. Simulation of GPP and Regional Adaptability Test

In this study, a point-by-point approach was used to simulate the GPP flux of 1138 typical vegetation grids from 2000 to 2021. And then the R^2 and RMSE values were calculated between the monthly simulated GPP values and GPP "observations" for each grid from 2012 to 2014. Through the statistical analysis, we tested the adaptive capacity of the Biome-BGC model at the regional scale.

2.3.4. Analysis of Spatiotemporal Change Trend of GPP

After completing the work of calibration, validation, and the adaptability test, we could determine the simulation accuracy of the Biome-BGC model in the YRB. Based on the GPP simulation data from 2000 to 2021 in the YRB, this study analyzed the GPP's spatiotemporal change trend and investigated its regular variations.

3. Results

3.1. The Optimized Parameters of Typical Vegetation Types

The optimized parameters of temperate grasses, alpine meadows, shrubs, evergreen needle leaf forests, deciduous broad leaf forests, one crop a year, and two crops a year are shown in the Table 1. Most parameters required calibration through the coupled differential evolution algorithm with the Biome-BGC model, while a few parameters used the model's default values.

Table 1. The parameter table for temperate grasses (GRA), alpine meadows (AM), shrubs, evergreen needle leaf forests (ENF), deciduous broad leaf forests (DBF), one crop a year (Crop1), wheat (Crop2-W), and maize (Crop2-M). * represents that this type of vegetation does not involve this parameter. The corresponding parameters are not working.

Parameter	GRA	AM	Shrub	ENF	DBF	Crop1	Crop2-W	Crop2-M	Units
Year day to start new growth	130	120	*	*	*	138	60	180	day
Year day to end litterfall	280	275	*	*	*	283	180	300	day
Transfer growth period as fraction of growing season	0.9912	0.6201	0.6217	*	0.0782	0.5021	0.6623	0.7976	/
Litterfall as fraction of growing season	0.2810	0.2959	0.6207	*	0.8964	0.5514	0.1985	0.2541	/
Annual leaf and fine root turnover fraction	1	1	0.8871	0.3023	1	0.4	1.0	0.1657	$1\ a^{-1}$
Annual live wood turnover fraction	0	0	0.7	0.2528	0.7	0	0	0	$1\ a^{-1}$
Annual whole-plant mortality fraction	0.01	0.06	0.005	0.0001	0.005	0.005	0.005	0.01	$1\ a^{-1}$
Annual fire mortality fraction	0.005	0.05	0.005	0.0001	0.0025	0.005	0.005	0.01	$1\ a^{-1}$
New fine root C: new leaf C	2.1644	1.6664	5.1482	5.6219	2.1620	0.1	1.9773	1.1619	ratio
New stem C: new leaf C	0	0	0.7497	0.5200	3.0301	0	0	0	ratio
New live wood C: new total wood C	0	0	0.2664	0.0867	0.1	0	0	0	ratio
New root C: new stem C	0	0	0.4017	0.3031	0.23	0	0	0	ratio
Current growth proportion	0.6744	0.7684	0.2370	0.5355	0.1865	0.05	0.1059	0.5937	/

Table 1. *Cont.*

Parameter	GRA	AM	Shrub	ENF	DBF	Crop1	Crop2-W	Crop2-M	Units
C:N of leaves	20.32	30.95	35.29	28.00	13.27	26.50	45.00	64.62	Kg C kg^{-1} N
C:N of leaf litter, after retranslocation	58.59	65.64	116.23	92.00	43.33	65.50	95.00	88.37	Kg C kg^{-1} N
C:N of fine roots	74.15	72.63	63.21	100.00	43.22	80.00	120.00	120.64	Kg C kg^{-1} N
C:N of live wood	0	0	89.37	80.00	50.00	0	0	0	Kg C kg^{-1} N
C:N of dead wood	0	0	1003.42	729.00	442.00	0	0	0	Kg C kg^{-1} N
Leaf litter labile proportion	0.39	0.39	0.38	0.32	0.39	0.37	0.68	0.68	DIM
Leaf litter cellulose proportion	0.44	0.44	0.45	0.44	0.44	0.46	0.23	0.23	DIM
Leaf litter lignin proportion	0.17	0.17	0.17	0.24	0.17	0.17	0.09	0.09	DIM
Fine root labile proportion	0.30	0.30	0.30	0.30	0.3	0.30	0.34	0.34	DIM
Fine root cellulose proportion	0.45	0.45	0.45	0.45	0.45	0.45	0.44	0.44	DIM
Fine root lignin proportion	0.25	0.25	0.25	0.25	0.25	0.25	0.22	0.22	DIM
Dead wood cellulose proportion	0.75	0.75	0.76	0.76	0.76	0.75	0.76	0.76	DIM
Dead wood lignin proportion	0.25	0.25	0.24	0.24	0.24	0.25	0.24	0.24	DIM
Canopy water interception coefficient	0.0180	0.0227	0.0354	0.0107	0.0164	0.001	0.0124	0.0772	1 $LAI^{-1}d^{-1}$
Canopy light extinction coefficient	0.4533	0.7359	0.5720	0.4744	0.6868	0.32	0.5431	0.4468	DIM
All-sided to projected leaf area ratio	2	2	2.3	3.4250	1.0815	0.325	2.0	2.0	DIM
Canopy average specific leaf area	30.4374	12.9271	35.9618	20.9445	37.6114	26.9	32.1638	25.8928	m^2 kg^{-1} C
Ratio of shaded SLA: sunlit SLA	2	2	2	2	2.3988	3.5	2.0	2.0	DIM
Fraction of leaf N in Rubisco	0.2637	0.2505	0.5210	0.3035	0.0845	0.34	0.2956	0.1937	DIM
Maximum stomatal conductance	0.0026	0.0057	0.0018	0.0010	0.0024	0.0081	0.0032	0.0043	m s^{-1}
Cuticular conductance	3.6×10^{-5}	7.5×10^{-5}	7.2×10^{-5}	6×10^{-7}	1.4×10^{-5}	1.1×10^{-4}	8.6×10^{-5}	7.2×10^{-5}	m s^{-1}
Boundary layer conductance	0.0758	0.0987	0.0752	0.1083	0.0265	0.048	0.0896	0.0370	m s^{-1}
Leaf water potential: start of conductance reduction	−0.73	−0.73	−0.6	−0.6	−0.6	−0.6	−0.73	−0.73	Mpa
Leaf water potential: complete conductance reduction	−2.7	−2.7	−2.3	−2.7	−2.3	−2.3	−2.7	−2.7	Mpa
Vapor pressure deficit: start of conductance reduction	953.71	712.63	1184.60	980.00	930.00	930.00	930.00	930.0	pa
Vapor pressure deficit: complete conductance reduction	3064.11	5647.09	2796.75	5000.00	4100.00	4100.00	4000.00	4000.0	pa

3.2. Calibration and Validation of the Model

This study selected seven research sites where the typical vegetation types were densely distributed (Figure 1c; 918, 78, 64, 701, 764, 945, and 276 grids). Because these seven research sites corresponded to the most densely distributed vegetation types and had the best validation data quality, they were selected. Based on the Biome-BGC-DE parameter optimization method, the uncertain parameters of the typical vegetation types in the YRB were optimized. Notably, wheat and maize used site number 276 together.

The curve graphs of the monthly simulated and observed GPP values for the typical vegetation types in the YRB (Figure 2) can be summarized into the following results: After parameter calibration, the GPP simulation curve of each typical vegetation type was very close to the GPP observation points, which not only reflected the GPP change trend but also accurately simulated the change in monthly GPP values. The parameter optimization with the differential evolution algorithm resulted in strong modeling performance for the various vegetation types.

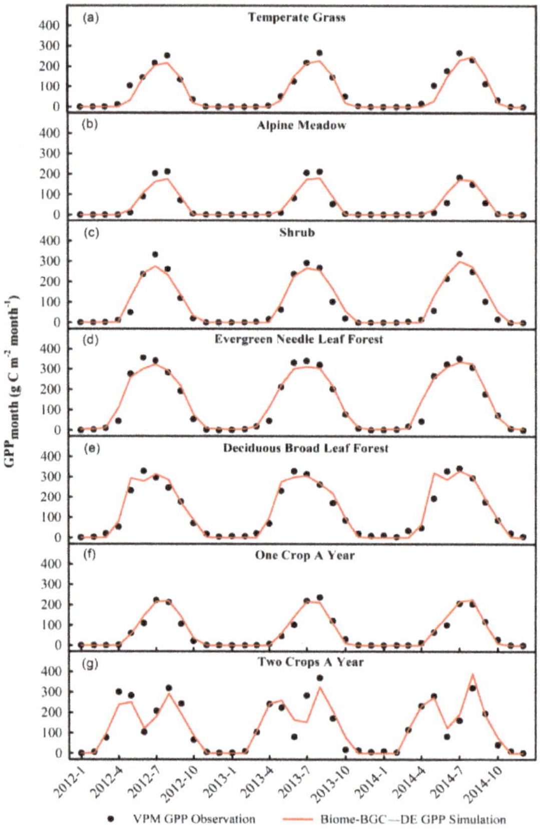

Figure 2. The curve graphs of monthly simulated and observed GPP values for typical vegetation types in the YRB.

The scatter plots of the monthly simulated and observed GPP values for the typical vegetation types in the YRB (Figure 3) can be summarized into the following results: Through the parameter optimization with the differential evolution algorithm, the R^2 values between the simulated GPP and the observed GPP at the seven optimal sites were above 0.9, and the RMSE values were below 40 g C m^{-2}. One crop a year and alpine meadows had better simulation performance, of which the R^2 and RMSE values were 0.96 and 0.94 and 16.19 g C m^{-2} and 18.13 g C m^{-2}, respectively. For temperate grasses, shrubs, evergreen needle leaf forests, and deciduous broad leaf forests, the R^2 values were 0.94, 0.94, 0.96, and 0.94, respectively, and the RMSE values were 24.33 g C m^{-2}, 29.20 g C m^{-2}, 27.88 g C m^{-2}, and 32.09 g C m^{-2}, respectively. In this study, winter wheat and summer maize were simulated separately and then superposed due to the lack of crop rotation for the Biome-BGC model. Hence, the simulation performance of two crops a year was slightly poor, with an R^2 value of 0.90 and RMSE value of 38.15 g C m^{-2}.

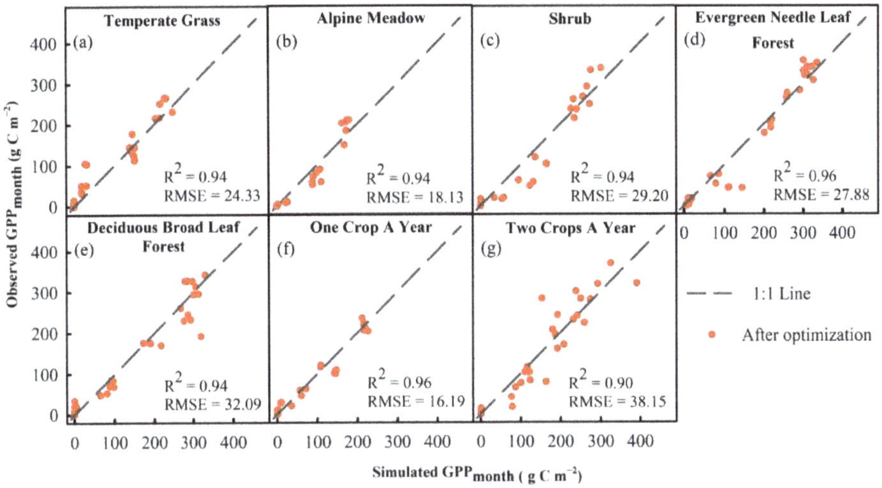

Figure 3. The scatter plots of monthly simulated and observed GPP values for typical vegetation types in the YRB.

3.3. Simulation and Regional Adaptability Test

After calibrating and validating the parameters of the typical vegetation types, this study simulated the monthly GPP from 2000 to 2021 at each grid with the meteorological, soil, initialization files, etc.

The adaptability of a model refers to its ability to cope with changes in the environment. Although Biome-BGC had high accuracy at the seven parameter optimization sites, the performance of the model at the regional scale was uncertain. The mismatch in spatial resolution among the input data of the model would lead to unavoidable errors, and the accumulation of these errors might weaken the simulation effectiveness at the regional scale. It was unknown how much the performance of the model would be reduced. Therefore, a further regional adaptability test was needed to judge the overall simulation performance of the optimized parameters of different typical vegetation types.

The optimized parameters exhibited great adaptability for the typical vegetation types in this study, as shown in Figure 4. For the temperate grass grids, the R^2 and RMSE values of the simulated results after parameter optimization were mostly centered (25–75%) in the range of 0.76–0.91 and 27–60 g C m^{-2}, respectively (Figure 4a). Similarly, for the alpine meadow grids, the R^2 and RMSE values were concentrated within 0.90–0.94 and 18–33 g C m^{-2}, respectively (Figure 4b). For the shrub grids, R^2: 0.84–0.94, RMSE: 34–66 g C m^{-2} (Figure 4c); for the evergreen needle leaf forests, R^2: 0.83–0.93, RMSE: 48–68 g C m^{-2} (Figure 4d); for the deciduous broadleaved forests, R2: 0.76–0.95, RMSE: 40–55 g C m^{-2} (Figure 4e); for one crop a year, R2: 0.83–0.93, RMSE: 44–77 g C m^{-2} (Figure 4f); for two crops a year, R2: 0.72–0.83, RMSE: 54–72 g C m^{-2} (Figure 4g). Overall, our results demonstrated the great simulation accuracy of the Biome-BGC model at the regional scale.

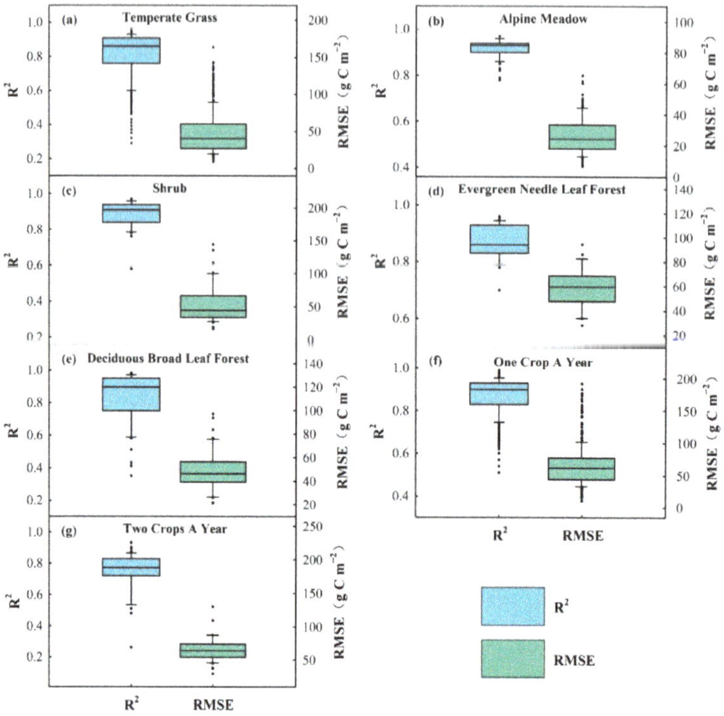

Figure 4. The R^2 and RMSE box plots of the GPP on a monthly scale simulated by the model (the quantiles are 90, 75, 50, 25, and 10 from top to bottom).

On the whole, simulating the GPP of the typical vegetation types with the optimized eco-physiological parameters, the average R^2 value between the simulated GPP values and observed GPP values in the Yellow River Basin from 2012 to 2014 was 0.85 and the average RMSE was 50.92 g C m^{-2}. Therefore, this model could well adapt to the Yellow River Basin (Table 2).

Table 2. Summary of average R^2 and RMSE values for monthly-scale GPP simulation of typical vegetation types in the YRB.

Vegetation Type	R^2	RMSE (g C m^{-2})
Temperate Grass	0.82	48.06
Alpine Meadow	0.92	26.57
Shrub	0.89	54.18
Evergreen Needle Leaf Forest	0.87	58.12
Deciduous Broad Leaf Forest	0.83	49.19
One Crop A Year	0.87	66.24
Two Crops A year	0.75	65.56
Yellow River Basin Vegetation	0.85	50.92

3.4. Analysis of Spatiotemporal Change Trend of GPP

The annual GPP variation in the YRB from 2000 to 2021 exhibited an obvious trend of fluctuating changes (Figure 5). The annual-scale GPP of evergreen needle leaf forests was the highest among all typical vegetation types, with a fluctuation range of 1596.64–1854.65 g C m^{-2} y^{-1} and an average of 1715.90 g C m^{-2} y^{-1} ($p < 0.01$); two crops a year (1432.33–1682.70 g C m^{-2} y^{-1}, 1541.31 g C m^{-2} y^{-1}); deciduous broad leaf forests (1107.36–1367.24 g C m^{-2} y^{-1}, 1267.77 g C m^{-2} y^{-1}); one crop a year (983.3912–1160.41 g C m^{-2} y^{-1}, 1068.85 g C m^{-2} y^{-1}); shrubs (777.03–974.29 g C m^{-2} y^{-1}, 869.34 g C m^{-2} y^{-1}); temperate grasses (575.12–729.48 g C m^{-2} y^{-1}, 658.10 g C m^{-2} y^{-1}); alpine meadows (410.36–541.69 g C m^{-2} y^{-1}, 471.80 g C m^{-2} y^{-1}); Yellow River Basin (681.61–785.51 g C m^{-2} y^{-1}, 741.90 g C m^{-2} y^{-1}).

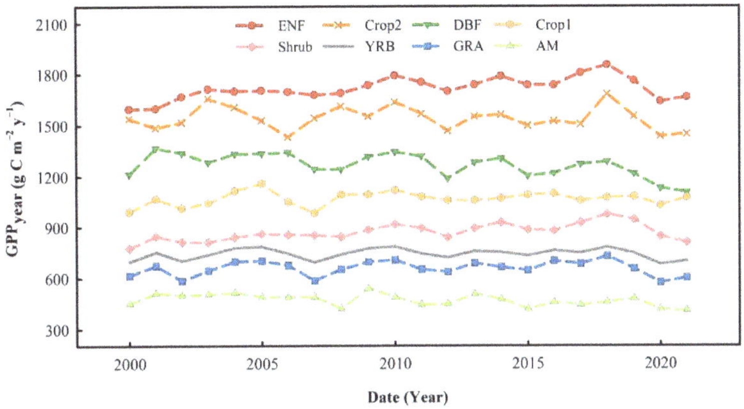

Figure 5. Curve graphs of GPP variation trends for the Yellow River Basin as a whole and different typical vegetation types during the current period (2000–2021).

The change rates of annual GPP variation for each typical vegetation type were obtained through one linear regression analysis and F significance testing: evergreen needle leaf forests (4.97 g C m^{-2} y^{-1}, $p < 0.05$); shrubs (4.24 g C m^{-2} y^{-1}, $p < 0.01$); one crop a year (1.11 g C m^{-2} y^{-1}, $p > 0.1$); temperate grasses (0.15 g C m^{-2} y^{-1}, $p > 0.1$); two crops a year (−1.80 g C m^{-2} y^{-1}, $p > 0.1$); alpine meadows (−3.18 g C m^{-2} y^{-1}, $p < 0.01$); deciduous broad leaf forests (−6.69 g C m^{-2} y^{-1}, $p < 0.01$); Yellow River Basin (−0.19 g C m^{-2} y^{-1}, $p > 0.1$).

The annual average GPP in the Yellow River Basin from 2000 to 2021 showed distinct spatial variations (Figure 6). In the western region of the YRB, the annual average GPP remained consistently very low throughout the year. In other regions of the YRB, there was a different distribution pattern of the annual average GPP. The southern areas exhibited higher values compared to the northern areas, while the eastern and western areas appeared to have relatively higher GPP values compared to the central region, which showed relatively lower GPP values. Due to the fact that the main vegetation types in the southeastern part of the Yellow River Basin were crops, temperate shrubs, and deciduous broad-leaved forests, while the northwestern part was mainly temperate meadows and other land uses (Figure 1c), the vegetation productivity in the southeastern part was significantly higher. From an east–west perspective, temperate grasslands and other land uses were mainly distributed in the central part of the Yellow River Basin (Figure 1c), so the GPP in the central part was slightly lower than that in the east–west part.

Through the distribution map of the annual GPP change rate from 2000 to 2021, we could find that (Figure 7a) the annual GPP in the YRB showed an overall increasing trend. The majority of the regions showed an upward trend, covering approximately 59.91% of the total area of the YRB. The other areas with a decreasing or unchanged trend in the annual average GPP occupied about 40.09% of the total area. But the significant area only accounted for 30.38% of the total area of the YRB (Figure 7a,b). In addition, it is important to note that only the annual GPP of forest vegetation, shrubs, and alpine meadows showed significant change trends. However, the annual GPP of temperate grass vegetation and crops showed non-significant trends.

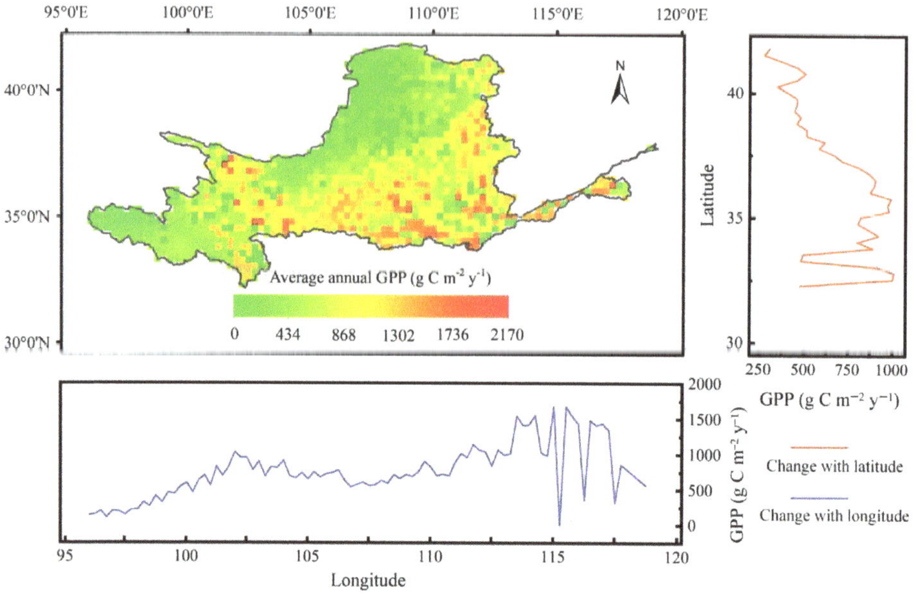

Figure 6. Spatial distribution map of annual average GPP in the Yellow River Basin during the current period (2000–2021).

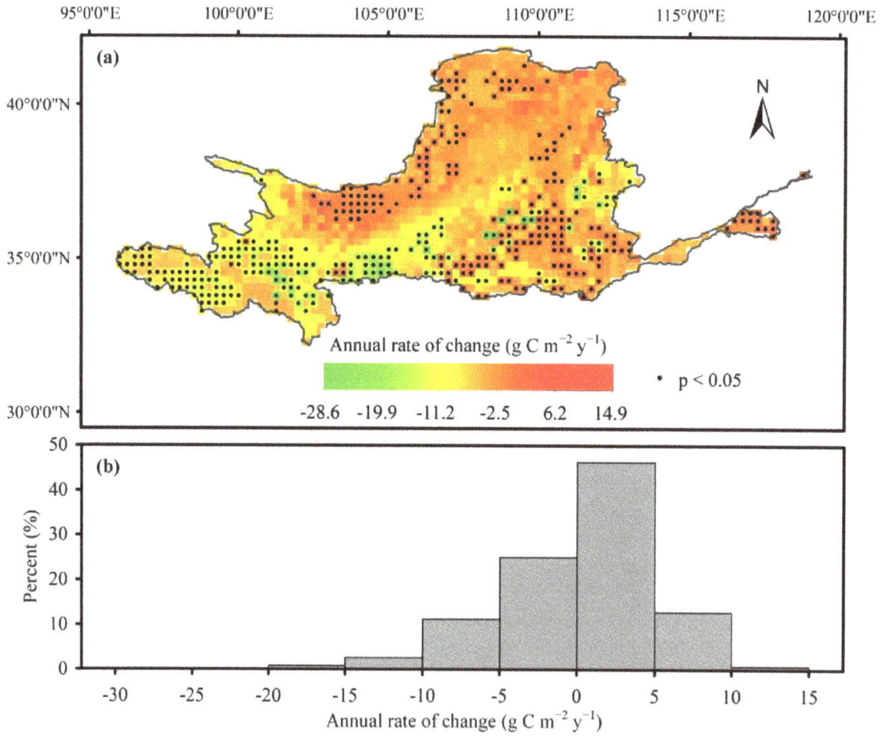

Figure 7. Spatial distribution map of annual GPP change rate in the Yellow River Basin during the current period (2000–2021).

4. Discussion

4.1. The Effectiveness of the Differential Evolution Optimization Method

It is difficult to optimize numerous eco-physiological parameters because of the strong correlations among them. Previous studies usually used a manual tuning method to calibrate the model based on EC flux data [23]. When there are so many parameters and vegetation types, it is not suitable to use the manual method. So, some scholars have used the PEST and Biome-BGC coupling method to optimize the sensitive parameters of the model [44]. But PEST has the disadvantages of easily falling into local optima and low efficiency of para38meter optimization. If we used the DE and Biome-BGC coupling method to calibrate the model, we were able to globally optimize over 20 parameters with a fast convergence rate at the same time.

We could determine that the optimization effect of Biome-BGC-DE was very great through the validation results of this study. The simulated GPP values of each typical vegetation type were close to the observed GPP values in trend and size. For two crops a year, previous studies usually simulated wheat and maize separately because of the lack of a crop rotation mechanism in the Biome-BGC model [43]. By contrast, this study adopted a method of separately simulating wheat and maize and then overlaying the simulation results. This study achieved an accurate simulation for two crops a year by optimizing the parameters of wheat and maize with DE ($R^2 = 0.90$, RMSE = 38.15 g C m^{-2}). In conclusion, if we have enough EC flux data and combine with the DE parameter optimization method, we can simulate the actual water and carbon fluxes of typical vegetation more accurately.

4.2. Adaptability of the Model

At the regional scale, the Biome-BGC model is widely used to study the variations in water and carbon fluxes within individual vegetation ecosystems [17,45,46]. Most studies usually conducted research on the carbon flux after optimizing the model parameters using data of a single EC site. And their research object most only included one kind of vegetation. Therefore, there was a lack of exploration into the following two questions: What is the model adaptability of optimized parameters for the same typical vegetation type in different environments? What are the differences in model adaptability among different typical vegetation types?

The GPP product data, as the observation data for this study, provided great conditions for the adaptability test. The average R^2 values for temperate grasses, alpine meadows, shrubs, evergreen needle leaf forests, deciduous broad leaf forests, one crop a year, and two crops a year were all above 0.8 (two crops a year was 0.75). The average RMSE values ranged from 26.57 to 66.24 g C m^{-2}. Moreover, there were relatively few outliers among the simulation of all typical vegetation types (R^2: Percentile < 10%, RMSE: Percentile > 90%). The overall average R^2 value for the Yellow River Basin was 0.85, with an average RMSE value of 50.92 g C m^{-2}. From the results of the adaptability test it was clear that the Biome-BGC model could achieve stable and accurate simulation of the carbon flux in complex environments across large regions after great calibration and validation. And, surely, Biome-BGC had an excellent adaptability test.

The model adaptability varied significantly between different typical vegetation types, and the reasons are as follows: (1) To carry out regional simulation, this study divided the YRB into grids with a resolution of 0.25°. This would unavoidably produce errors due to the scaling-up of vegetation coverage data. As a result, this led to the confusion of vegetation coverage information in areas with complex vegetation type distributions, whereas regions with uniform vegetation distribution were less affected. Alpine meadows are primarily distributed in the eastern part of the Tibetan Plateau, characterized by a single type of vegetation. Therefore, the adaptability test of the alpine meadow ecosystem presented the best results. On the other hand, other vegetation types are affected to some extent by errors in upscaling the data. (2) Some temperate grass and crop areas need human management, such as mowing grass, crop rotation, and so on,.

The Biome-BGC model can only reflect the influence of climatic factors and cannot account for human activities. Despite achieving good simulations for crop rotation in this study, there will still be some errors in simulating temperate grass and crops. In summary, following great parameter calibration and validation, the Biome-BGC model can be used for long-term carbon flux simulations in large-scale study areas.

4.3. *Analysis of the Results of GPP Simulations of Typical Vegetation Types*

The interannual GPP spatiotemporal trends of change for typical vegetation types in the YRB were in the following order: evergreen needle leaf forests > two crops a year > deciduous broad leaf forests > one crop a year > shrubs > temperate grasses > alpine meadows. Forest vegetation generally exhibited a higher annual average GPP, having a stronger capacity for absorbing atmospheric CO_2. The GPP values of crop vegetation were mostly larger than those of shrubs and alpine meadows. And the GPP values of two crops a year were only smaller than those of evergreen needle leaf forests. The main reason is the cultivation of maize (a C4 crop) during the 180 to 300-day period. Compared to C3 plants, C4 plants (using the Hatch and Slack pathway) exhibit higher photosynthetic rates and greater water, nitrogen, and radiation use efficiency (WUE, NUE, and RUE, respectively) [47,48]. Furthermore, C4 plants are better adapted to warm and arid climates than C3 plants. [49]. As a result, the annual average GPP of two crops a year was so high. The annual average GPP of shrubs was slightly higher than that of grassland vegetation. Due to climatic conditions, alpine meadows had consistently low GPP levels over multiple years. The vegetation in the Yellow River Basin was predominantly composed of temperate grasses, one crop a year, and alpine meadows, resulting in a relatively low average GPP for the entire region.

By comparing the annual average simulated GPP of various vegetation types in this study with the results of previous research (Table 3), it could be found that the simulated values of temperate grass in this study were slightly higher than the annual average GPP of this vegetation type obtained by [50] based on Flux-net data. Ref. [51] indicated that the annual average GPP of the Tibetan Plateau was 312.3 g Cm^{-2} y^{-1}. The spatial distribution pattern of the GPP on the Tibetan Plateau showed a gradual increase from west to east. Therefore, the annual average value of alpine meadows in the Yellow River Basin should be more than 312.3 g Cm^{-2} y^{-1}. For other vegetation types in this study, such as shrubs, evergreen needle leaf forests, and deciduous broad leaf forests, the simulated annual average GPP values generally fell within the range of results from previous studies. The multi-year observational data from Flux-net sites on one crop a year and two crops a year in the study by [50] closely agreed with the simulated annual average GPP values of this study. By combining the separately simulated annual average GPP values for wheat and maize from Wang et al. [43], it could be observed that their results also closely aligned with the results of this study. In summary, this paper argues that the simulated GPP values for various representative vegetation types from 2000 to 2021 in this study were reasonably accurate. The calibrated and validated vegetation parameters could effectively simulate the GPP of multiple typical vegetation types in the Yellow River Basin.

Table 3. The comparison of results between this study and previous studies.

Vegetation Type	Multi-Year Average GPP (g C m^{-2} y^{-1}) Simulation of This Study	Multi-Year Average GPP (g C m^{-2} y^{-1}) Simulation of Other Studies	Reference
Temperate Grass	658.10	542.0	[50]
Alpine Meadow	471.80	312.3 461.6 712.0	[51] [52] [53]
Shrub	869.34	744.0–867.0	[54]

Table 3. *Cont.*

Vegetation Type	Multi-Year Average GPP (g C m^{-2} y^{-1}) Simulation of This Study	Simulation of Other Studies	Reference
Evergreen Needle Leaf Forest	1715.90	1325.0 1319.0–1924.0	[45] [50]
Deciduous Broad Leaf Forest	1267.77	1355.0 1122.0–1507.0	[45] [50]
One Crop A year	1068.85	1101–1396	[50]
Two Crops A year	1541.31	1471 1604	[50] [43]

4.4. The Change of GPP Under the Influence of Natural Factors

Due to the lack of consideration for human activities, such as land use changes, mowing grass, and so on, the Biome-BGC model can only reflect the influence of natural factors on the water and carbon fluxes in the study area. In terms of the research results, we found that the annual GPP showed a change trend of fluctuation in the YRB, without the influence of human activities. Even the annual GPP values of the deciduous broad leaf forests and alpine meadows showed a significant decreasing trend.

Therefore, to cope with climate change, we must take positive measures to control these adverse trends and ensure that the GPP in the Yellow River Basin shows a growth trajectory. To ensure ecological integrity and promote high-quality development in the Yellow River Basin, we need to hold back the declining trends in certain regions and enhance the carbon sequestration capacity of terrestrial ecosystems.

5. Conclusions

(1) The parameter optimization method of coupling Biome-BGC with the differential evolution (DE) algorithm was used for model parameter calibration and validation. Through validation and the adaptability test, it was demonstrated that the Biome-BGC model had great adaptability and high model accuracy in simulating the gross primary productivity of typical vegetation types in the Yellow River Basin. This model can be applied to carbon flux simulation studies for historical periods in the Yellow River Basin.

(2) From 2000 to 2021, the average gross primary productivity in the Yellow River Basin for the past 22 years was 741.90 g C m^{-2} y^{-1}, with a range of 681.61 g C m^{-2} y^{-1}–785.51 g C m^{-2} y^{-1}. On the whole, the change trend of the GPP was not significant, showing a fluctuating trend. The annual GPP of evergreen needle leaf forests and shrubs showed a significant increasing trend, while the annual GPP of deciduous broad leaf forests and alpine meadows exhibited a significant decreasing trend. On the other hand, the annual GPP variations for crops and temperate grasses were not significant. The annual change in the GPP in the Yellow River Basin showed mainly a single peak change pattern, while only the GPP of one crop a year showed a double peak change pattern.

(3) From 2000 to 2021, the spatial distribution of the annual average GPP in the Yellow River Basin had the following characteristics: The annual average GPP of the Yellow River Basin had a great spatial difference because of different vegetation coverage. The annual average GPP values in the westernmost region were comparatively the lowest among the other areas. Beyond this region, the distribution of the annual average GPP presented a distinct pattern with higher values in the southern parts, lower values in the northern parts, higher values in the eastern and western parts, and lower values in the central area. The GPP in most regions of the Yellow River Basin showed an increasing trend, accounting for approximately 59.91% of the total area.

Author Contributions: Conceptualization, L.J. and B.Z.; methodology, B.Z.; software, L.J.; investigation, L.J. and B.Z.; writing—original draft preparation, L.J.; writing—review and editing, B.Z.; visualization, L.J. and B.Z.; funding acquisition, B.Z. All authors have read and agreed to the published version of the manuscript.

Funding: This research was funded by the Fundamental Research Funds for the Inner Mongolia Normal University (2022JBQN065).

Data Availability Statement: Data is contained within the article.

Acknowledgments: We are very grateful to the relevant programs for funding, and we are very grateful to the editors and reviewers for their constructive comments on this study.

Conflicts of Interest: The authors declare no conflicts of interest.

References

1. Piao, S.; Yue, C.; Ding, J.; Guo, Z. Perspectives on the role of terrestrial ecosystems in the 'carbon neutrality' strategy. *Sci. China Earth Sci.* **2022**, *65*, 1178–1186. [CrossRef]
2. World Meteorological Organization. *WMO Greenhouse Gas Bulletin No. 17: The State of Greenhouse Gases in the Atmosphere Based on Global Observations Through 2020*; World Meteorological Organization: Geneva, Switzerland, 2021.
3. Piao, S.; He, Y.; Wang, X.; Chen, F. Estimation of China's terrestrial ecosystem carbon sink: Methods, progress and prospects. *Sci. China Earth Sci.* **2022**, *65*, 641–651. [CrossRef]
4. Chapin, F.S.; Matson, P.A.; Vitousek, P.M. The Ecosystem Concept. In *Principles of Terrestrial Ecosystem Ecology*; Springer: New York, NY, USA, 2011; pp. 3–22.
5. Le Quéré, C.; Andrew, R.M.; Friedlingstein, P.; Sitch, S.; Hauck, J.; Pongratz, J.; Pickers, P.A.; Korsbakken, J.I.; Peters, G.P.; Canadell, J.G.; et al. Global Carbon Budget 2018. *Earth Syst. Sci. Data* **2018**, *10*, 2141–2194. [CrossRef]
6. Shevliakova, E.; Stouffer, R.J.; Malyshev, S.; Krasting, J.P.; Hurtt, G.C.; Pacala, S.W. Historical warming reduced due to enhanced land carbon uptake. *Proc. Natl. Acad. Sci. USA* **2013**, *110*, 16730–16735. [CrossRef] [PubMed]
7. Beer, C.; Reichstein, M.; Tomelleri, E.; Ciais, P.; Jung, M.; Carvalhais, N.; Rödenbeck, C.; Arain, M.A.; Baldocchi, D.; Bonan, G.B.; et al. Terrestrial Gross Carbon Dioxide Uptake: Global Distribution and Covariation with Climate. *Science* **2010**, *329*, 834–838. [CrossRef]
8. Chapin, F.S.; Matson, P.A.; Mooney, H.A.; Vitousek, P.M. *Principles of Terrestrial Ecosystem Ecology*; Springer: Berlin/Heidelberg, Germany, 2002.
9. Verma, M.; Friedl, M.A.; Richardson, A.D.; Kiely, G.; Cescatti, A.; Law, B.E.; Wohlfahrt, G.; Gielen, B.; Roupsard, O.; Moors, E.J.; et al. Remote sensing of annual terrestrial gross primary productivity from MODIS: An assessment using the FLUXNET La Thuile data set. *Biogeosciences* **2014**, *11*, 2185–2200. [CrossRef]
10. Bai, J.; Zhang, H.; Sun, R.; Li, X.; Xiao, J.; Wang, Y. Estimation of global GPP from GOME-2 and OCO-2 SIF by considering the dynamic variations of GPP-SIF relationship. *Agric. For. Meteorol.* **2022**, *326*, 109180. [CrossRef]
11. Liu, F.; Wang, C.; Wang, X. Can vegetation index track the interannual variation in gross primary production of temperate deciduous forests? *Ecol. Process.* **2021**, *10*, 51. [CrossRef]
12. Lv, Y.; Liu, J.; He, W.; Zhou, Y.; Nguyen, N.T.; Bi, W.; Wei, X.; Chen, H. How well do light-use efficiency models capture large-scale drought impacts on vegetation productivity compared with data-driven estimates? *Ecol. Indic.* **2023**, *146*, 109739. [CrossRef]
13. Ogutu, B.O.; Dash, J. Assessing the capacity of three production efficiency models in simulating gross carbon uptake across multiple biomes in conterminous USA. *Agric. For. Meteorol.* **2013**, *174–175*, 158–169. [CrossRef]
14. Law, B.; Thornton, P.; Irvine, J.; Anthoni, P.; Van Tuyl, S. Carbon storage and fluxes in ponderosa pine forests at different developmental stages. *Glob. Chang. Biol.* **2001**, *7*, 755–777. [CrossRef]
15. Liu, J.; Chen, J.M.; Cihlar, J.; Park, W.M. A process-based boreal ecosystem productivity simulator using remote sensing inputs. *Remote Sens. Environ.* **1997**, *62*, 158–175. [CrossRef]
16. Running, S.W.; Hunt, E.R. *8—Generalization of a Forest Ecosystem Process Model for Other Biomes, BIOME-BGC, and an Application for Global-Scale Models*; Academic Press: Cambridge, MA, USA, 1993.
17. Sun, Q.; Li, B.; Zhang, T.; Yuan, Y.; Gao, X.; Ge, J.; Li, F.; Zhang, Z. An improved Biome-BGC model for estimating net primary productivity of alpine meadow on the Qinghai-Tibet Plateau. *Ecol. Model.* **2017**, *350*, 55–68. [CrossRef]
18. White, M.A.; Thornton, P.E.; Running, S.W.; Nemani, R.R. Parameterization and Sensitivity Analysis of the BIOME–BGC Terrestrial Ecosystem Model: Net Primary Production Controls. *Earth Interact.* **2000**, *4*, 1–85. [CrossRef]
19. Bai, J.; Chen, X.; Li, L.; Luo, G.; Yu, Q. Quantifying the contributions of agricultural oasis expansion, management practices and climate change to net primary production and evapotranspiration in croplands in arid northwest China. *J. Arid. Environ.* **2013**, *100–101*, 31–41. [CrossRef]
20. Nunes, L.; Gower, S.T.; Peckham, S.D.; Magalhaes, M.; Lopes, D.; Rego, F.C. Estimation of productivity in pine and oak forests in northern Portugal using Biome-BGC. *For. Int. J. For. Res.* **2014**, *88*, 200–212. [CrossRef]
21. Liu, J.; Wu, Z.; Yang, S.; Yang, C. Sensitivity Analysis of Biome-BGC for Gross Primary Production of a Rubber Plantation Ecosystem: A Case Study of Hainan Island, China. *Int. J. Environ. Res. Public Health* **2022**, *19*, 14068. [CrossRef]

22. Schmid, S.; Zierl, B.; Bugmann, H. Analyzing the carbon dynamics of central European forests: Comparison of Biome-BGC simulations with measurements. *Reg. Environ. Chang.* **2006**, *6*, 167–180. [CrossRef]
23. Srinet, R.; Nandy, S.; Patel, N.; Padalia, H.; Watham, T.; Singh, S.K.; Chauhan, P. Simulation of forest carbon fluxes by integrating remote sensing data into biome-BGC model. *Ecol. Model.* **2023**, *475*, 110185. [CrossRef]
24. Running, S.W.; Coughlan, J.C. A general model of forest ecosystem processes for regional applications I. Hydrologic balance, canopy gas exchange and primary production processes. *Ecol. Model.* **1988**, *42*, 125–154. [CrossRef]
25. Ma, S.; Churkina, G.; Wieland, R.; Gessler, A. Optimization and evaluation of the ANTHRO-BGC model for winter crops in Europe. *Ecol. Model.* **2011**, *222*, 3662–3679. [CrossRef]
26. Keller, E.D.; Baisden, W.T.; Timar, L.; Mullan, B.; Clark, A. Grassland production under global change scenarios for New Zealand pastoral agri-culture. *Geosci. Model. Dev.* **2014**, *7*, 2359–2391. [CrossRef]
27. Xue, S.; Jian, H.; Yang, F.; Liu, Q.; Yao, Q. Impact of water-sediment regulation on the concentration and transport of dissolved heavy metals in the middle and lower reaches of the Yellow River. *Sci. Total Environ.* **2022**, *806*, 150535. [CrossRef]
28. Zhang, X.; Liu, K.; Wang, S.; Wu, T.; Li, X.; Wang, J.; Wang, D.; Zhu, H.; Tan, C.; Ji, Y. Spatiotemporal evolution of ecological vulnerability in the Yellow River Basin under ecological restoration initiatives. *Ecol. Indic.* **2022**, *135*, 108586. [CrossRef]
29. Liu, C.; Zhang, X.; Wang, T.; Chen, G.; Zhu, K.; Wang, Q.; Wang, J. Detection of vegetation coverage changes in the Yellow River Basin from 2003 to 2020. *Ecol. Indic.* **2022**, *138*, 108818. [CrossRef]
30. Zhang, Q.; Xu, C.-Y.; Chen, Y.D.; Ren, L. Comparison of evapotranspiration variations between the Yellow River and Pearl River basin, China. *Stoch. Environ. Res. Risk Assess.* **2011**, *25*, 139–150. [CrossRef]
31. Liu, S.; Shao, Q.; Ning, J.; Niu, L.; Zhang, X.; Liu, G.; Huang, H. Remote-Sensing-Based Assessment of the Ecological Restoration Degree and Restoration Potential of Ecosystems in the Upper Yellow River over the Past 20 Years. *Remote Sens.* **2022**, *14*, 3550. [CrossRef]
32. Ran, Y.; Ma, H. 1 km Plant Functional Types Map over China in 2000. *Remote Sens. Technol. Appl.* **2016**, *31*, 827–832.
33. Wu, J.; Gao, X.-J. A gridded daily observation dataset over China region and comparison with the other datasets. *Chin. J. Geophys.* **2013**, *56*, 1102–1111.
34. Hungerford, R.D.; Nemani, R.R.; Running, S.W.; Coughlan, J.C. *MTCLIM: A Mountain Microclimate Simulation Model*; U.S. Department of Agriculture, Forest Service, Intermountain Forest and Range Experiment Station: Washington, DC, USA, 1989.
35. FAO/IIASA/ISRIC/ISSCAS/JRC. *Harmonized World Soil Database (Version 1.2)*; FAO: Rome, Italy; IIASA: Laxenburg, Austria, 2012.
36. Zhang, Y.; Xiao, X.; Wu, X.; Zhou, S.; Zhang, G.; Qin, Y.; Dong, J. A global moderate resolution dataset of gross primary production of vegetation for 2000–2016. *Sci. Data* **2017**, *4*, 170165. [CrossRef]
37. Running, S.W.; Gower, S.T. FOREST-BGC, A general model of forest ecosystem processes for regional applications. II. Dynamic carbon allocation and nitrogen budgets. *Tree Physiol.* **1991**, *9*, 147–160. [CrossRef] [PubMed]
38. Storn, R.; Price, K. Differential evolution—A simple and efficient heuristic for global optimization over continuous spaces. *J. Glob. Optim.* **1997**, *11*, 341–359. [CrossRef]
39. Slowik, A. Application of an Adaptive Differential Evolution Algorithm with Multiple Trial Vectors to Artificial Neural Network Training. *IEEE Trans. Ind. Electron.* **2010**, *58*, 3160–3167. [CrossRef]
40. Raj, R.; Hamm, N.A.; van der Tol, C.; Stein, A. Variance-based sensitivity analysis of BIOME-BGC for gross and net primary production. *Ecol. Model.* **2014**, *292*, 26–36. [CrossRef]
41. Hunt, N.D.; Gower, S.T.; Nadelhoffer, K.; Lajtha, K.; Townsend, K.; Brye, K.R. Validation of an agroecosystem process model (AGRO-BGC) on annual and per-ennial bioenergy feedstocks. *Ecol. Model.* **2016**, *321*, 23–34. [CrossRef]
42. Li, C.; Sun, H.; Wu, X.; Han, H. Dataset of the net primary production on the Qinghai-Tibetan Plateau using a soil water content improved Biome-BGC model. *Data Brief* **2019**, *27*, 104740. [CrossRef]
43. Wang, Q.; Watanabe, M.; Ouyang, Z. Simulation of water and carbon fluxes using BIOME-BGC model over crops in China. *Agric. For. Meteorol.* **2005**, *131*, 209–224. [CrossRef]
44. Wu, Z.; Li, Q.; Xu, S.; Chen, Q.; Chen, Y. Investigation into climate change effects on carbon and water fluxes, and water use efficiency of the temperate grassland ecosystems in Inner Mongolia of China. *Water Supply* **2023**, *23*, 2147–2161. [CrossRef]
45. Chiesi, M.; Maselli, F.; Moriondo, M.; Fibbi, L.; Bindi, M.; Running, S. Application of BIOME-BGC to simulate Mediterranean forest processes. *Ecol. Model.* **2007**, *206*, 179–190. [CrossRef]
46. Ueyama, M.; Ichii, K.; Hirata, R.; Takagi, K.; Asanuma, J.; Machimura, T.; Nakai, Y.; Ohta, T.; Saigusa, N.; Takahashi, Y.; et al. Simulating carbon and water cycles of larch forests in East Asia by the BIOME-BGC model with AsiaFlux data. *Biogeosciences* **2010**, *7*, 959–977. [CrossRef]
47. Hatch, M.D.; Slack, C. Photosynthesis by sugar-cane leaves. A new carboxylation reaction and the pathway of sugar formation. *Biochem. J.* **1966**, *101*, 103–111. [CrossRef]
48. Sheehy, J.E.; Mitchell, P.L.; Hardy, B. *Charting New Pathways to C4 Rice*; World Scientific: Singapore, 2008.
49. Schuler, M.L.; Mantegazza, O.; Weber, A.P. Engineering C4 photosynthesis into C3 chassis in the synthetic biology age. *Plant J.* **2016**, *87*, 51–65. [CrossRef] [PubMed]
50. Falge, E.; Baldocchi, D.; Tenhunen, J.; Aubinet, M.; Bakwin, P.; Berbigier, P.; Bernhofer, C.; Burba, G.; Clement, R.; Davis, K.J.; et al. Seasonality of ecosystem respiration and gross primary production as derived from FLUXNET measurements. *Agric. For. Meteorol.* **2002**, *113*, 53–74. [CrossRef]

51. He, H.; Liu, M.; Xiao, X.; Ren, X.; Zhang, L.; Sun, X.; Yang, Y.; Li, Y.; Zhao, L.; Shi, P.; et al. Large-scale estimation and uncertainty analysis of gross primary production in Tibetan alpine grasslands. *J. Geophys. Res. Biogeosci.* **2014**, *119*, 466–486. [CrossRef]
52. You, Y.; Wang, S.; Ma, Y.; Wang, X.; Liu, W. Improved Modeling of Gross Primary Productivity of Alpine Grasslands on the Tibetan Plateau Using the Biome-BGC Model. *Remote Sens.* **2019**, *11*, 1287. [CrossRef]
53. Zhuang, Q.; He, J.; Lu, Y.; Ji, L.; Xiao, J.; Luo, T. Carbon dynamics of terrestrial ecosystems on the Tibetan Plateau during the 20th century: An analysis with a process-based biogeochemical model. *Glob. Ecol. Biogeogr.* **2010**, *19*, 649–662. [CrossRef]
54. Hinojo-Hinojo, C.; Castellanos, A.E.; Huxman, T.; Rodriguez, J.C.; Vargas, R.; Romo-León, J.R.; Biederman, J.A. Native shrubland and managed buffelgrass savanna in drylands: Impli-cations for ecosystem carbon and water fluxes. *Agric. For. Meteorol.* **2019**, *268*, 269–278. [CrossRef]

water

MDPI

Article

Spatiotemporal Drought Analysis Using the Composite Drought Index (CDI) over Dobrogea, Romania

Cristina Serban [1] and Carmen Maftei [2,*]

1 Faculty of Mathematics and Informatics, Ovidius University of Constanta, 900527 Constanta, Romania; serban.cristina@365.univ-ovidius.ro
2 Civil Engineering Faculty, Transilvania University of Brasov, 500036 Brasov, Romania
* Correspondence: carmen.maftei@unitbv.ro

Abstract: This paper discusses a study that examined the severity of droughts and their changes in the Dobrogea region in southeastern Romania between 2001 and 2021 and develops a high-resolution (1 km) Composite Drought Index (CDI) dataset. To explore the effectiveness of the index, we carried out a correlation analysis between the CDI, the Standardized Precipitation Index (SPI), and the Standardized Precipitation Evapotranspiration Index (SPEI), which shows a strong positive relationship among these indices. Analysis of the CDI time series reveals an increase in drought frequency for the study period, due to high temperature and below-normal rainfall. Most parts of the region were affected by moderate, severe, or extreme droughts, except for the years 2002–2005 and 2013. The worst drought events were in 2011, 2012, and 2020, when the region was under severe land surface temperature stress, with values up to 39.13 °C. The central and northern areas of the region had the longest period of drought, at 22 months, which started in 2018 and culminated in 2020 when extreme drought covered over 70% of the region. Another major event was in 2015 when 95% of the region experienced severe drought. These results show the potential of the CDI as one of the significant indices in the assessment of drought and provide useful insights into drought monitoring in the future. More than that, we consider that the GPM IMERG satellite product can be used in the implementation of Drought Management Plans in Dobrogea in order to calculate drought indices derived from remote sensing data.

Keywords: CDI; CHIRPS; Dobrogea; drought; IMERG; LST; NDVI; SPI

1. Introduction

Drought is a prevalent natural calamity that has a substantial impact on human life, the environment, and agricultural productivity. According to an investigation launched by the World Meteorological Organization (WMO), the economic impact of drought in Europe over the past 30 years has been around EUR 30 to 100 billion [1–3]. The escalating frequency and intensity of global drought events, intensified by climate change, necessitates a re-evaluation of current drought management strategies. Facing such challenges and realizing the importance of prognosis for the upcoming period, the Communication "Addressing the challenge of water scarcity and droughts" was adopted by the European Commission in 2007 [4]. As part of the Water Blueprint, which was agreed earlier in November 2012, the European Commission was asked by the Environmental Council to evaluate and develop the "European policy concerning the water scarcity and drought", which ended in November 2012 [5,6]. The EU Commission suggested including drought risk management in future RBMPs (River Basin Management Plans) beginning in 2013, acknowledging the little progress made in putting the 2007 instrument into practice. In this regard, the Integrated Drought Management Programme (IDMP) was initiated in 2013 by the World Meteorological Organization (WMO) and the Global Water Partnership (GWP). An examination of drought in the Central and East European (CEE) region and its inclusion in the initial RBMPs revealed that the region under investigation had a very poor Drought Management Plan (DMP) development [7]. The fifth report on the implementation of the Water Framework Directive (WFD) and the Floods Directive was released by the EU Commission in 2019. Among the suggestions it made to the Member States was the third RBMP, which was scheduled for adoption by the end of 2021. It should better address climate change's effects on water availability, as well as make greater use of DMPs.

According to the United Nations Office for Disaster Risk Reduction (UNDRR) [8], drought as a natural hazard is a hydrometeorological hazard. From an operational point of view, in 1997, the American Meteorological Society [9] classified the definitions into four categories: hydrological, agricultural, meteorological, and socio-economic. For each of them, there are numerous indices. In 2023, more than 150 indices were used in drought analysis [10], and approximately half of them were investigated [11]. A manual of indices was realized by the WMO and GWP in 2016 [12]. The PDSI—Palmer Drought Severity Index introduced in [13] may be the easiest to use. Calculated using a water budget and on homogeneous soil regions, PDSI is recommended for use only in the USA [9]. At the same time, this index is not the best solution for use in arid and subhumid climates. To eliminate these disadvantages, Mc Kee introduced the SPI—Standardized Precipitation Index [14]. Using the Gamma distribution function applied to a long series of observed rainfall, followed by normalization, this index has the following advantages: it requires fewer data, helps determine the severity of drought, and offers early warning of it. However, this index does not take into consideration other climate parameters. To overcome this disadvantage, Vicente-Serrano proposed a new index, the SPEI—Standardized Precipitation Evapotranspiration Index [15], which followed the identical algorithm as the SPI, but added temperature and based its computation on the water deficit (difference between precipitation and potential evapotranspiration). Due to the large number of drought characterization indices and the generation of new ones, participants in the 2009 Inter-regional Workshop on Drought Early Warning Systems in Lincoln, Nebraska, USA, recognized that no single drought index fits every scenario. Additionally, they concluded that the SPI should be used to describe worldwide meteorological dryness [16].

Drought can be assessed through drought indices using meteorological stations or remote sensing (RS)-based datasets. With the advent of satellite imagery and RS, a new class of drought indices can be developed since the new technique provides real-time information, with precision from a few hundred meters to a few kilometers. The importance of RS technologies in real-time or near-real-time drought monitoring has been emphasized in recent decades of research, underlining their critical role in providing timely and accurate data for drought management [17]. RS-based drought indices can be classified into several groups. First, there are indices used for monitoring soil water content (e.g., Normalized Multiband Drought Index—NMDI); next, the indices specifically designed to capture changes in vegetation associated with drought stress (e.g., Normalized Difference Vegetation Index—NDVI, Normalized Difference Drought Index—NDDI, Leaf Area Index—LAI). The indices within the third group allow for estimating soil moisture using surface and crop temperature (e.g., Crop Water Stress Index—CWSI). Indicators based on the land surface's thermal stress are part of a different category (e.g., Temperature Condition Index—TCI). While no single indicator can definitively predict droughts, several factors and indicators can be considered together to assess the risk, especially since several types of droughts can occur simultaneously in an area. Therefore, several other indices were developed based on the NDVI-LST (Land Surface Temperature) relationship, which, in general, expresses a negative correlation (e.g., Vegetation Health Index—VHI, Drought Severity Index—DSI). The latest techniques used to develop remote sensing drought indices are due to advancements in microwave sensors, which allow for the assessment of global climate parameters as a substitute to ground measurements, such as rainfall (e.g., Climate Hazards Group InfraRed Precipitation with Station data—CHIRPS). Thus, NDVI, LST, and satellite rainfall data have been used to create a new generation of remote sensing indices (e.g., Scaled Drought Condition Index—SDCI). To provide an overall assessment of ecological drought conditions across various ecosystems, composite indices, like the Combined Drought Indicator (CDI) [18], have been developed by mixing various drought indices and data sources, such as precipitation, soil moisture, and vegetation indices. In its original solution, the CDI uses the SPI, Soil Moisture Anomaly (SMA), and fAPAR Anomaly ("fraction of incident solar radiation that is absorbed by land vegetation for photosynthesis"). The SPI is calculated based on daily or monthly precipitation registered to meteorological stations, SMA values are based on hydrological models, and fAPAR is a satellite-derived product. This index is used currently by Copernicus European Drought Observation (EDO) to estimate agricultural drought over Europe [18]. The spatial resolution is 5 km, and the temporal resolution is 10 days.

Romania is situated in the Central Southeastern part of the EU. The Romanian climate is temperate continental with different influences (e.g., the Atlantic in the western part and the Black Sea in the southeastern part of Romania). Mountains, hills, and plains are organized concentrically in tiers that make up our country's landscape. Approximately 30% of Romania's land area is considered desertified, having a climate that is arid, semiarid, or subhumid–dry [19]. The most impacted areas of Romania are in the south, southeast, and eastern parts, where the precipitation's multiannual rate is 600 mm or less (the multiannual precipitation in the Dobrogea region is approximately 400 mm) and the temperature is increasing [20]. In recent works, the authors note a temperature increase from 0.3 °C [20] to 0.8 °C [21], and even 1.7 °C after 1997 [22]. As we mentioned in the previous paragraphs, even if the EU Commission strongly recommends introducing the DMP in the RBMPs, Romania, like other countries in the Eastern European area, failed to achieve this. Therefore, in 2015, the GWP for Central and Eastern Europe (GWPCEE) published a set of guidelines for developing DMPs. Additional documents were also released, e.g., the Handbook of Drought Indicators and Indices [8]. The important sub-steps in developing a DMP are (i) quantifying the effects of drought and identifying and assessing drought conditions and (ii) establishing a set of indicators appropriate for different types of droughts. The most important sub-step is to set the best appropriate drought index. To achieve this, important studies must be carried out in different areas, and, in Romania, different indices were tested in different regions [2,23–27]. Most studies use the SPI or de Martonne as a drought index. Hydrological drought is investigated using a comparison between precipitation and river discharge [27], analysis of low flow characteristics [28], or using the Standardized Flow Index (similar to the SPI) [22]. In recent work [29], the authors tested the "Synthetic Drought Hydrograph" to detect parameters related to hydrological drought characteristics.

As shown, the SPI requires a minimum 30-year precipitation dataset, which is not usually available. Moreover, not all regions have enough meteorological stations to provide the necessary data. Dobrogea is in this situation, with only 10 meteorological stations managed by the National Meteorological Administration. To overcome this disadvantage, indices derived from satellite data can be used.

Remote sensing data were used in Romania for different purposes such as the investigation of forest changes [30,31], to determine heat islands in Bucharest and other regions [32], and to estimate the risk of desertification in the southeastern part of Romania [33]. Recently, Angearu et al. [34] examined the DSI (Drought Severity Index) in Baragan Plain. For the areas located in the southeast part of Europe (including Romania), few bibliographic resources demonstrate the importance of satellite data for drought research [35]. Using the SPEI and NDVI, the relationship between drought occurrences and vegetative stress in Romania and the Republic of Moldova from 1998 to 2014 was investigated in [36]. The study focused on the growing season (April to October) and on the severe drought event of 2000–2001. The study in [37] analyzed the drought at the country level (including Dobrogea) during the period 1901–2021, using the SPEI. The results indicated an increased frequency and duration of moderate, severe, and extreme drought events over the last 120 years. Using the NDDI and fAPAR, spanning the years 2000–2015, the authors of [38] analyzed the severity and degree of drought in Dobrogea. The results showed that the whole region experienced a type of drought during this period, which affected agricultural areas more than the semi-natural areas or forests.

This paper's major goal is to use high-resolution (1 km) and long-term (2001–2021) datasets to evaluate the performance of a Composite Drought Index (CDI) over the Dobrogea region. To compute the CDI, we used the algorithm proposed by Abdourahamane et al. [39]. The specific objectives are (i) to evaluate the performance of two global rainfall products (CHIRPS and GPM IMERG), used in the computation of the CDI to replace the precipitation measurements from meteorological stations; (ii) to estimate the NDVI and LST to calculate the CDI; (iii) to perform a complex correlation analysis between the CDI, SPI, and SPEI, and (iv) to carry out a spatiotemporal analysis of the CDI during the dry season in Dobrogea.

2. Materials and Methods

As was explained in the previous section, the WMO recommends using the SPI to estimate drought. Since ground measurements are not always available, we decided to use the CDI, which is one of the indices derived solely from RS data. The main disadvantage of the SPI is that it does not consider other climatic parameters besides precipitation, such as temperature. Unlike the SPI, the SPEI employs both precipitation and temperature. As the CDI is built on temperature and the NDVI, we decided to use the SPEI as well to inspect the reliability of the seasonal CDI.

The following diagram provides a description of the methodology that was proposed in this research (Figure 1).

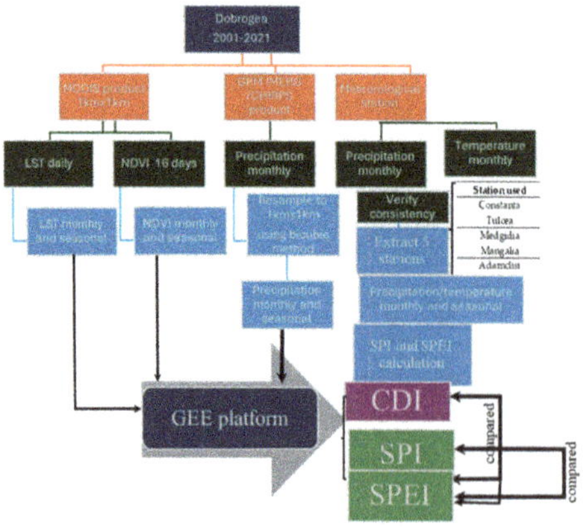

Figure 1. Schematization of the methodology.

The current study was carried out in the Dobrogea region (see Section 2.1); we used MODIS and global gridded precipitation products (GPM IMERG and CHIRPS) to extract LST, NDVI (see Section 2.2.1), and precipitation (see Section 2.2.2) data, respectively. Precipitation data measured at several meteorological stations (Section 2.2.4) in the Dobrogea region (Figure 2), provided by the National Agency of Meteorology, were used to verify the reliability of satellite precipitation. Then, monthly and seasonal (July to September) drought indices (CDI, SPI, and SPEI) were estimated for 20 years. For the SPI and SPEI, we computed two variants, derived from (a) satellite rainfall products and (b) ground measurements of precipitation and temperature, which were then used to examine the reliability of the CDI in different years.

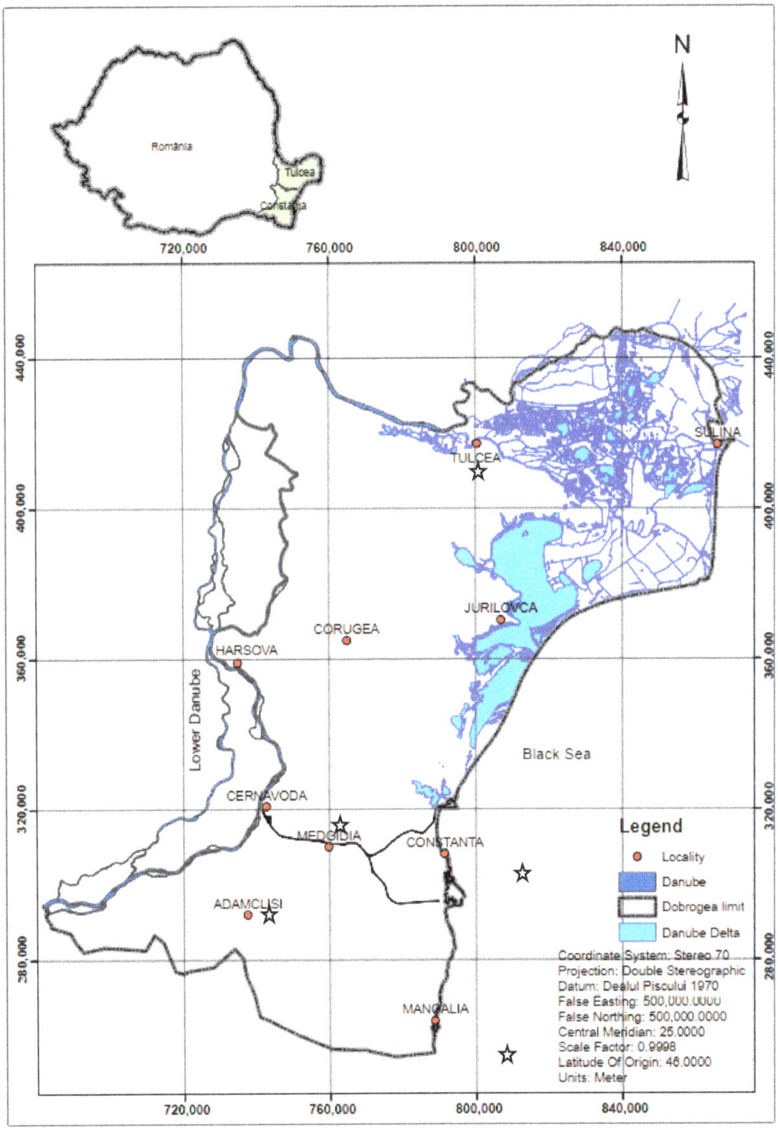

Figure 2. Study area; the stations used in this study are marked with stars (see also Table 1).

2.1. Study Area

The approach illustrated in Figure 1 was implemented in the Dobrogea region. This region is located in the southeast of Romania and is bordered by the lower Danube to the west and north, the Black Sea to the east (Figure 2), and Bulgaria to the south.

The Romanian coastline stretches 275 km, dominated by the Danube Delta and the coastal lakes. The Danube Delta is situated in the northeastern part of Dobrogea, a unique ecosystem introduced in the UNESCO Word Heritage List (December 1991) after the area was recognized as a Biosphere Reserve (in September 1991). Dobrogea also benefits from the most varied relief, which rises from the Danube and the coast inland (467 m—Greci—Macin Mountain). The Macin Mountains, located in the National Park in the north of Dobrogea, represent the oldest geological structure in Romania, developed during the Hercynian orogeny. Hydrologically speaking, in Dobrogea, there are 2 large catchments, one of which is a tributary of the Danube River, and the other is connected to the Black Sea (the littoral basin) via a network of lakes along the coast. The two river basins include several small river systems, with low average flows (below 0.7 m^3/s). Coastal lakes are divided into two parts: the North Unit (Danube Delta and Razim–Sinoe Lake) and the South Unit (the littoral). Generally, the lakes are used for fisheries (4 lakes from the Danube Delta and 6 lakes from the South Unit), leisure, and therapeutic activity (e.g., Techirghiol Lake). Some of these lakes are hypersaline (Techirghiol Lake) and/or brackish. The main characteristic of Razim–Sinoe Lake is (or rather was) the existence of "gates" that connected it to the Black Sea, which contributed to maintaining a unique ecosystem and a balanced water budget. The temperate climate, the water with low salinity, and the orientation of the beach to the east create excellent conditions for tourism; so, today, the coast is considered the Romanian Riviera due to investments in tourist infrastructure. The climate of Dobrogea is temperate, influenced by the Black Sea, the precipitation is under 450 mm/year, and the average temperature is around 11–12 °C [40,41]. According to [42,43], these weather features increase the risk of drought, all the more so as Dobrogea represents one of the Romanian regions most impacted by global warming [44,45].

Dobrogea (Scythia Minor in antiquity) has been an inhabited territory for over 2000 years; it still preserves the vestiges of the Greek colonies Histria, Callatis, and Tomis, and also the architectural monument, Tropaeum Traiani, built by the emperor Traian to mark the victory against the Dacians (104–106 AD). Due to the geographical particularities and the great potential of history, the Dobrogea region's economy is based on tourism (including therapeutic tourism), fisheries, industry (especially energy and electricity), transportation (fluvial and maritime), and agriculture [39].

From hydraulic works used to improve fisheries in the Danube Delta, lakes, and canals to major landscape transformations, the territory of Dobrogea has been subject to major anthropogenic influences. With a total surface area of 15,485 m^2 (including the Danube Delta), more than 80% of Dobrogea is used for agriculture. In 1850, the agronomic researcher Ion Ionescu de la Brad stated that "the plants suffer more of water than nourishment". To improve the situation, in the 1945–1947 period, about 116,000 hectares of Dobrogea land were planned to be planted with wind-break forest protection. Unfortunately, only 3800 hectares were successfully planted and, after 10 years, massive deforestation was undertaken to increase the agricultural territory [46]. In the early 1970s, an extensive irrigation system was built to improve the conditions for agricultural development. It operated at full capacity until 1989 when the irrigated area had reached 581,500 hectares. After 1989, the irrigation capacity decreased enormously (under 20,000 hectares) as a result of the government policy of the 1990s, followed by abandonment and destruction of the irrigation infrastructure [46]. Nowadays, the drought phenomenon is a reality in the

Dobrogea region [2,3,22,46–49] and a major risk to diminish part of natural resources such as lakes and the production of fisheries, agriculture, and electricity.

2.2. Datasets

The datasets for this research study were selected between January 2001 and December 2021, this being the period in which both types of satellite rainfall data, CHIRPS and GPM IMERG, are available (see Section 2.2.2). The effectiveness of satellite-based drought and precipitation indices was assessed using meteorological observations from ground stations. We focused our study on the dry season (July–September) [50] due to the low precipitation rates, which have a significant impact on drought in the Dobrogea area.

There are 10 main meteorological stations (Figure 2) in Dobrogea, administered by the National Administration of Meteorology (NAM), and they are grouped as follows: 4 on the Black Sea coast (Mangalia, Constanta, Jurilovca, and Sulina), 3 on the Danube banks (Cernavoda, Harsova, and Tulcea), and 3 inland (Adamclisi, Medgidia, and Corugea). For these stations, the NAM provided data starting from 1965 to 2021. There are other stations (such as Sf Gheorghe and Chilia), but they were set up later, and their data series does not cover the period mentioned above. However, it should be noted that Sulina station is located on a dam that extends 13 km into the sea. It is practically surrounded by water (the Danube and the Black Sea); therefore, we discarded this station for our study since the NDVI values are negative. The same situation can be found at Jurilovca weather station, which is located in the proximity of Razim–Sinoe Lake. Ultimately, for our study, we chose 5 stations out of 10 (Table 1) because, after a detailed analysis of all station datasets, we observed that some stations did not have the complete data series for the analyzed period (2001–2021). The stations selected cover all the areas mentioned above: 2 on the seacoast, 1 on the Danube banks, and 2 inland.

Table 1. Selected meteorological stations across Dobrogea.

Station Used	Longitude/Latitude
Constanta	28.64/44.21
Tulcea	28.82/45.19
Medgidia	28.25/44.24
Mangalia	28.58/43.82
Adamclisi	27.96/44.09

2.2.1. NDVI and LST Datasets

The present study used MODIS satellite imagery to retrieve the data needed for further computations. Having a more frequent repeat cycle (1–2 days) than Landsat and high spatial resolutions (250 m, 500 m, and 1000 m), MODIS data are well-suited for large-scale drought monitoring. We resorted to the MOD21A1D.061 Terra product with 1000 m resolution to acquire daily LST, which was estimated as an average from all observations that were cloud-free and had good LST&Emissivity accuracy. The MOD13A2.061 Terra product, with a 1000 m spatial resolution, provided the vegetation index NDVI, its values being composited using the finest valid pixel value (low view angle, low clouds, the highest NDVI value) from the whole of the acquisitions within a 16-day interval. These NDVI and LST products can be used on the Google Earth Engine (GEE) platform [51].

NDVI values vary from +1.0 to −1.0. Here, we computed the average values to establish the monthly and seasonal (July to September) NDVI and LST for the 2001–2021 period (datasets named NDVI-1, LST-1 and NDVI-3, LST-3, respectively).

2.2.2. Precipitation Datasets

Two high-resolution satellite precipitation products were utilized—GPM IMERG and CHIRPS—in order to rate their accuracy against data from ground stations and then select the one with the best performance.

Climate Hazards Group InfraRed Precipitation with Station data (CHIRPS) is a daily gridded precipitation dataset [52] that provides high-resolution (0.05°), quasi-global (50° S–50° N) satellite, and observation-based precipitation data, over land, from 1981 to up until now. This consistent, long time series dataset is extremely useful in trend investigation and drought monitoring. The daily CHIRPS data (CHIRPS v.2.0 Final, Dataset Availability: 1 January 1981–1 January 2024) can be obtained from the GEE platform.

Global Precipitation Measurement (GPM) was launched in 2014 by NASA and JAXA, while Integrated Multi-Satellite Retrievals for GPM (GPM IMERG) is a precipitation algorithm developed by NASA that uses inter-calibrated estimates from the GPM constellation of precipitation-related satellites and other data (such as precipitation gauge analyses) to create a long (20+ years), global-scale, high-temporal (30 min) and high-spatial (0.1°) resolution satellite-based precipitation dataset. In this study, monthly GPM IMERG images were employed, spanning two decades (2001–2021), provided by GPM: Monthly Global Precipitation Measurement (GPM) vRelease 07 (Dataset Availability: 1 June 2000–17 January 2025), which is available on the GEE platform.

In the present study, to make all satellite datasets consistent, the CHIRPS and GPM IMERG images were spatially resampled to 1 km × 1 km (the spatial resolution of MODIS LST and NDVI products) using the bicubic method. Also, the daily CHIRPS and monthly GPM IMERG precipitation data were organized in monthly and seasonal data (July–September) by the arithmetic mean (datasets named CHIRPS-1, CHIRPS-3, IMERG-1, and IMERG-3, respectively), for a time span ranging from 2001 to 2021.

2.2.3. SPEI Dataset

The SPEI is an important drought indicator that uses precipitation and evapotranspiration data [15], thus being able to assess the end results of temperature increase on water demand. In the context of global warming, the SPEI enables monitoring and analysis of distinct drought types. The Global SPEI database offers information about drought conditions at a global scale, with a spatial resolution of 0.5°, a monthly temporal resolution, and 1- to 48-month timescales. The computation of the SPEI is based on the FAO Penman–Monteith estimation of potential evapotranspiration, which is considered a better method than the Thornthwaite PET one; this is why the SPEIbase is appropriate for most short-, medium-, or long-term analyses. In our study, we used the SPEI on a 3-month timescale (dataset named SPEI-3) for the dry season of the 2001–2021 period, after first spatially resampling the satellite images to 1 km × 1 km using the bicubic method. The SPEIbase Standardized Precipitation Evapotranspiration Index database, Version 2.9 is available through the GEE platform.

2.2.4. Ground Stations Datasets

As mentioned in Section 2.2, there is a limited number of meteorological stations across Dobrogea, and many of them have large gaps in the time series, which makes them unsuitable for further analysis. Hence, for the study period (2001–2021), the first step in the selection of ground stations was to ensure no data gaps in the time series of monthly precipitation. As a result of this selection criteria, 5 stations were retained (Figure 2 and Table 1), and their monthly precipitation time series were adopted for validation. The calculation of the in situ SPEI requires temperature value as well; therefore, another condition placed on the ground data was to have no gaps in the temperature dataset. As a result, of the 5 stations with complete precipitation datasets, 3 stations (Constanta, Tulcea, and Mangalia), with full precipitation and temperature datasets, were selected to be used in the in situ SPEI calculation.

2.3. *Methods*

2.3.1. Composite Drought Index (CDI)

The Composite Drought Index proposed in [39] was used in this study to understand the relationship between rainfall, temperature, and vegetation in the monitoring and evaluation of agricultural drought in Dobrogea area. Therefore, the satellite rainfall data, the MODIS NDVI, and LST were all used to calculate the CDI with a method that combines entropy and weighted Euclidian distance (Equation (1))

$$CDI = \frac{S_i^-}{S_i^- + S_i^+} \tag{1}$$

where S_i^- and S_i^+ are the weighted Euclidian distance between the current condition and MDC (Maximum Driest Condition) and MWC (Maximum Wettest Condition), respectively (Equations (2)–(4)):

$$S_i^- = \sqrt{\sum_k Ew_k (r_{ki} - MDC_k)^2} \tag{2}$$

$$S_i^+ = \sqrt{\sum_k Ew_k (r_{ki} - MWC_k)^2} \tag{3}$$

$$\begin{aligned} MDC_k &= \{\max(r_{ki}), k \in \{LST\}; \min(r_{ki}), k \in \{NDVI, precipitation\}\} \\ MWC_k &= \{\min(r_{ki}), k \in \{LST\}; \max(r_{ki}), k \in \{NDVI, precipitation\}\} \end{aligned} \tag{4}$$

$$r_{ki} = \frac{x_{ki}}{\sum_{i=1}^{m} x_{ki}} \tag{5}$$

where x_{ki} is the value of the kth input variable (LST, NDVI, and precipitation), r_{ki} is the normalized value (Equation (5)), and i is the time index ($i = 1, 2, \ldots, m$).

Ew_k is the entropy weight of each input variable and varies from 0 to 1 (Equation (6))

$$Ew_k = \frac{D_k}{\sum_{j=1}^{k} D_j} \tag{6}$$

where

$$D_k = 1 - e_k \tag{7}$$

For each variable, the entropy measure, e_k, was computed by the following equation:

$$e_k = \frac{-\sum_{i=1}^{m} r_{ki} \ln(r_{ki})}{\ln m} \tag{8}$$

The CDI, built only with satellite data, proved to be suitable for meteorological and agricultural drought monitoring in other regions, being useful, in particular, in the identification of moderate and mild drought events, which usually are not detected by the in situ drought indicators [39].

In this study, the computation of the CDI was performed with JavaScript on GEE platform. The obtained CDI range is between 0 and 1. CDI was assessed at 1- and 3-month (seasonal: July to September) timescales (CDI-1 and CDI-3, respectively), as they reveal moisture conditions on short- and medium-term and allow for the estimation of seasonal drought.

Based on the CDI, we classified the drought severity into seven levels: extreme, critical, severe, moderate, mild, near normal, and no drought (Table 2). According to the color identifier for each level defined in this table, we generated drought spatial distribution maps.

Table 2. Drought severity levels by the CDI.

CDI	Drought Severity Categories
>0.5	No drought
0.4–0.5	Near Normal
0.3–0.4	Mild
0.2–0.3	Moderate
0.1–0.2	Severe
0.03–0.1	Critical
≤0.03	Extreme

2.3.2. Standardized Precipitation Index (SPI)

The SPI, as proposed by [14], depicts the probability of rain in a certain period of time in a region. Since 1993, the SPI has been used to a great extent in recording and estimating the aspects of various drought types (meteorological, agricultural, hydrological) for 1-, 3-, 6-, 12-month, and longer timescales [53–55]. Compared to other drought indices, a large number of studies have attained robust results and greater performance [56,57], leading to the SPI being endorsed by the World Meteorological Organization as a worldwide tool to monitor different kinds of drought [16].

The SPI is able to detect the outset and end of drought events, and to measure the severity of drought, using its intensity. Table 3 shows the drought severity types, based on the interpretation of SPI values [14].

Table 3. Drought severity levels by the SPI.

SPI	Drought Severity Level
≥2	Extremely Wet
1.50–1.99	Severely Wet
1.00–1.49	Moderately Wet
−0.99–0.99	Nearly Normal
−1.49–−1.0	Moderate Drought
−1.99–−1.5	Severe Drought
≤−2	Extreme Drought

This study calculated 1- and 3-month timescales of the SPI, named SPI-1 and SPI-3, respectively, to monitor meteorological drought based on precipitation anomalies from 2001 to 2021, for the Dobrogea area. The input of the JavaScript code that runs on the GEE platform was satellite daily rainfall data, which were summed up to monthly (or 3-month) precipitation data. The SPI was estimated by using the following equation:

$$SPI = (P - P_m)/\sigma \quad (9)$$

where P represents the total rainfall of a period (mm); P_m is the historical mean precipitation of the period (mm); and σ is the historical standard deviation of precipitation of the period (mm).

2.3.3. Drought Severity, Duration, Frequency, Intensity, and Probability of Occurrence

Drought features include duration, severity, intensity, frequency, and probability of occurrence. Here, we used the runs theory proposed by [58] to estimate the drought characteristics. The runs theory defines a slice of the drought index set in which every single value is lower or higher than a designated threshold (K).

Drought duration (T) is the time interval between the outset (t_o) and end (t_e) of a drought occurrence based on the CDI ($CDI \leq 0.4$, $K = 0.4$). In the present study, we chose to express drought severity by the longest drought duration [59]; therefore, severity (S) was computed as the sum of S_p parameters during the longest drought duration (Equations (10) and (11))

$$S = \sum_{t_o}^{t_e} S_p \tag{10}$$

$$S_p = K - CDI \tag{11}$$

Drought frequency is the number of drought occurrences at a particular level within a specific period. Drought intensity (I) was obtained by dividing S by the maximum drought duration, T (Equation (12)).

$$I = \frac{S}{T} \tag{12}$$

To determine the probability of occurrence, P (%), of a specific drought event, we divided the number of classified drought month occurrences, m_c, by the total number of drought events taking place, m_t [60] (Equation (13)).

$$P = \frac{m_c}{m_t} \times 100\% \tag{13}$$

2.3.4. Validation of High-Resolution Rainfall Products

We checked the accuracy of the satellite precipitation datasets from 2001 to 2021, at annual timescales, using data from 5 independent weather stations in Dobrogea. Ground measurements and satellite rainfall estimations were averaged for each year of the study period. We applied the "nearest method" for spatial data extraction, namely, for each satellite precipitation product, we extracted the mean precipitation value at each weather station for each year using the spatial coordinates of that station; then, we used the mean precipitation value of the grid cell in which the meteorological station resides to compare with the corresponding in situ mean precipitation value for the corresponding year. Afterward, we applied several pairwise comparison statistics methods, e.g., Pearson's correlation coefficient (r), root mean square error (RMSE), normalized root mean square error (NRMSE), and agreement, to quantitatively validate the satellite precipitation estimates against all in situ datasets (Table 4).

Table 4. Statistical evaluation metrics used to validate satellite products.

Metric	Equation	Interpretation
Pearson Correlation Coefficient (r)	$r = \frac{\sum_{i=1}^{n}(x_i-\overline{x})(y_i-\overline{y})}{\sqrt{\sum_{i=1}^{n}(x_i-\overline{x})^2}\sqrt{\sum_{i=1}^{n}(y_i-\overline{y})^2}}$ (14) where n represents the number of samples, x_i/y_i are the ith value of the measured/estimated samples, and $\overline{x}$ and $\overline{y}$ are the average of the measured and estimated values of all samples. Strong correlations have the following values: - Positive (0.5 to 1.0) and negative (−0.5 to −1.0). Medium correlations have the following values: - Positive (0.3 to 0.5) and negative (−0.3 to −0.5). Small correlations have the following values: - Positive (0.1 to 0.3) and negative (−0.1 to −0.3).	r takes values from 1.0 to −1.0, meaning a positive to negative strong linear correlation; a zero or near-zero value indicates no correlation among the variables [61].
Root Mean Square Error (RMSE)	$RMSE = \sqrt{\frac{1}{n}\sum_{i=1}^{n}(y_i - x_i)^2}$ (15)	The lower the value, the better the central tendencies and the smaller the errors. A value of zero is the perfect outcome [61].
Normalized Root Mean Square Error (NRMSE)	$NRMSE = 100 \times \frac{RMSE}{\overline{x}}$	For a model, smaller values show reduced residual variance [62].
Agreement (%)	$Agreement = 100 \times (1 - (RMSE/\overline{x}))$	Values close to 100% suggest strong agreement; values closer to 0% show the least agreement [63].

2.3.5. Performance Evaluation of Satellite Precipitation Products

In the present study, the SPI, attained from satellite rainfall time series, was used to verify the reliability of the CDI in different years; therefore, we validated it with in situ SPI estimation. We used the monthly precipitation datasets from 5 weather stations in Dobrogea, from 2001 to 2021, to calculate the 1-month SPI dataset (in situ SPI-1) by using the SPIGenerator [64]. These data were used as reference data. Then, we computed two variants of the satellite-based 1-month SPI, namely, SPI-1c and SPI-1i, using the gridded data provided by the high-resolution precipitation products CHIRPS and GPM IMERG, respectively. By performing a correlation analysis between these indices and the corresponding reference data (in situ SPI-1), we were able to investigate the applicability of each global precipitation product to the computation of the CDI. The product with the best performance was selected for the computation of the Composite Drought Index considered in this study.

2.3.6. Correlation Between the SPI and CDI

To validate composite drought indices, the SPI has often been used as a benchmark [65]. We used the monthly precipitation datasets from weather stations to calculate the seasonal (July–September) SPI dataset (in situ SPI-3), from 2001 to 2021, by running the SPIGenerator. Then, we conducted a Pearson correlation analysis and comparison of scatter plots, between CDI-3 and both the satellite-based and in situ SPI-3. Afterward, we analyzed the Pearson correlation between the monthly averaged CDI-1 and satellite-based SPI-1 during the driest and wettest years of the study period. Furthermore, we expanded our analysis with the correlation between CDI-3 and satellite-based SPI-3 for years corresponding to each type of drought event established by the CDI.

2.3.7. Correlation Between the SPEI and CDI

Apart from the SPI, the CDI was also correlated with the SPEI to analyze its robustness. We used the monthly precipitation and temperature datasets from the weather stations to calculate the seasonal SPEI dataset (in situ SPEI-3), from 2001 to 2021, by running the SPEI Calculator [66]. Then, in situ and satellite-based estimates of SPEI-3 were validated against the corresponding variant of SPI-3. After that, we carried out a correlation analysis for both the in situ and satellite-based CDI-3 with the appropriate SPEI-3 index by using scatter plots and the Pearson correlation coefficient.

3. Results

3.1. Annual Precipitation Evaluation of Two High-Resolution Precipitation Products at the Regional Scale

We compared the satellite precipitation with that recorded by the weather stations through the assessment of *r*, RMSE, NRMSE, and agreement (Table 5). The results were derived for annual precipitation from January 2001 to December 2021.

Table 5. Statistical assessment of annual satellite precipitation for the period 2001–2021.

Weather Station	Constanta	Tulcea	Medgidia	Mangalia	Adamclisi
Index			**GPM IMERG**		
r	0.90	0.95	0.70	0.77	0.70
RMSE	7.97	6.20	10.20	11.73	10.36
NRMSE	19.7	14.6	25.0	27.7	22.20
Agreement	80.3	85.4	75.0	72.3	77.8
			CHIRPS		
r	0.85	0.84	0.58	0.71	0.55
RMSE	6.16	6.12	10.45	8.66	10.73
NRMSE	15.2	14.4	25.4	20.4	23.0
Agreement	84.8	85.6	74.6	79.6	77.0

According to statistical results, the correlation coefficient (*r*) took values between 0.55 and 0.95, and the greatest values of 0.95 and 0.90, were attained with GPM IMERG, followed by CHIRPS (0.85). The lowest *r* of 0.55 was obtained with CHIRPS at Adamclisi station.

The lowest annual RMSE, a value of 6.12, was recorded with CHIRPS, followed closely by GPM IMERG, which attained an RMSE of 6.20 for the same station, i.e., Tulcea. GPM IMERG registered the highest RMSE of 11.73 next to CHIRPS (10.73). The agreement of both precipitation products with measured values was more than 70%, denoting the high correlation between both products and ground truth.

The *r* values for IMERG varied from 0.70 to 0.95, with a median of 0.80, and the average agreement value was 78%. As for CHIRPS, the median correlation coefficient and agreement were 0.70 and 80%, respectively.

3.2. Correlation Between 1-Month SPI-1 and In Situ SPI-1

This study carried out a Pearson correlation analysis between SPI-1c and SPI-1i, computed with gridded data provided by CHIRPS and GPM IMERG, respectively, and in situ SPI-1, to measure the performance of the satellite data. The Pearson correlation coefficient between SPI-1c and in situ SPI took values between 0.46 and 0.85, with a median of 0.65. Regarding SPI-1i and in situ SPI, the minimum, maximum, and median values of r were 0.62, 0.87, and 0.74, respectively. The coefficient of determination (Figure 3) values were close to a best-fit correlation at both ground stations, $R^2 = 0.75$ (p-value = 0.0001), for SPI-1i. A strong positive relationship was also observed for SPI-1c, with R^2 values of 0.72 and 0.63 (p-value = 0.0001).

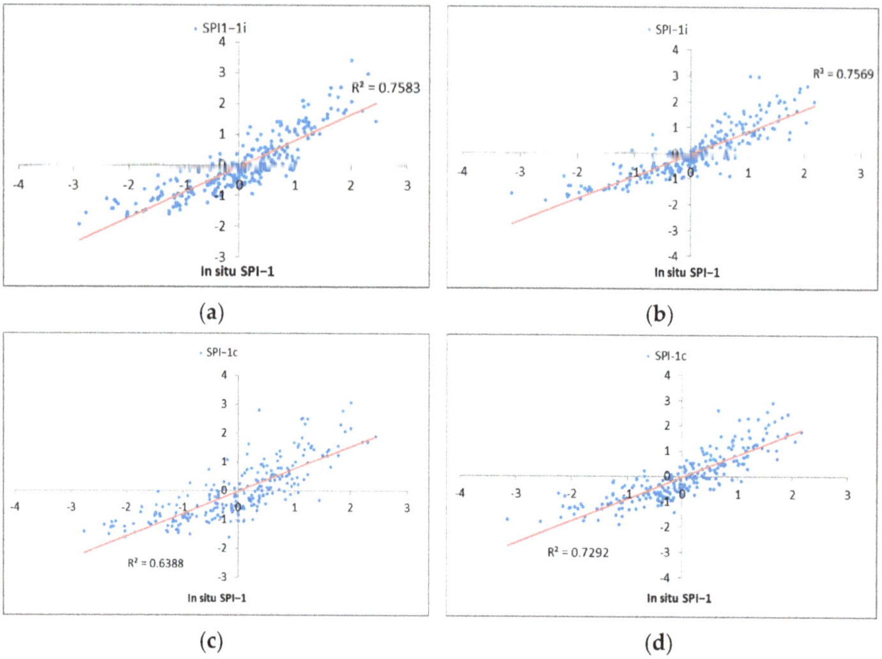

Figure 3. Scatterplots of in situ SPI and estimated SPI: (**a**) SPI-1i, Constanta, (**b**) SPI-1i, Tulcea, (**c**) SPI-1c, Constanta, and (**d**) SPI-1c, Tulcea.

The statistics indicate that SPI-1i exhibited a significant positive correlation with in situ SPI-1 from 2001 to 2021. These results, along with those from the validation of high-resolution precipitation products, encouraged us to select the GPM IMERG product for further computation, as its performance was higher, indicating that is dependable for use during low rainfall conditions [67].

3.3. Temporal Dynamics of LST and the NDVI During the Dry Season

Generally, dry soil shows a high level of LST due to acute temperature and diminished soil moisture content. Figure 4 shows the long-term dry season mean LST over Dobrogea and reveals an ascending trend of LST during the study period, with a high average temperature of the dry season (>34 °C).

Seasonal changes in vegetation conditions over a vast region can be monitored with MODIS NDVI data [68]. Furthermore, these data can be used to identify drought phenomena due to their ability to sense soil surface moisture. The temporal analysis of mean seasonal NDVI for Dobrogea (Figure 5) shows good vegetative cover, with relatively high NDVI values (>0.43), and a descending trend for the period from 2001 to 2021. This could indicate that agricultural lands were under great stress due to high-level temperatures and irregular and below-normal rainfall.

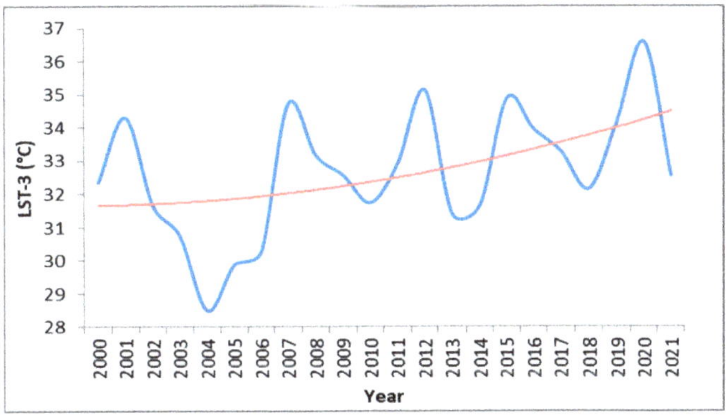

Figure 4. Long-term dry season mean LST-3 (°C) over Dobrogea.

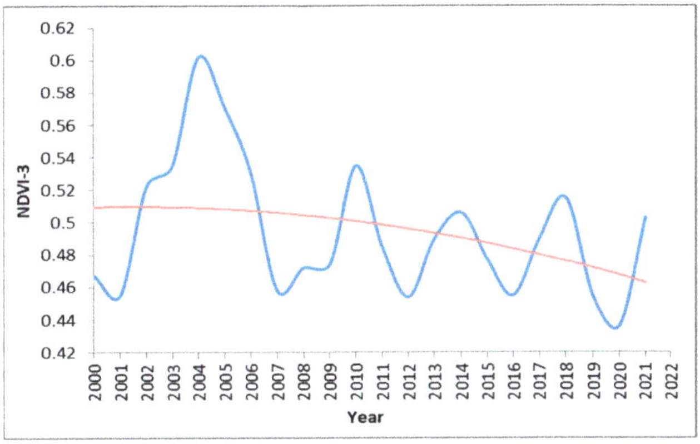

Figure 5. Long-term dry season mean NDVI-3 over Dobrogea.

3.4. Spatiotemporal Dynamics of 3-Month SPI During the Dry Season

The spatiotemporal analysis of the seasonal (3-month) SPI for the dry season shows that drought and near-normal conditions were recurrent in most of the years of the study period. The time series of SPI-3 (Figure 6) during the 2001–2021 period had a significant downward trend (R^2 = 0.33, p-value = 0.006). Thus, the Dobrogea area became less humid over the period of 2001 to 2021.

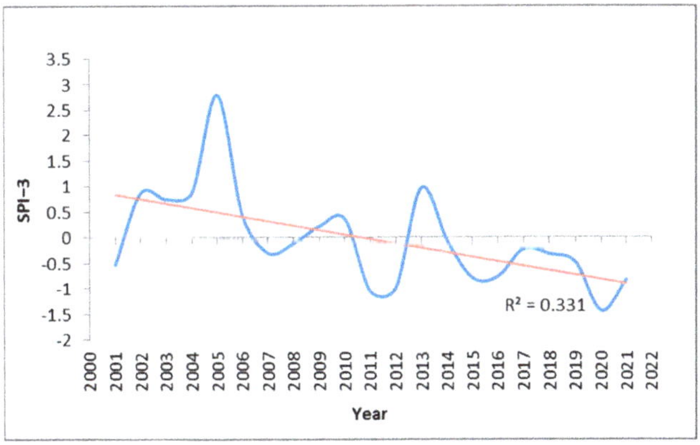

Figure 6. Long-term dry season average SPI-3 over Dobrogea.

Due to the deficit rainfall in the years 2011, 2012, 2015, 2016, 2019, and 2021, different parts of Dobrogea (Figure 7) experienced severe to moderate dry conditions (−1.99 < SPI < −1). In the year 2020, a vast area of the region was exposed to severe dry conditions (−1.99 < SPI < −1.5), even extreme drought in the east (SPI < −2). During the years 2001, 2007, and 2008, moderate to severe droughts were indicated, whereas in 2017 and 2018, there were moderate to mild drought conditions in most of the area. In the year 2014, there were mild drought conditions in most parts of the zone, and near-normal to moderately wet conditions were met in 2006, 2009, 2010, and 2013. Extreme wet conditions (SPI > 2) were met only in 2005, whereas, for the rest of the period, the region experienced near-normal conditions (−0.99 to 0.99) to moderately wet conditions (1.0 to 1.49).

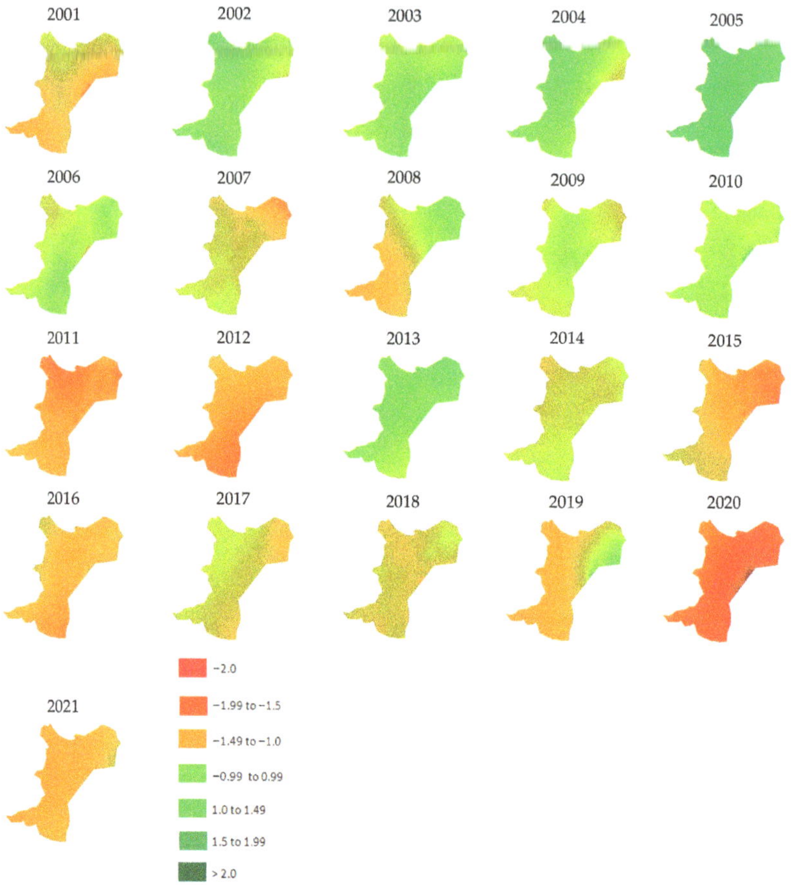

Figure 7. Spatiotemporal patterns of SPI-3 over the study area during the period from 2001 to 2021. Identified dry conditions: severe to extreme (2020), severe (2011, 2012, 2015, 2016, 2019, and 2021), moderate to severe (2001, 2007, and 2008), mild to moderate (2017 and 2018), and mild (2014).

3.5. Spatiotemporal Dynamics of the CDI During the Dry Season

The analysis of the spatiotemporal dynamics of the CDI for the dry season revealed that the region experienced frequent agricultural droughts. The time series of CDI-3 (Figure 8) during the 2001–2021 period had a significant downward trend (R^2 = 0.35, *p*-value = 0.004). These results indicated that during the study period, most parts of the region were exposed to some type of drought, except for the years 2002–2005 and 2013.

These findings are consistent with other research on Dobrogea's major drought events. As indicated in [37], in Romania, the period 1991–2021 was the driest since 1901, with an increase in the duration of moderate, severe, and extreme droughts being identified throughout the country, including Dobrogea. Normal wet conditions characterized the period 1997–2009 [35], which was followed by a dry period, from 2010 to 2021 [25,37]; the drought for this period had the highest degree of severity [37].

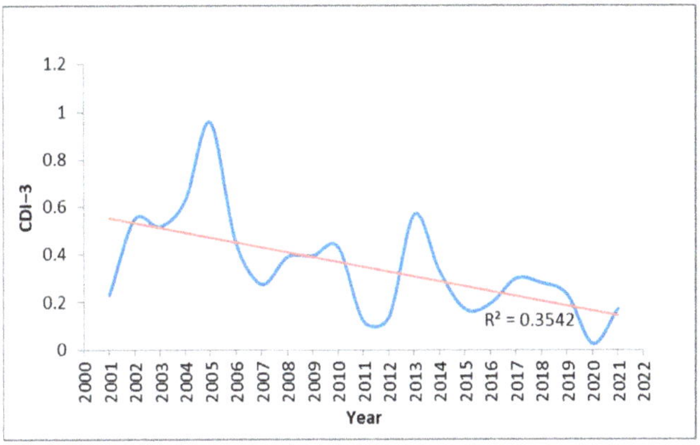

Figure 8. Long-term dry season average CDI-3 over Dobrogea.

In the years 2011, 2012, and 2020, the region encountered the worst drought events (CDI < 0.1), and these years were followed by 2007, 2015, 2016, 2019, and 2021 (Table 6). The authors of [37] identified similar results at the country level, and those of [38] determined that Dobrogea experienced extreme drought in 2007, 2012, and 2015. As indicated in [69], the year 2019 was considered one of the warmest years of all the measurements made in Romania from 1900 to the present, with the drought developing rapidly and attaining the utmost intensity and spatial expanse by 2020. This was proven by the authors of [70], who showed that Nuntasi–Tuzla Lake completely dried up in August 2020 and partially in 2012. Based on the present study, critical to extreme drought conditions were met in large areas across the region (about 95% in 2020, and 35% in 2011 and 2012), whereas severe drought was encountered across 62% and 41% during the years 2011 and 2012, respectively. During these years, the region recorded high LST values (33.5 °C < LST < 39.13 °C), which shows that it was under severe temperature stress (Figure 9). In 2007, more than 77% of the region was affected by moderate drought, and in the years 2015, 2016, 2019, and 2021, an extensive part of the area (43% to 95%) experienced severe drought conditions (Figure 10). In Europe, the summer of 2015 was among the warmest and driest summers of the last seven decades [71], caused by high evapotranspiration rates and a great shortage of precipitation. The drought affected large areas of central and eastern Europe, including Dobrogea [38].

In the years 2017 and 2018, there were moderate to mild drought conditions in most of the region, whereas in 2001 and 2008, there were moderate to severe dry conditions in the southern, western, and central areas. In the year 2014, there were mild drought conditions in most of the region (61%); near-normal conditions were met in 2006, 2009, and 2010, except for a few areas in the west and south of the region (mild drought in about 23%, 35%, and 30% of the area, respectively). Unlike all other years, in 2005, the whole area benefited from extreme wet conditions (CDI = 0.95 on 90% of the region), and this was also clearly indicated through the SPI analysis (extreme wet conditions). Normal wet conditions were indicated in the years 2002–2004 and 2013, suggesting that the region was free from drought.

The analysis of spatial drought distribution, based on SPI and CDI data, reveals that the southern and central parts of Dobrogea are most exposed to drought; here, severe to extreme droughts occur most often. A similar observation was made in [38].

Table 6. The area under different categories of droughts during the years 2001 to 2021.

Year/Drought	Mild	Moderate	Severe	Critical	Extreme
2001	24.89%	33.88%	26.67%	11.00%	0.77%
2002	0.18%	0.08%	0.02%	0.02%	0.00%
2003	3.66%	0.08%	0.03%	0.02%	0.01%
2004	0.11%	0.05%	0.03%	0.00%	0.00%
2005	0.08%	0.05%	0.04%	0.00%	0.00%
2006	22.87%	4.85%	0.05%	0.03%	0.01%
2007	16.77%	77.95%	4.66%	0.02%	0.02%
2008	12.58%	19.88%	36.10%	3.59%	0.01%
2009	34.49%	14.20%	0.09%	0.02%	0.01%
2010	30.01%	0.21%	0.05%	0.03%	0.01%
2011	0.42%	2.86%	62.14%	34.07%	0.22%
2012	1.28%	20.19%	41.04%	34.74%	2.38%
2013	9.23%	0.20%	0.00%	0.00%	0.01%
2014	60.99%	24.33%	0.75%	0.01%	0.00%
2015	0.54%	4.42%	94.56%	0.06%	0.00%
2016	11.03%	25.09%	50.05%	13.24%	0.01%
2017	43.42%	43.66%	7.70%	0.02%	0.00%
2018	14.25%	62.00%	11.23%	0.05%	0.01%
2019	8.97%	14.77%	43.02%	15.79%	0.13%
2020	0.30%	0.71%	2.68%	21.83%	74.22%
2021	7.07%	16.88%	66.46%	9.13%	0.05%

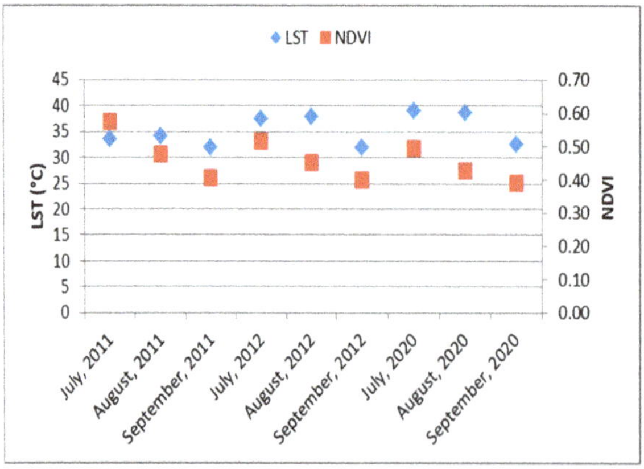

Figure 9. LST-1 and NDVI-1 for the dry season (July–September) in 2011, 2012, and 2020.

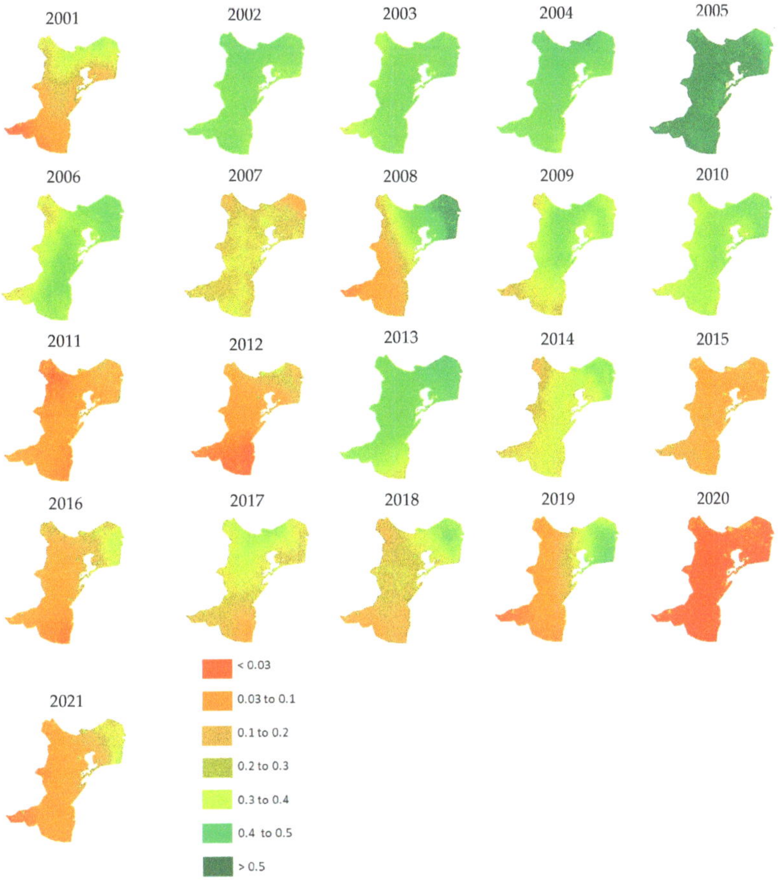

Figure 10. Spatiotemporal patterns of CDI-3 over the study area during the period 2001–2021. Identified dry conditions: critical to extreme (2011, 2012, 2020), severe (2015, 2016, 2019, and 2021), moderate to severe (2001, 2007, and 2008), mild to moderate (2017 and 2018), and mild (2014).

3.6. Correlation Between the CDI and SPI

The Pearson correlation analysis was implemented between CDI-3 and in situ SPI-3 for the dry season of the study period to establish their relationship and seasonal fluctuations in drought and precipitation. The results show that, for the years considered, CDI-3 had a strong positive correlation with in situ SPI-3 (Figure 11). The correlation coefficient (*r*) values varied from 0.55 to 0.85, with a median of 0.70.

The coefficient of determination (Figure 12) denoted an almost best-fit correlation at Tulcea ground station, R^2 = 0.75 (*p*-value = 0.0001), and a positive correlation was also obtained in Constanta (R^2 = 0.55, *p*-value = 0.0001), Mangalia (R^2 = 0.51, *p*-value = 0.0002), and Medgidia (R^2 = 0.40, *p*-value = 0.002), but the results are not a significantly linear fit. This could be explained by the fact that the short term droughts in Dobrogea not only arise due to a lack of precipitation but are also induced by high temperatures and a high level of evapotranspiration.

Furthermore, we conducted a Pearson correlation analysis between CDI-1 and satellite-based SPI-1, considering their average for each month of the dry season, from July to September, during the driest year (2020) and the wettest year (2005). The results showed a significant positive correlation between CDI-1 and SPI-1 for the years and months considered, with a correlation coefficient of 0.81 and 0.71, respectively. In the dry year, poorly distributed rainfall affected healthy vegetation growth and enhanced the intensity of drought in the area. The strong positive correlation during the wet year indicated that the CDI responds well to vegetation conditions and the soil moisture content of a region.

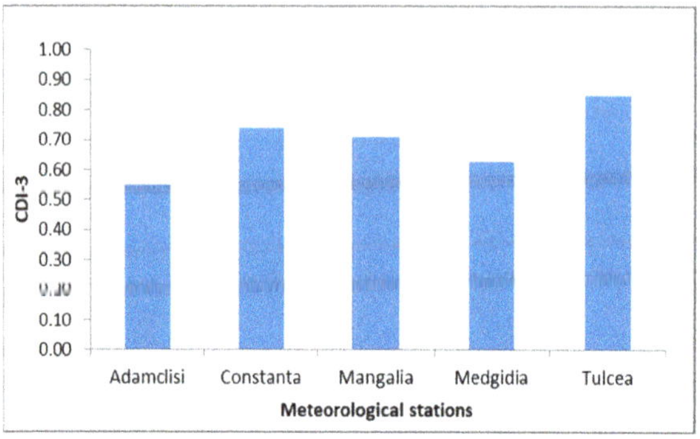

Figure 11. The correlation coefficient between CDI-3 and in situ SPI-3 at meteorological stations during the dry season (July–September) from 2001 to 2021.

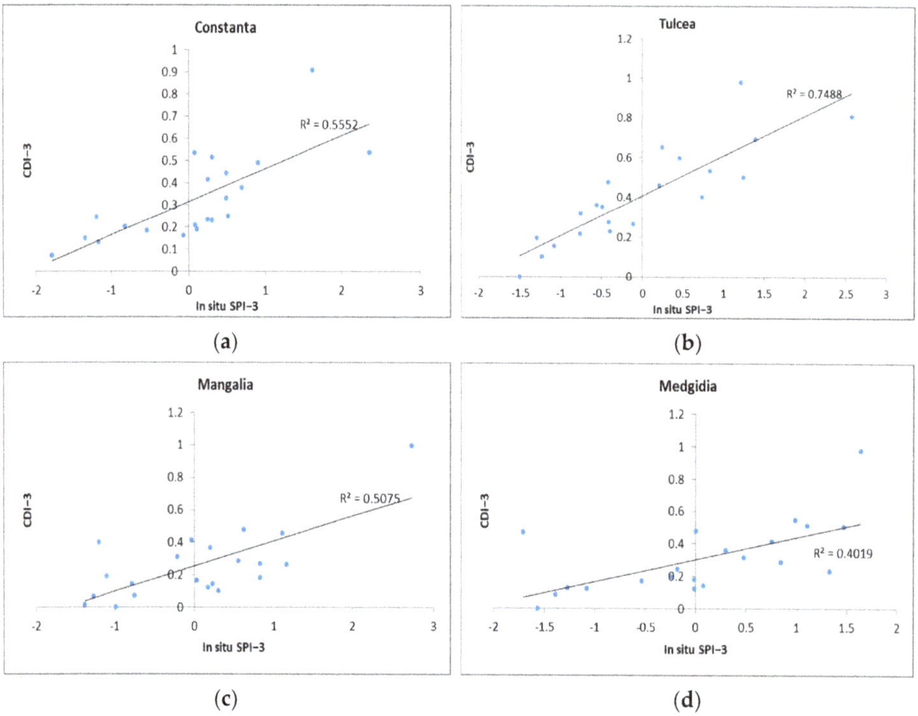

Figure 12. Scatterplots of seasonal CDI-3 and in situ SPI-3 from 2001 to 2021 for weather stations (**a**) Constanta, (**b**) Tulcea, (**c**) Mangalia, and (**d**) Medgidia.

We extended the analysis by computing the Pearson correlation between seasonal CDI-3 and satellite-based SPI-3 for the years corresponding to each drought class established for the CDI (Table 2): no drought (2002), mild (2006, 2009, 2013, 2014), moderate (2001, 2007, 2017), severe (2008, 2019, 2021), and critical to extreme (2012). The results (Figure 13) indicated a strong positive correlation between CDI-3 and SPI-3 during the selected years in most parts of Dobrogea, and the correlation values were close to 1 in some areas. Despite that, a strong negative correlation was detected in some parts of the north, northeast, and southeast.

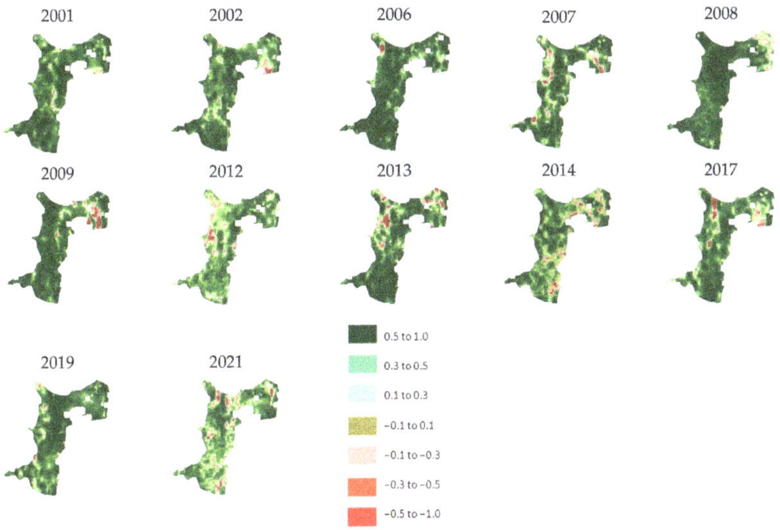

Figure 13. Spatial distribution of Pearson's correlation between CDI-3 and SPI-3 during the dry season (July–September) in Dobrogea.

The spatiotemporal analysis of seasonal CDI-3 and SPI-3 during the period from 2001 to 2021 indicated that the entire area of the Dobrogea was vulnerable to some type of drought in most of the years due to below-normal rainfall. The correlation analysis between the seasonal 1- and 3-month CDI and SPI, both in situ and satellite-based, for the study period revealed a strong positive linear relationship in most parts of Dobrogea between below-normal precipitation and drought intensity.

3.7. Correlation Between the SPEI and SPI

Before its further use, in situ SPEI-3 was compared with in situ SPI-3 data, both datasets being built from ground measurements from three weather stations (see Section 2.2.4). The Pearson correlation analysis showed that in situ SPEI-3 had a very strong positive correlation with in situ SPI-3, with r being 0.98, 0.76, and 0.97 at Constanta, Tulcea, and Mangalia stations, respectively. Likewise, the coefficient of determination (Figure 14a–c) indicated a very strong linear association among these indices at Constanta and Mangalia ($0.94 \leq R^2 \leq 0.96$, p-value = 0.0001). The dispersion of data points was larger at Tulcea (R^2 = 0.57, p-value = 0.0001), which is a sign of a good relationship between indices in that area, but with a greater degree of variability. Further on, we quantified the relationship between satellite-based SPEI-3 and SPI-3, considering their average for each year of the study period (Figure 14d). The results showed that there was a very strong correlation (r = 0.85) and a significant linear association between the SPEI-3 and SPI-3 indices at the 0.0001 significance level (Figure 14d). Overall, we found that there is a good level of agreement between both the in situ and satellite-based SPEI and SPI estimations in the study area at the 3-month timescale.

The total number of months for the study period is 252, of which we identified the number of drought months for each station, as shown in Table 8. Medgidia had the highest number of months of drought (76.19%), followed by Constanta and Tulcea. The longest drought event was in Tulcea and Medgidia, spanning 22 months, between 2018 and 2020 (Figure 17). The high number of months of drought recorded during the study period is consistent with the results of [35], which indicate an increase in the duration of moderate, severe, and extreme droughts in Dobrogea for 1991–2021 compared to the previous periods.

Table 8. Drought characteristics for ground stations under CDI-1 during the years 2001 to 2021.

Station	Total Months	No. of Drought Months	Drought (%)	Longest Drought Period	Longest Drought Duration (Months)	Severity	Intensity
Constanta	252	179	71.03%	July 2018 to August 2019	14	1.36	0.10
Tulcea	252	178	70.63%	February 2019 to November 2020	22	2.81	0.13
Medgidia	252	192	76.19%	August 2018 to June 2020	22	3.41	0.16

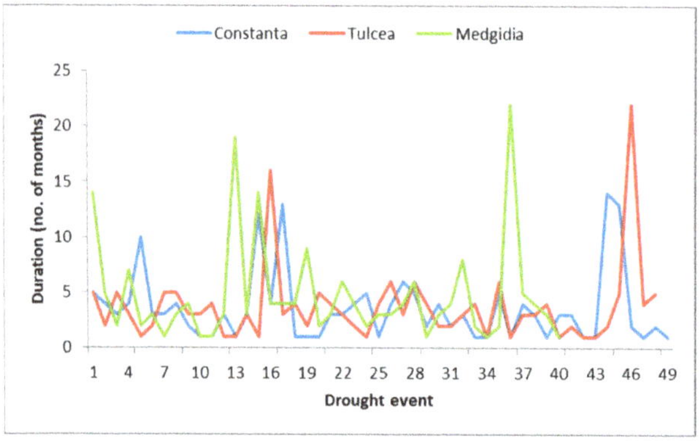

Figure 17. Drought duration (in months) for ground stations during the years 2001 to 2021.

For less severe drought events, the drought severity values must be closer to zero. Table 8 shows that the severity of drought ranges from moderate in Constanta to severe in Tulcea and extreme in Medgidia.

Most of the drought event intensities were mild to moderate, with more than 40 drought events for each station over the entire study period (Figure 17). One extreme drought occurred in August 2020 in Medgidia. A significant drought incident that was recorded by all stations took place between February 2019 and August 2019, which may indicate that most parts of Dobrogea suffered drought phenomena during this period. This event was extended until November 2020, for all stations, with a month of normal precipitation conditions in Constanta (September 2019) and Medgidia (June 2020). Another long drought event, registered by all stations, was between April 2006 and October 2007, which was interrupted by 3 months of normal conditions in Constanta and 2 months in Tulcea, unevenly distributed throughout the period. It had the longest duration in Medgidia (19 months). Generally, a drought incident was not necessarily followed by a wet period; there were most often near-normal months ($0.4 < CDI \leq 0.5$: 33 (13%), 36 (14%) and 55 (22%) months in Medgidia, Tulcea, and Constanta, respectively) or short moist breaks.

This study computed the frequencies of the drought classes identified by CDI-1 and in situ SPI-1. Figure 18 shows that CDI-1 and in -situ SPI-1 have their closest similarity for severe and no drought conditions; there is a dissimilarity for the other types of drought classes, which may be due to the CDI's ability to identify moderate and mild drought events, which were not identified by the in situ SPI or were classified as near-normal conditions.

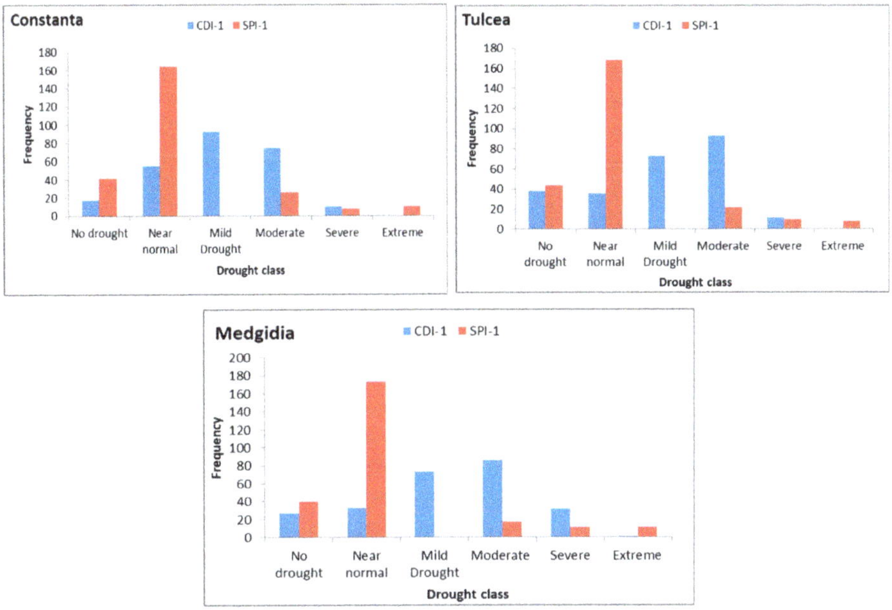

Figure 18. Comparison of the frequency of drought classes between CDI-1 and in situ SPI-1.

4. Discussion

4.1. Assessment of Drought Based on the CDI in Dobrogea Area and Practical Implication

As explained in Sections 1 and 2.1, Dobrogea is one of the driest regions of Romania, with precipitation that does not exceed 500 mm/year, temperatures that have increased by over 0.8 °C in the last 3 decades, and the presence of dry winds [72]. Studies conducted by [72] show that the coastal area has the longest periods of drought. In order to estimate drought in Dobrogea, a composite drought indicator proposed by Abdourahamane [39] was selected. This type of indicator was used for the first time in a research study in this region, where the density of meteorological stations is small (see Section 2.2). This index was chosen for the following reasons: (i) it integrates data obtained from satellite images, (ii) it uses precipitation, a climatic parameter that governs the wet and dry periods in the region, along with temperature (LST) and vegetation conditions (NDVI), and (iii) it has a spatial resolution of 1 km^2. The results are promising and consistent with other drought research studies in Dobrogea, which will allow us to overcome the limits described in Section 3.8. This study cannot be extended to the entire period for which data measurements are available at meteorological stations (1965–2000) since the GPM IMERG product cannot provide data for this period, and, although the CHIRPS product provides information for the period 1981–2000, the errors are greater than those obtained with the GPM IMERG product, as described in Section 3.1.

Using the proposed methodology, we can highlight the following results: (i) the LST value results indicate a temperature rise, consistent with the previous Dobrogea region results [22]; (ii) the NDVI values decreased, which could indicate that agricultural lands were under considerable stress due to high temperatures and irregular and below normal rainfall; (iii) the time series of CDI-3 during 2001–2021 had a significant downward trend, signaling that during the study period, most parts of the study area experienced drought conditions; (iv) the spatiotemporal analysis of the CDI reveals the periods and areas prone to drought, i.e., 2001, 2007–2008, 2011–2012, and 2014–2021, in mostly southern and central areas; (v) there is an acceptable level of agreement between the CDI, SPI and SPEI estimations in the study area, at different timescales; and (vi) the CDI can provide consistent information regarding duration, severity, intensity, and frequency of drought.

These results can be used as part of the DMP in Dobrogea. Still, the DMP must also take into account the climatic particularities of the area, and especially the consequences of using irrigation. The irrigation systems used in Dobrogea were extremely expensive. The water source was mainly the Danube and less often the Danube–Black Sea Canal or the freshwater lakes situated along the coastline. Because of the relief that rises from the Danube and the sea inland, it was necessary to design irrigation systems on pumping energy stages, which led to high energy costs for transporting water from the water source to the highest stage [41]. This is the reason why, nowadays, only irrigation systems located in the proximity of water sources still function. Sometimes, these irrigation systems come with enormous costs for other industries, such as tourism and energy. For example, the studies carried out in Techirghiol and Nuntasi revealed ecological catastrophes, which consisted of the transformation of the lake water from salty or brackish into freshwater [22,70]. This led to a decrease in salinity and the therapeutic quality of the sapropelic mud and, finally, to its disappearance. In Techirghiol, measures were taken in time because the area is well known for its therapeutic center [49]; however, in Nuntasi, no measures were seized, and the mud completely disappeared. However, in Dobrogea, due to climatic conditions, in order to grow large water-consuming plants (such as corn, sunflower, etc.), irrigation is absolutely necessary; therefore, they should be planted in areas that do not influence aquatic ecosystems (some of them being protected areas). For these types of areas, small water-consuming plants combined with a drip irrigation system should be used. To conclude, in Dobrogea, projects are needed that combine the cultivation of small water-consuming plants and irrigation systems with water sources that do not lead to energy consumption for transportation, which will aid in the protection of lakes and rivers. Otherwise, natural ecosystems will be destroyed.

4.2. Limitations

This research facilitates a deeper comprehension of the changes in the Drobrogea area in recent decades (2001–2021), but it is subject to certain limitations. For a study of extreme events such as meteorological drought, at least 30 years of records of climatic parameters (precipitation, temperatures) are needed. Unfortunately, these data are not always available, and not all ground stations have data without any gaps. Also, given the fact that only a few meteorological stations operated by the National Administration of Meteorology (NAM) cover the Dobrogea area, eventually, a small number of meteorological stations were used in this research to collect data for validation, and this could alter the precision of our tests. Nevertheless, the selected stations largely cover the measurement areas proposed by the NAM (the coast, the Danube bank, and the interior area of Dobrogea), being representative (50%) for each area.

This study aims to estimate the CDI using two satellite products for precipitation: CHIRPS and GPM IMERG. Considering the availability of GPM IMERG data (starting with June 2000), for a coherent study, we chose the period 1 January 2001–31 December 2021, this being the period in which satellite data for both products are available, which also corresponds to the availability of test data from weather stations (until 2021). Therefore, this research is conducted in a period not long enough for our results to be significant for understanding long-term drought conditions [73], but rather short- and medium-term trends.

It is well known that the NDVI reflects the health of vegetation during drought periods, but it is difficult to interpret when vegetation is sparse or when areas are analyzed in proximity to water (high absorption of the near-infrared spectrum). For this reason, it will be necessary to integrate other more robust vegetation indices into the CDI calculation.

5. Conclusions

Understanding how drought affects agriculture and the development of the Dobrogea region is an essential component of this study, which can provide useful insights into land and environmental planning in this area. This study's primary goal was to identify, apply, and evaluate a method to calculate a drought index based only on the RS data for the Dobrogea region, given that other indices such as the SPI or SPEI reveal their limitations, due, in particular, to strong dependency on ground datasets provided by meteorological stations. The prime European platform that provides information about drought, the EDO web portal [74], offers several products related to drought indicators, like the SPI, CDI, SMA, etc., the resolution of these products being around 5 km × 5 km. The technique proposed in this study gives the possibility to calculate a drought hazard index in near real time, based only on RS data, and with a better spatiotemporal resolution, using MODIS satellite imagery (with a spatial resolution of 1 km × 1 km and a frequent repeat cycle of 1–2 days) to retrieve the data related to the NDVI and LST, and a high-resolution satellite rainfall product—GPM IMERG, to estimate precipitation. The research in this paper also reveals that there is a reasonable level of agreement between the CDI, SPI, and SPEI as drought evaluation tools in the study area. To conclude, the Dobrogea area is vulnerable to drought, and the CDI could be used for drought detection as part of the DMP. In the future, we will extend our study to other stations. As the interpretation of the NDVI is quite difficult, we plan to introduce other indices that express vegetation conditions much more robustly.

Author Contributions: Conceptualization, C.S. and C.M.; methodology, C.S. and C.M.; software, C.S.; validation, C.S.; writing—original draft preparation, C.S. and C.M.; writing—review and editing, C.S. and C.M. All authors have read and agreed to the published version of this manuscript.

Funding: The APC was funded by Transilvania University of Brasov.

Data Availability Statement: Dataset available on request from the authors.

Conflicts of Interest: The authors declare no conflict of interest.

References

1. Bokal, S.; Müller, R. *Integrated Drought Management in Central and Eastern Europe*; WMO: Geneva, Switzerland, 2018.
2. Maftei, C.E.; Bărbulescu, A.; Osman, A. Assessment of the Drought Risk in Constanta County, Romania. *Atmosphere* **2024**, *15*, 1281. [CrossRef]
3. Maftei, C.; Dobrica, G.; Cerneaga, C.; Buzgaru, N. Drought Land Degradation and Desertification—Case Study of Nuntasi-Tuzla Lake in Romania. In *Water Safety, Security and Sustainability: Threat Detection and Mitigation*; Vaseashta, A., Maftei, C., Eds.; Advanced Sciences and Technologies for Security Applications; Springer International Publishing: Cham, Switzerland, 2021; pp. 583–597, ISBN 978-3-030-76008-3.
4. CE. *COM (2007) 414—Addressing the Challenge of Water Scarcity and Droughts in the EU—EU Monitor*; CE: Brussels, Belgium, 2007.
5. CE. *COM (2012) 673—Blueprint to Safeguard Europe's Water Resources—EU Monitor*; CE: Brussels, Belgium, 2012.
6. Stein, U.; Özerol, G.; Tröltzsch, J.; Landgrebe, R.; Szendrenyi, A.; Vidaurre, R. European Drought and Water Scarcity Policies. In *Governance for Drought Resilience: Land and Water Drought Management in Europe*; Bressers, H., Bressers, N., Larrue, C., Eds.; Springer International Publishing: Cham, Switzerland, 2016; pp. 17–43, ISBN 978-3-319-29671-5.

7. Fatulová, E.; Majerčáková, O.; Houšková, B.; Bardarska, G.; Alexandrov, V.; Kulířová, P.; Gayer, J.; Molnár, P.; Fiala, K.; Tamás, J.; et al. *Guidelines for Preparation of the Drought Management Plans: Development and Implementation of Risk-Based Drought Management Plans in the Context of the EU Water Framework Directive—As Part of the River Basin Management Plans*; Global Water Partnership Central and Eastern Europe: Bratislava, Slovakia, 2015; ISBN 978-80-972060-1-7.
8. United Nations Office for Disaster Risk Reduction (UNDRR). Available online: https://www.undrr.org/terminology/hazard (accessed on 15 October 2024).
9. Kogan, F.N. Droughts of the Late 1980s in the United States as Derived from NOAA Polar-Orbiting Satellite Data. *Bull. Am. Meteorol. Soc.* **1995**, *76*, 655–668. [CrossRef]
10. Haied, N.; Foufou, A.; Khadri, S.; Boussaid, A.; Azlaoui, M.; Bougherira, N. Spatial and Temporal Assessment of Drought Hazard, Vulnerability and Risk in Three Different Climatic Zones in Algeria Using Two Commonly Used Meteorological Indices. *Sustainability* **2023**, *15*, 10. [CrossRef]
11. Zargar, A.; Sadiq, R.; Naser, B.; Khan, F.I. A review of drought indices. *Environ. Rev.* **2011**, *19*, 333–349. [CrossRef]
12. WMO; GWP. *Handbook of Drought Indicators and Indices*; Svoboda, M., Fuchs, B., Eds.; WMO: Geneva, Switzerland, 2016; ISBN 978-92-63-11173-9.
13. Palmer, W.C. Meteorological Drought. *Weather. Bur. Res. Pap.* **1965**, *45*, 1–58.
14. McKee, T.B.; Doesken, N.J.; Kleist, J. The relationship of drought frequency and duration to time scales. *Eighth Conf. Appl. Climatol.* **1993**, *22*, 174–184.
15. Vicente-Serrano, S.M.; Beguería, S.; López-Moreno, J.I. A Multiscalar Drought Index Sensitive to Global Warming: The Standardized Precipitation Evapotranspiration Index. *J. Clim.* **2010**, *23*, 1696–1718. [CrossRef]
16. Hayes, M.; Svoboda, M.; Wall, N.; Widhalm, M. The Lincoln Declaration on Drought Indices: Universal Meteorological Drought Index Recommended. *Bull. Am. Meteorol. Soc.* **2011**, *92*, 485–488. [CrossRef]
17. Alahacoon, N.; Edirisinghe, M.A. Comprehensive Assessment of Remote Sensing and Traditional Based Drought Monitoring Indices at Global and Regional Scale. *Geomat. Nat. Hazards Risk* **2022**, *13*, 762–799. [CrossRef]
18. European Drought Observatory | Copernicu. Available online: https://www.copernicus.eu/en/european-drought-observatory (accessed on 23 October 2024).
19. EDO. Available online: https://www.copernicus.eu/ro/node/8534 (accessed on 15 October 2024).
20. Mateescu, E.; Smarandache, M.; Jeler, N.; Apostol, V. Drought Conditions and Management Strategies in Romania (Capacity Development to Support National Drought Management Policy) [Country Report]. Available online: https://www.droughtmanagement.info/literature/UNW-DPC_NDMP_Country_Report_Romania_2013.pdf (accessed on 1 July 2015).
21. Maftei, C.; Barbulescu, A. Statistical Analysis of Climate Evolution in Dobrudja Region. In Proceedings of the World Congress on Engineering, WCE, London, UK, 2–4 July2008; Volume 2, pp. 2–4.
22. Dobrica, G. Evaluarea Secetei Hidrologice în Bazinul Hidrografic al Lacului Nuntași-Tuzla, Județul Constanța (Assessment of the Hydrological Drought in the Nuntași-Tuzla Lake Watershed, Constanța County). Ph.D. Thesis, Unpublished. Ovidius University of Constanta, Constanta, Romania, 2023.
23. Croitoru, A.-E.; Piticar, A.; Imbroane, A.M.; Burada, D.C. Spatiotemporal distribution of aridity indices based on temperature and precipitation in the extra-Carpathian regions of Romania. *Theor. Appl. Climatol.* **2013**, *112*, 597–607. [CrossRef]
24. Dascălu, S.I.; Gothard, M.; Bojariu, R.; Birsan, M.-V.; Cică, R.; Vintilă, R.; Adler, M.-J.; Chendeș, V.; Mic, R.-P. Drought-related variables over the Bârlad basin (Eastern Romania) under climate change scenarios. *CATENA* **2016**, *141*, 92–99. [CrossRef]
25. Ionita, M.; Scholz, P.; Chelcea, S. Assessment of droughts in Romania using the Standardized Precipitation Index. *Nat. Hazards* **2016**, *81*, 1483–1498. [CrossRef]
26. Paltineanu, C.; Mihailescu, I.F.; Prefac, Z.; Dragota, C.; Vasenciuc, F.; Claudia, N. Combining the standardized precipitation index and climatic water deficit in characterizing droughts: A case study in Romania. *Theor. Appl. Climatol.* **2009**, *97*, 219–233. [CrossRef]
27. Stefan, S.; Ghioca, M.; Rimbu, N.; Boroneant, C. Study of meteorological and hydrological drought in southern Romania from observational data. *Int. J. Climatol.* **2004**, *24*, 871–881. [CrossRef]
28. Chelcea, S.M.; Aldea, A.A.; Trifu, M.C. An Assessment of Low Flow and Water Deficits on the Danube and Romanian Rivers During 1980–2020. In *Modeling and Monitoring Extreme Hydrometeorological Events*; IGI Global Scientific Publishing: Hershey, PA, USA, 2024; pp. 185–229. [CrossRef]
29. Drobot, R.; Draghia, A.F.; Sîrbu, N.; Dinu, C. Synthetic Drought Hydrograph. *Hydrology* **2023**, *10*, 10. [CrossRef]
30. Pravalie, R.; Sîrodoev, I.; Peptenatu, D. Changes in the forest ecosystems in areas impacted by aridization in south-western Romania. *J. Environ. Health Sci. Eng.* **2014**, *12*, 2. [CrossRef]
31. Nasiri, V.; Sadeghi, S.M.M.; Moradi, F.; Afshari, S.; Deljouei, A.; Griess, V.C.; Maftei, C.; Borz, S.A. The Influence of Data Density and Integration on Forest Canopy Cover Mapping Using Sentinel-1 and Sentinel-2 Time Series in Mediterranean Oak Forests. *ISPRS Int. J. Geo-Inf.* **2022**, *11*, 423. [CrossRef]

32. Cheval, S.; Dumitrescu, A. The summer surface urban heat island of Bucharest (Romania) retrieved from MODIS images. *Theor. Appl. Climatol.* **2015**, *121*, 631–640. [CrossRef]
33. Vorovencii, I. Assessing and monitoring the risk of desertification in Dobrogea, Romania, using Landsat data and decision tree classifier. *Environ. Monit. Assess.* **2015**, *187*, 204. [CrossRef]
34. Angearu, C.-V.; Ontel, I.; Boldeanu, G.; Mihailescu, D.; Nertan, A.; Craciunescu, V.; Catana, S.; Irimescu, A. Multi-Temporal Analysis and Trends of the Drought Based on MODIS Data in Agricultural Areas, Romania. *Remote Sens.* **2020**, *12*, 23. [CrossRef]
35. Vicente-Serrano, S.M.; Juez, C.; Potopová, V.; Boincean, B.; Murphy, C.; Domínguez-Castro, F.; Eklundh, L.; Peña-Angulo, D.; Noguera, I.; Jin, H.; et al. Drought Risk in Moldova under Global Warming and Possible Crop Adaptation Strategies. *Ann. N. Y. Acad. Sci.* **2024**, *1538*, 144–161. [CrossRef]
36. Páscoa, P.; Gouveia, C.M.; Russo, A.C.; Bojariu, R.; Vicente-Serrano, S.M.; Trigo, R.M. Drought Impacts on Vegetation in Southeastern Europe. *Remote Sens.* **2020**, *12*, 2156. [CrossRef]
37. Badaluta, C.A.; Haliuc, A.; Badaluta, G.; Scriban, R.E. Spatiotemporal variability of drought in Romania during 1901-2021 using the Standardized Precipitation Evapotranspiration Index (SPEI). *Analele Univ. Din Oradea Ser. Geogr.* **2024**, *1*, 33–43. [CrossRef]
38. Angearu, C.V.; Irimescu, A.; Mihailescu, D.; Virsta, A. Evaluation of droughts and fires in the Dobrogea region, using Modis satellite data, Agriculture for Life Life for Agriculture. *Conf. Proc.* **2018**, *1*, 336–345.
39. Abdourahamane, Z.S.; Garba, I.; Gambo Boukary, A.; Mirzabaev, A. Spatiotemporal Characterization of Agricultural Drought in the Sahel Region Using a Composite Drought Index. *J. Arid. Environ.* **2022**, *204*, 104789. [CrossRef]
40. Maftei, C.; Barbulescu, A. Statistical Analysis of Precipitation Time Series in the Dobrudja Region. *Mausam* **2012**, *63*, 553–564. [CrossRef]
41. Roşu, L.; Macarov, L.I. Management of Drought and Floods in the Dobrogea Region. In *Extreme Weather and Impacts of Climate Change on Water Resources in the Dobrogea Region*; IGI Global: Hershey, PA, USA, 2015; pp. 403–443, ISBN 978-1-4666-8438-6.
42. Ciulache, S.; Torica, V. Dobrudja 's climate (in Romanian), Annals of the University of Bucharest. *Geogr. Ser.* **2003**, *17*, 83–105.
43. Posea, G.; Bogdan, O.; Zăvoianu, I. (Eds.) Geography of Romania. In *The Romanian Plain, Danube, Dobrudja Platou, The Romanian Black Sea Seaside and the Continental Platform (in Romanian)*; Romanian Academy Printing House: Bucharest, Romania, 2005; Volume 5.
44. Bandoc, G.; Pravalie, R. Climatic water balancedynamics over the last five decades in Romania'smost arid region, Dobrogea. *J. Geogr. Sci.* **2015**, *25*, 1307–1327. [CrossRef]
45. Mitrică, B.; Mateescu, E.; Dragotă, C.S.; Grigorescu, I.; Dumitraşcu, M.; Popovici, E.A. Climatechange impacts on agricultural crops in the Timiş Plain, Romania. *Rom. Agric. Res.* **2015**, *32*, 93–101.
46. Lup, A.; Miron, L. *Drought Management in the Agriculture of Dobrogea Province, MPRA Paper 53403*; University Library of Munich: Munich, Germany, 2013.
47. Chelcea, S.; Ionita, M.; Adler, M.-J. Identification of Dry Periods in the Dobrogea Region. In *Extreme Weather and Impacts of Climate Change on Water Resources in the Dobrogea Region*; IGI Global: Hershey, PA, USA, 2015; pp. 52–72, ISBN 978-1-4666-8438-6.
48. Ionita, M.; Chelcea, S. Spatio-Temporal Variability of Seasonal Drought over the Dobrogea Region. In *Extreme Weather and Impacts of Climate Change on Water Resources in the Dobrogea Region*; IGI Global: Hershey, PA, USA, 2015; pp. 17–51, ISBN 978-1-4666-8438-6.
49. Maftei, C.; Buta, C.; Carazeanu Popovici, I. The Impact of Human Interventions and Changes in Climate on the Hydro-Chemical Composition of Techirghiol Lake (Romania). *Water* **2020**, *12*, 2261. [CrossRef]
50. Balteanu, D.; Șerban, M. Global environmental changes. In *An Interdisciplinary Evaluation of Uncertainties*; Coresi Printing House: Bucharest, Romania, 2005. (In Romanian)
51. Google Earth Engine (GEE) Platform. Available online: https://developers.google.com/earth-engine/datasets/ (accessed on 10 May 2024).
52. Funk, C.; Peterson, P.; Landsfeld, M.; Pedreros, D.; Verdin, J.; Shukla, S.; Husak, G.; Rowland, J.; Harrison, L.; Hoell, A. The climate hazards infrared precipitation with stations—A new environmental record for monitoring extremes. *Sci. Data* **2015**, *2*, 150066. [CrossRef] [PubMed]
53. Javed, T.; Li, Y.; Rashid, S.; Li, F.; Hu, Q.; Feng, H.; Chen, X.; Ahmad, S.; Liu, F.; Pulatov, B. Performance and relationship of four different agricultural drought indices for drought monitoring in China's mainland using remote sensing data. *Sci. Total Environ* **2021**, *759*, 143530. [CrossRef] [PubMed]
54. Won, J.; Choi, J.; Lee, O.; Kim, S. Copula-based Joint Drought Index using SPI and EDDI and its application to climate change. *Sci. Total Environ* **2020**, *744*, 140701. [CrossRef]
55. Achour, K.; Meddi, M.; Zeroual, A.; Bouabdelli, S.; Maccioni, P.; Moramarco, T. Spatio-temporal analysis and forecasting of drought in the plains of northwestern Algeria using the standardized precipitation index. *J. Earth Syst. Sci.* **2020**, *129*, 42. [CrossRef]
56. Guttman, N.B. Comparing the palmer drought index and the standardized precipitation index 1. *JAWRA J. Am. Water Resour. Assoc.* **1998**, *34*, 113–121. [CrossRef]

57. Mpelasoka, F.; Hennessy, K.; Jones, R.; Bates, B. Comparison of suitable drought indices for climate change impacts assessment over Australia towards resource management. *Int. J. Climatol.* **2008**, *28*, 1283–1292. [CrossRef]
58. Yevjevich, V.M. An Objective Approach to Definitions and Investigations of Continental Hydrologic Droughts. In *Hydrology Papers*; Colorado State University: Denver, CO, USA, 1967.
59. Malik, A.; Kumar, A.; Salih, S.Q.; Yaseen, Z.M. Hydrological Drought Investigation Using Streamflow Drought Index. In *Intelligent Data Analytics for Decision-Support Systems in Hazard Mitigation: Theory and Practice of Hazard Mitigation*; Springer: Berlin/Heidelberg, Germany, 2021; pp. 63–88.
60. Aiyelokun, O.; Ogunsanwo, G.; Ojelabi, A.; Agbede, O. Gaussian Naïve Bayes Classification Algorithm for Drought and Flood Risk Reduction. In *Intelligent Data Analytics for Decision-Support Systems in Hazard Mitigation: Theory and Practice of Hazard Mitigation*; Springer: Berlin/Heidelberg, Germany, 2021; pp. 49–62.
61. Chen, S.; Li, Q.; Zhong, W.; Wang, R.; Chen, D.; Pan, S. Improved monitoring and assessment of meteorological drought based on multi-source fused precipitation data. *Int. J. Environ. Res. Public Health* **2022**, *19*, 1542. [CrossRef]
62. Yang, T.; Zhang, W.; Zhou, T.; Wu, W.; Liu, T.; Sun, C. Plant phenomics & precision agriculture simulation of winter wheat growth by the assimilation of unmanned aerial vehicle imagery into the WOFOST model. *PLoS ONE* **2021**, *16*, e0246874.
63. Araujo, J.; Born, D.G. Calculating percentage agreement correctly but writing its formula incorrectly. *Behav. Anal.* **1985**, *8*, 207. [CrossRef]
64. SPI Generator. Available online: https://drought.unl.edu/monitoring/SPI/SPIProgram.aspx (accessed on 10 July 2024).
65. Bayissa, Y.; Tadesse, T.; Demisse, G. Building a High-Resolution Vegetation Outlook Model to Monitor Agricultural Drought for the Upper Blue Nile Basin, Ethiopia. *Remote Sens.* **2019**, *11*, 371. [CrossRef]
66. Beguería, S.; Vicente Serrano, S.M. *SPEI Calculator*; Digital.CSIC: Madrid, Spain, 2009. [CrossRef]
67. Pazhanivelan, S.; Geethalakshmi, V.; Samykannu, V.; Kumaraperumal, R.; Kancheti, M.; Kaliaperumal, R.; Raju, M.; Yadav, M.K. Evaluation of SPI and Rainfall Departure Based on Multi-Satellite Precipitation Products for Meteorological Drought Monitoring in Tamil Nadu. *Water* **2023**, *15*, 1435. [CrossRef]
68. Gulacsi, A.; Kovacs, F. Drought monitoring with spectral indices calculated from MODIS satellite images in Hungary. *J. Environ. Geog.* **2015**, *8*, 11–20. [CrossRef]
69. Popescu, A.; Dinu, T.A.; Stoian, E.; Serban, V. Variation of the main agricultural crops yield due to drought in Romania and Dobrogea region in the period 2000–2019, Scientific Papers Series Management. *Econ. Eng. Agric. Rural. Dev.* **2020**, *20*, 397–416.
70. Serban, C.; Maftei, C.; Dobrică, G. Surface Water Change Detection via Water Indices and Predictive Modeling Using Remote Sensing Imagery: A Case Study of Nuntasi-Tuzla Lake, Romania. *Water* **2022**, *14*, 556. [CrossRef]
71. Ionita, M.; Tallaksen, L.M.; Kingston, D.G.; Stagge, J.H.; Laaha, G.; Van Lanen, H.A.J.; Scholz, P.; Chelcea, S.M.; Haslinger, K. The European 2015 drought from a climatological perspective. *Hydrol. Earth Syst. Sci.* **2017**, *21*, 1397–1419. [CrossRef]
72. Păltineanu, C.; Mihăilescu, I.F.; Seceleanu, I. *Dobrogea: Pedoclimatic Conditions, Consumption and Irrigation Water Requirements of the Main Agricultural Crops (in Romanian)*; Ex Ponto: Constanta, Romania, 2000.
73. Pedro-Monzonís, M.; Solera, A.; Ferrer, J.; Estrela, T.; Paredes-Arquiola, J. A review of water scarcity and drought indexes in water resources planning and management. *J. Hydrol.* **2015**, *527*, 482–493. [CrossRef]
74. EDO Web Portal. Available online: https://edo.jrc.ec.europa.eu/edov2/php/index.php?id=1051 (accessed on 15 October 2024).

water

MDPI

Article

Hydrochemical Characterization, Source Identification, and Irrigation Water Quality Assessment in the Voghji River Catchment Area, Southern Armenia

Gor Gevorgyan [1,*], Gor Khachatryan [1], Anita Varagyan [2], Vahagn Varagyan [2] and Ashok Vaseashta [3,4,5,*]

1 Scientific Center of Zoology and Hydroecology, National Academy of Sciences of the Republic of Armenia, Yerevan 0014, Armenia; gor.khachatryan2@edu.ysu.am
2 Faculty of Biology, Yerevan State University, Yerevan 0025, Armenia; avaragyan@ysu.am (A.V.); varagyan@ysu.am (V.V.)
3 International Clean Water Institute, Applied Research, Manassas, VA 20110, USA
4 Ghitu IEEN, Academy of Sciences of Moldova, MD-2001 Chisinau, Moldova
5 Institutul de Cercetare al Universităţii din Bucureşti, 050663 Bucharest, Romania
* Correspondence: gor.gevorgyan@sczhe.sci.am (G.G.); prof.vaseashta@ieee.org (A.V.)

Abstract: Water quality is a fundamental parameter for assessing the suitability of surface waters. Likewise, the hydrochemical behavior is critically important to understand for rivers used in irrigation. This study aims to evaluate and characterize the surface water quality of the Voghji River catchment basin for irrigation, as it reveals the hydrochemical origins in the catchment basin. Nine key parameters, including EC, Cl^-, SO_4^{2-}, Ca^{2+}, Mg^{2+}, Na^+, K^+, CO_3^{2-}, and HCO_3^-, were measured at seven sampling points in July and September 2017. The ion concentration patterns in July followed the sequence: $Ca^{2+} > Na^+ > K^+ > Mg^{2+}$ and $HCO_3^- > SO_4^{2-} > Cl^- > CO_3^{2-}$, while in September, they were $Ca^{2+} > Na^+ > Mg^{2+} > K^+$ and $HCO_3^- > SO_4^{2-} > Cl^- > CO_3^{2-}$. The sequences were almost similar between the two months, with minor differences in cation distribution, particularly between Mg^{2+} and K^+. Overall, Ca^{2+} and HCO_3^- were the dominant ions in the studied surface water samples. The concentrations of K^+, Na^+, Mg^{2+}, Ca^{2+}, Cl^-, SO_4^{2-}, and HCO_3^- were found to be well below the FAO irrigation water standards, indicating that the waters of the Voghji River and its tributaries (Achanan, Vachagan, and Geghi) were generally safe for irrigation. However, the FAO threshold value was exceeded only for CO_3^{2-} in the Vachagan River in Kapan Town. The chemical analysis of surface waters in the Voghji River catchment basin revealed dominant Ca^{2+}-HCO_3^- and mixed Ca^{2+}-K^+-SO_4^{2-}-Cl^- facies, with key geochemical processes including carbonate and gypsum dissolution, silicate weathering, and cation exchange. Ionic correlations indicated that Na^+ and Cl^- sources were influenced by both natural (e.g., halite dissolution, weathering) and anthropogenic inputs, while Ca^{2+} and Mg^{2+} primarily originated from carbonate dissolution. The Gibbs diagram suggested that rock–water interactions were the primary natural mechanism controlling the water chemistry, with evaporation also playing a significant role. Various indices, including the Kelly index, magnesium adsorption ratio, sodium percentage, sodium adsorption ratio, permeability index, potential salinity, residual sodium carbonate, soluble sodium percentage, and irrigation water quality index, were applied, along with US Salinity Laboratory diagram and Wilcox diagram, to further assess the irrigation suitability. Most indices confirmed the suitability of the waters for irrigation; however, the Achanan River near the mouth and the Voghji River downstream of Kapan Town exhibited moderate salinity levels, underscoring the need for water management to prevent potential soil degradation.

 https://doi.org/10.3390/w17060854

Keywords: surface water; Voghji river catchment basin; hydrogeochemical characteristics; irrigation properties; agroecological risks

1. Introduction

Agriculture plays a key role in sustaining both the economy and the quality of life (QoL) for the population. As the primary source of irrigation water, rivers are essential for socio-economic development and environmental protection [1–3]. Moreover, water security has become a critical global challenge [4], further complicating sustainable development in the agricultural sector. Limited water resources and pollution not only hinder yield growth but also impact the quality and quantity of agricultural products [5,6]. Insufficient water security also affects livelihoods, leading to food shortages and economic risks, which hinder the sustainable development of rural communities [3,7]. The availability of freshwater constrains sustainable development opportunities worldwide, particularly in semi-arid and arid regions [8]. As a mountainous country with an arid climate, Armenia, located on the northern edge of the subtropical zone, faces similar challenges. Under these conditions, surface waters become a critical resource, playing a key role in ecosystem maintenance. They not only support biodiversity but also provide essential ecosystem services vital for the country's agricultural sector and sustainable regional development [9,10]. Surface water resources face significant challenges, including water quality degradation and freshwater depletion. These issues arise from both human activities and natural geological processes [11,12]. Key anthropogenic impacts include industrial and agricultural pollution, inadequate wastewater treatment, and inefficient water resource management [13]. Meanwhile, geological processes such as erosion, soil salinization, and the introduction of natural pollutants into water systems also contribute to the deterioration of surface water quality [14,15]. Consequently, these challenges threaten ecosystem stability, human health, and regional socio-economic development, highlighting the need for comprehensive and effective solutions for sustainable water resource management [16].

Sustainable agriculture plays a key role in managing Armenia's water resources. By promoting the use of modern and efficient irrigation methods, environmentally friendly agricultural approaches, and the use of non-traditional water sources for agriculture, such as treated wastewater and brackish water, it is possible to significantly reduce water consumption in agriculture while maintaining or increasing crop yields. These approaches can reduce pressure on freshwater resources, ensuring their more sustainable and efficient use [17,18]. Such solutions not only enhance agricultural productivity but also mitigate the risks associated with inefficient water resource management, thus supporting Armenia's long-term economic and environmental sustainability. The adoption of sustainable agricultural practices can improve water quality, ensure more reliable water supplies for agricultural communities, and contribute to the health of both agriculture and ecosystems [19]. Overall, the agricultural sector plays a crucial role in sustainable water resources management, ensuring that Armenia's limited water resources are preserved for future generations. The approaches presented here are a significant contribution to the long-term strategic development of the country's agriculture and water resources [20].

Water quality assessment is vital for monitoring and conserving surface water resources, which are highly vulnerable to physicochemical stresses caused by human activities [21]. Thus, an in-depth study of the physicochemical variables and geochemical processes that characterize surface water features is essential for advancing the hydrochemical understanding of ecosystems. This, in turn, can enhance sustainability and improve water quality management practices [22,23]. Integrated water quality indices have been developed by combining the physicochemical properties of rivers and streams to provide a more accurate assessment of water quality variations at different locations. These indices are useful for effectively describing water quality [24–26]. Therefore, they serve as key metrics for evaluating the overall water quality based on various variables. Among the most important water quality indicators for agricultural suitability are the irrigation water quality index, sodium adsorption ratio, sodium percentage, soluble sodium percentage, permeability index, potential salinity, residual sodium carbonate, and magnesium adsorption ratio, all of which collectively define the irrigation water quality status [27–29].

Recent advancements in hydrochemical research on irrigation water have increasingly focused on integrating traditional hydrogeochemical techniques with advanced modeling approaches to enhance water quality assessments and management strategies [30]. Despite these advancements, significant challenges persist. Unsustainable water management practices have exacerbated soil salinity and sodicity, leading to substantial agricultural productivity losses [31], highlighting the need for adaptable tools in irrigation water management. Addressing these challenges requires a systematic and multidimensional approach to hydrochemical assessments and monitoring frameworks. This study employed a comprehensive methodology, integrating graphical, mapping, and statistical methods to evaluate surface irrigation water quality in an anthropogenically influenced region. A detailed analysis of physicochemical parameters was conducted to assess the suitability of water for irrigation. By integrating multiple water quality indices, the study established a robust framework for capturing seasonal variations and spatial heterogeneity across different sites. Furthermore, geochemical relationship analysis was utilized to trace the evolution of water chemistry, offering critical insights into key hydrogeochemical processes such as mineral dissolution, ion exchange, and weathering dynamics. Beyond the results and conclusions, the study demonstrates that a comprehensive assessment of irrigation water quality can be effectively conducted at a regional scale with relatively limited effort and resources. There has been limited research on the irrigation properties of surface waters in the Republic of Armenia. While a few studies have reported on irrigation water quality in the Voghji River catchment basin [32–34], detailed information on surface water facies, hydrochemical origins, geochemical regulating mechanisms, and agroecological risks remains unpublished.

2. Materials and Methods

The materials and methods include site description, water sampling, sample and data analysis, and processing.

2.1. Sites Description

The Voghji River catchment basin is located between the high mountain ranges of the Bargushat and Meghri ranges in the southern part of the Republic of Armenia. The geological structure of the basin is primarily composed of volcanic-sedimentary rocks. The landscape features noticeable fragmentation, with numerous V-shaped valleys and gorges. Contemporary ecogenesis is largely influenced by fluvial processes. The soils in the region are classified as brown forest soils with a carbonate component [35]. The basin's primary watercourse is the Voghji River, which originates on the slopes of Mount Kaputjugh and is fed by several tributaries. Within Armenia's borders, the river spans a length of 56 km, with a catchment area of 1175 km^2. The river has an average slope of 0.058, and its annual discharge is approximately 9.6 m^3/s. The river network's density coefficient varies from 1.5 to 2.0 km/km^2, while the flow coefficient is 0.58. The river's hydrological regime is primarily influenced by meltwater, followed by rainwater and groundwater contributions. Due to the mountainous terrain of the basin, there is a distinct vertical climatic variation. The region exhibits the following climate types: dry subtropical, moderately warm semi-arid, moderately cold forested, cold mountainous, and alpine tundra [36–38]. Surface waters in the Voghji River catchment area are used for irrigation, energy, industrial, and other purposes. As shown in Figure 1, the land use within the catchment basin is predominantly oriented towards agricultural activities, with a substantial portion of the area allocated for crop production and related farming practices. This highlights the critical role of agriculture in the region's land management and irrigation water demand.

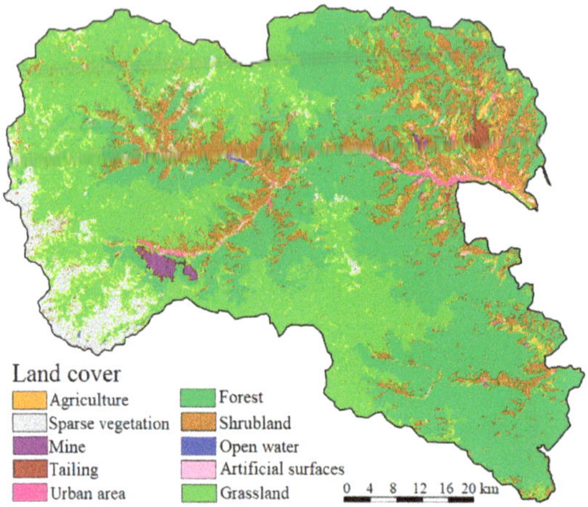

Figure 1. Land cover and land use classification within the Voghji River catchment area.

2.2. Water Sampling

Water samples were collected from seven sites across four rivers in July and September 2017 (Figure 2). River site V-1, located in an area with minimal anthropogenic influence, was used as a reference point for the river catchment basin, while the other sites were subject to noticeable human impact. Detailed information about the sampling sites and their corresponding names is provided in Table 1.

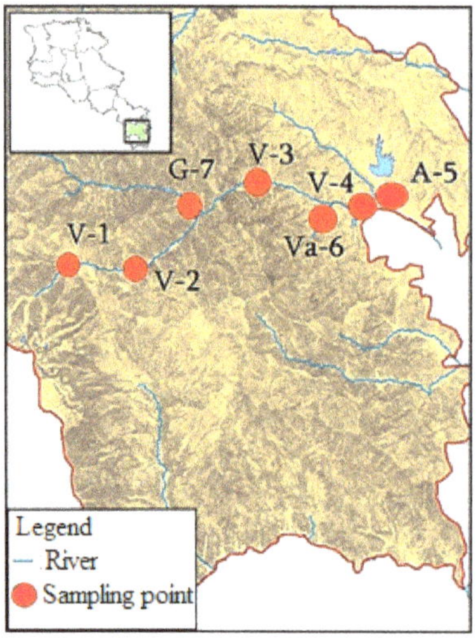

Figure 2. Location of the Voghji River and its catchment area in Armenia, along with the river observation sites as listed in Table 1.

Table 1. Sampling point details for rivers.

River	Sampling Point	The Location of the Sampling Point	Geographic Coordinates
Voghji	V-1	Upstream of Kajaran Town	39°09′26.9″ N 46°06′47.4″ E
Voghji	V-2	Downstream of Kajaran Town	39°09′01.8″ N 46°11′34.3″ E
Voghji	V-3	Upstream of Kapan Town	39°13′27.7″ N 46°20′19.4″ E
Voghji	V-4	Downstream of Kapan Town	39°11′53.0″ N 46°28′05.9″ E
Achanan	A-5	Near the river mouth	39°11′54.9″ N 46°28′02.8″ E
Vachagan	Va-6	In Kapan Town	39°11′53.5″ N 46°23′43.8″ E
Geghi	G-7	Near the river mouth	39°11′58.3″ N 46°15′31.8″ E

The water samples were collected in sterile sample containers and transported in a cool box under low-temperature conditions (4–8 °C). Electrical conductivity (EC) measurements were taken in the field using a conductometer (HI98129, Hanna Instruments, Nusfalau, Romania). The concentrations of total dissolved solids (TDS) were calculated from EC using a conversion factor of 0.65, following the approach described by Rusydi (2018) [39].

2.3. Sample Analysis

The quantitative analysis of Ca^{2+} and Mg^{2+} ions in water samples was performed using the ISO 7980:1986 method with an atomic absorption spectrometer (novAA 350, Analytik Jena, Jena, Germany) [40]. The concentrations of K^+ and Na^+ ions were measured using the same instrument but following the emission method [41].

The determination of CO_3^{2-} and HCO_3^- ions in water samples was conducted according to the GOST 31957-2012 method [42], while SO_4^{2-} and Cl^- ions were analyzed using the ISO 9297:2011 and GOST 4389-72 methods, respectively [43,44].

To ensure the accuracy and reliability of the surface water sample testing, several quality assurance measures were implemented. High-purity water was used throughout the analyses to avoid contamination. All glassware used was thoroughly pre-washed with a 10% nitric acid (HNO_3) solution to remove any potential residues or contaminants. Afterward, the glassware was carefully rinsed with distilled water to ensure it was free from any remaining chemicals before use in the sample analysis. Additionally, reference measurements were conducted alongside the sample testing to validate the results and confirm the reliability of the data. These precautions were taken to maintain the highest standards of analytical quality and ensure the credibility of the findings.

2.4. Data Analysis and Processing

All statistical analyses were conducted using Excel, ver. 2016 (Microsoft, Redmond, WA, USA) and Origin, ver. 2018 (OriginLab Corporation, Northampton, MA, USA). Mapping was performed using ArcGIS, ver. 10.3 (Esri, Redlands, CA, USA). Graphs were generated using Origin, ver. 2018 (OriginLab Corporation, Northampton, MA, USA). Freely accessible satellite datasets were utilized for land cover and land use classification within the Voghji River catchment area. Data from the Sentinel-1 and Sentinel-2 missions were selected due to their superior spatial resolution. To minimize cloud cover, enhance reflectance in the near-infrared (NIR) spectrum from green vegetation, and prevent misclassification caused by snow, all scenes were acquired between July and early October 2021 (https://apps.sentinel-hub.com/eo-browser, accessed on 5 March 2025).

3. Assessment of Irrigation Water Quality and Agroecological Hazards

EC values and the concentrations of major ions were compared with the FAO irrigation water standards [45]. The suitability of water for irrigation was evaluated using the Kelly index (KI) [46], magnesium adsorption ratio (MAR) [47], sodium percentage (Na%) [48], sodium adsorption ratio (SAR) [49], potential salinity (PS) [50], permeability index (PI) [50,51], residual sodium carbonate (RSC) [52], soluble sodium percentage (SSP) [53], Gibbs ratio [54], irrigation water quality index (IWQI) [55], US Salinity Laboratory diagram [56], and Wilcox diagram [57]. Additionally, GIS zoning maps were used to visualize the data for each parameter, providing a spatial understanding of the water quality variability across the study area.

The alkaline hazard of irrigation water for soil was assessed using the KI, calculated with the following equation [46]:

$$KI = \frac{Na^{+}}{Ca^{2+}\ Mg^{2+}}, \quad (1)$$

A KI greater than 1 indicates an excessive concentration of Na^{+}, rendering the water unsuitable for irrigation.

The magnesium hazard of irrigation water for soil was evaluated using the MAR, calculated with the following equation [47]:

$$MAR = \frac{Mg^{2+}}{Ca^{2+}\ Mg^{2+}} \times 100, \quad (2)$$

For irrigation water to be considered suitable, MAR values should be below 50%. The salinity levels of irrigation water were estimated using Na%, calculated with the following equation [48]:

$$Na\% = \frac{Na^{+} + K^{+}}{Ca^{2+}\ Mg^{2+} + Na^{+} + K^{+}} \times 100, \quad (3)$$

Based on Na% values, irrigation water can be classified into five categories: excellent (Na% < 20), good (Na% = 20–40%), acceptable (Na% = 40–60%), questionable (Na% = 60–80%), and unsuitable (Na% > 80%).

The sodium hazard of irrigation water for crops was assessed using the SAR, calculated with the following equation [49]:

$$SAR = \frac{Na^{+}}{\sqrt{\frac{Ca^{2+} + Mg^{2+}}{2}}}, \quad (4)$$

The SAR categorizes water into four hazard levels:

- Low sodium hazard (S1): SAR < 10. Water in this category is generally safe for irrigation, posing minimal risk to soil structure.
- Medium sodium hazard (S2): SAR = 10–18. This water may lead to moderate sodium accumulation, particularly in soils with low permeability, such as clay-rich soils.
- High sodium hazard (S3): SAR = 18–26. Water in this range can significantly affect the soil structure by causing clay dispersion, reducing permeability, and restricting aeration.
- Very high sodium hazard (S4): SAR > 26. Irrigation with such water is generally unsuitable without prior treatment, as it can lead to severe soil degradation, reduced infiltration rates, and long-term agricultural unsustainability.

The combined effect of SAR and EC determines the overall suitability of water for irrigation, as high sodium levels (SAR) can amplify the effects of salinity (EC), leading to reduced soil permeability and an increased risk of soil degradation. Based on this relationship, irrigation water is classified into four categories according to EC [56]:

- Low salinity risk (C1): EC < 250 μS/cm. Water in this category is suitable for most crops and soil types, posing minimal risk to plant growth.
- Medium salinity risk (C2): EC = 250–750 μS/cm. This water is generally suitable for soils with moderate drainage.
- High salinity risk (C3): EC = 750–2250 μS/cm. Water in this range may be problematic for soils with poor drainage, as excessive salt buildup can reduce crop yields.
- Very high salinity risk (C4): EC > 2250 μS/cm. This water is generally unsuitable for irrigation under normal conditions.

The risk of high salt concentrations, caused by Cl^- and $SO_4{}^{2-}$, in irrigation water was determined using the PS, calculated with the following equation [47]:

$$PS = Cl^- + \frac{SO_4^{2-}}{2}, \tag{5}$$

For soils with low permeability, irrigation water is classified based on PS (meq/L) into three risk categories: class I (low, <3 meq/L), suitable for most crops with minimal risk of salt accumulation; class II (moderate, 3–5 meq/L), requiring careful management to prevent salinity buildup; and class III (high, >5 meq/L), posing a significant risk that necessitates mitigation measures such as leaching and improved drainage to sustain soil productivity.

The permeability of irrigation water in soil was evaluated using the PI, calculated with the following equation [50,51]:

$$PI = \frac{(Na^+ + \sqrt{HCO_3^-}) \times 100}{(Ca^{2+} + Mg^{2+} + Na^+)}, \tag{6}$$

PI levels are classified into three categories: class I (excellent) and class II (good), with values above 75% and between 25% and 75%, respectively, both considered suitable for irrigation. Class III, with values below 25%, is considered unsuitable for irrigation.

The sodicity risk of irrigation water for soil was estimated using the RSC, calculated with the following equation [52]:

$$RSC = (HCO_3^- + CO_3^{2-}) - (Mg^{2+} + Ca^{2+}), \tag{7}$$

When RSC < 1.25 meq/L, the water is safe for irrigation with minimal sodicity risk. RSC = 1.25–2.50 meq/L indicates marginal quality, and RSC > 2.5 meq/L poses a high sodicity risk.

The salinity of irrigation water was assessed using the SSP, calculated with the following equation [53]:

$$SSP = \frac{Na^+}{Ca^{2+} + Mg^{2+} + Na^+ + K^+} \times 100, \tag{8}$$

SSP values above 50% indicate unsuitable water quality for irrigation, as they reduce soil permeability.

The chemical influence in irrigation water due to natural mechanisms was examined using Gibbs ratios, calculated with the following equations [54]:

$$\text{Gibbs ratio (I)} = \frac{Cl^-}{Cl^- + HCO_3^-}, \tag{9}$$

$$\text{Gibbs ratio (II)} = \frac{Na^+ + K^+}{Na^+ + K^+ + Ca^+}, \quad (10)$$

Irrigation water quality was evaluated using the IWQI, calculated with the following equation [55]:

$$IWQI = \sum_{i=1}^{n} q_i w_i, \quad (11)$$

where q_i represents the quality measurement value within the permissible range (Table 2), and w_i denotes the specified weight of each data set (Table 3). The IWQI is classified into five categories, as shown in Table 4.

Table 2. Irrigation water quality parameters and their limiting values.

	q_i	EC (μS/cm)	SAR (meq/L)	Na^+ (meq/L)	Cl^- (meq/L)	HCO_3^- (meq/L)
High	85–100	200–750	<3	2–3	<4	1.0–1.5
Medium	60–85	750–1500	3–6	3–6	4–7	1.5–4.5
Low	35–60	1500–3000	6–12	6–9	7–10	4.5–8.5
Very low	0–35	<200 or >3000	>12	<2 or >9	>10	<1.0 or >8.5

Table 3. The weights of the IWQI parameters.

Parameters	w_i
EC	0.211
Na^+	0.204
HCO_3^-	0.202
Cl^-	0.194
SAR	0.189
Total	1.000

Table 4. Classification of water quality for the investigated sites based on IWQI.

IWQI Values and Type of Restriction	Recommendations for Crops and Soil	
	Type of Plants	Soil
85–100 (no restriction)	No toxicity	Water can be used for all types of soils as a low risk of soil salinity and sodicity prevails.
70–85 (low restriction)	Avoid the use of salt-sensitive plants	Water can be used for light soil texture with high sand content and moderate to high permeability.
55–70 (moderate restriction)	Moderate salt-tolerance plants	Water can be used for moderate to high permeable soil taking into consideration moderate soil leaching processes.
40–55 (high restriction)	Moderate to high salt-tolerance plants	Water can be used for permeable soil without compact layers and taking into consideration the high frequency of the irrigation schedule for irrigation water with EC > 2000 μS/cm and SAR > 7.
0–40 (severe restriction)	High salt-tolerance plants only	Water cannot be used to irrigate soil under normal conditions.

The q_i values for the water quality parameters (qEC, $qSAR$, qNa^+, qCl^-, and $qHCO^{3-}$) were determined using the equation below:

$$q_i = q_{max} - \left(\frac{(x_{ij} - x_{inf}) \times q_{imap}}{x_{amp}}\right) \tag{12}$$

where q_{max} is the upper value of the corresponding class of q_i, x_{ij} represents the data points of the parameters (observed value of each parameter), x_{inf} refers to the lower limit value of the class to which the observed parameter belongs, q_{imap} represents the class amplitude for q_i classes, and x_{amp} corresponds to the class amplitude to which the parameter belongs.

4. Results and Discussion

4.1. Assessment of EC and Major Ions in Surface Waters

Figure 3 presents a summary of the results related to the physicochemical characteristics examined in this study. The average concentrations of K^+, Na^+, Mg^{2+}, Ca^{2+}, Cl^-, $SO_4{}^{2-}$, $HCO_3{}^-$, and $CO_3{}^{2-}$ in the Voghji River catchment area in July were 8.65, 19.43, 6.79, 29.18, 6.03, 85.51, 91.13, and 5.10 mg/L, respectively. In September, the concentrations were 9.26, 30.06, 12.13, 41.20, 11.93, 116.70, and 179.34 mg/L (Figure 3). In September, $CO_3{}^{2-}$ was not detected at any study site, while in July, it was recorded only at Va-6 and G-7 sites, with concentrations of 9 mg/L and 1.2 mg/L, respectively (Figure 3).

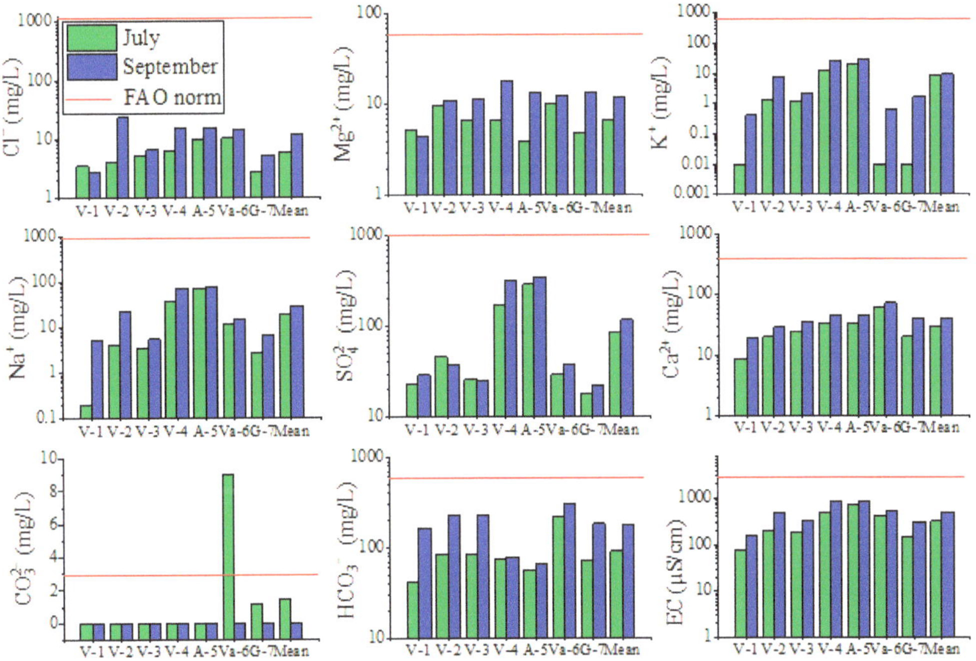

Figure 3. Quantitative values of EC and major ions at observation points in the Voghji River catchment basin in comparison with FAO regulatory thresholds [45].

The ion concentrations in July followed the sequences: $Ca^{2+} > Na^+ > K^+ > Mg^{2+}$ and $HCO_3{}^- > SO_4{}^{2-} > Cl^- > CO_3{}^{2-}$. In September, the sequences were: $Ca^{2+} > Na^+ > Mg^{2+} > K^+$ and $HCO_3{}^- > SO_4{}^{2-} > Cl^- > CO_3{}^{2-}$. The ion sequences in July and September were almost identical, with a slight difference observed in the cations, particularly between Mg^{2+}

and K^+. Overall, the findings indicated that Ca^{2+} and HCO_3^- were the dominant ions in the surface water samples studied.

EC values measured at seven locations in the Voghji River catchment basin ranged from 76 μS/cm to 834 μS/cm, with average values of 313 μS/cm in July and 491 μS/cm in September. The most abundant cation in the catchment basin was Ca^{2+}, with concentrations ranging from 8.8 mg/L to 72.9 mg/L. The average concentrations were 29.18 mg/L in July and 41.20 mg/L in September. The second most dominant cation was Na^+, with concentrations ranging from 0.20 mg/L to 80.09 mg/L, and average values of 19.43 mg/L in July and 30.06 mg/L in September. Mg^{2+} concentrations in the collected samples ranged from 3.9 mg/L to 17.9 mg/L, with average values of 6.79 mg/L in July and 12.13 mg/L in September. K^+ concentrations ranged from 0.42 mg/L to 27.94 mg/L, with average values of 8.65 mg/L in July and 9.26 mg/L in September. HCO_3^- was the predominant anion in the surface waters of the Voghji River catchment basin, with concentrations ranging from 42.7 mg/L to 298.0 mg/L. The mean concentrations were 91.13 mg/L in July and 179.34 mg/L in September. The second most abundant anion was SO_4^{2-}, with concentrations ranging from 17.9 mg/L to 350.4 mg/L and average values of 85.51 mg/L in July and 116.70 mg/L in September, followed by Cl^-, which ranged from 2.8 mg/L to 34.4 mg/L, with average values of 6.03 mg/L in July and 11.93 mg/L in September. These values indicate that the surface waters in the Voghji River catchment basin were suitable for irrigation, as they were well below the FAO threshold values (Figure 3). However, the CO_3^{2-} levels between 0 mg/L and 9 mg/L, with an average of 1.46 mg/L in July and 0 mg/L in September, exceeded the FAO threshold value at the Va-6 site in July (Figure 3). The CO_3^{2-} concentration was only detected at Va-6 and G-7 sites in July.

4.2. Hydrochemical Facies and Source Identification in Surface Waters

In the Voghji River catchment basin, surface water facies and geochemical factors were studied, revealing the dominance of HCO_3^- in cations and a more uniform distribution of Ca^{2+}, Na^+, and K^+ in anions. The hydrochemical properties of the surface water samples indicated the presence of Ca^{2+}-HCO_3^- and mixed Ca^{2+}-K^+- SO_4^{2-}-Cl^- water facies.

Ionic relationships were used to track the evolution of geochemical factors in the surface waters of the Voghji River catchment basin. In particular, the relationship between Na^+ and Cl^- is crucial for understanding the processes and mechanisms affecting water properties in the studied region. Surface water samples showed a moderate positive correlation (Figure 4a), suggesting that both ions likely originated from common sources such as halite dissolution or anthropogenic inputs. Most of the samples fell near or below the equiline, indicating that Cl^- levels in the surface waters were not primarily due to salt-bearing chlorides but may have been influenced by other sources, such as anthropogenic inputs or minor saline water intrusion. However, a subset of the samples plots above the equiline, pointing to additional geochemical processes such as silicate weathering and cation exchange. These processes resulted in Na^+ enrichment, where Na^+ ions were released through the weathering of silicate minerals or exchanged with Ca^{2+} and Mg^{2+} in soils and sediments. This deviation from the equiline further supports the role of these processes in influencing the water chemistry. The plot of HCO_3^- + SO_4^{2-} vs. Ca^{2+} + Mg^{2+} revealed a strong positive correlation (Figure 4b), indicating that carbonate rock weathering (e.g., calcite, dolomite) and gypsum dissolution were likely the dominant geochemical processes in the region. This suggests that the release of Ca^{2+} and Mg^{2+} ions into the surface waters were primarily driven by the dissolution of these minerals. A nearly perfect correlation between Na^+ and SO_4^{2-} further strengthens the geochemical link between these two ions (Figure 4c), likely resulting from the dissolution of sodium sulfate minerals or from agricultural and industrial inputs that contributed Na^+ and SO_4^{2-} ions to the water system. The Na^+ vs. (Ca^{2+} + Mg^{2+}) plot showed a weak relationship (Figure 4d), suggesting that Na^+ likely originated from different processes compared to Ca^{2+} and Mg^{2+}. While Na^+ may have derived from silicate weathering or anthropogenic sources, Ca^{2+} and Mg^{2+} were more likely released from carbonate dissolution processes. The weak correlation between Cl^- and SO_4^{2-} suggests that these two ions had distinct sources (Figure 4e). SO_4^{2-} was likely influenced by gypsum dissolution or human activities, while Cl^- may reflect saline water intrusion into the catchment basin. Finally, the negative correlation in the plot of (Ca^{2+} + Mg^{2+}) − (HCO_3^- + SO_4^{2-}) vs. (Na^+ − Cl^-) highlights the role of cation exchange processes (Figure 4f). In this case, Na^+ was likely replacing Ca^{2+} and Mg^{2+} in soils or clays, leading to the observed negative correlation between these variables.

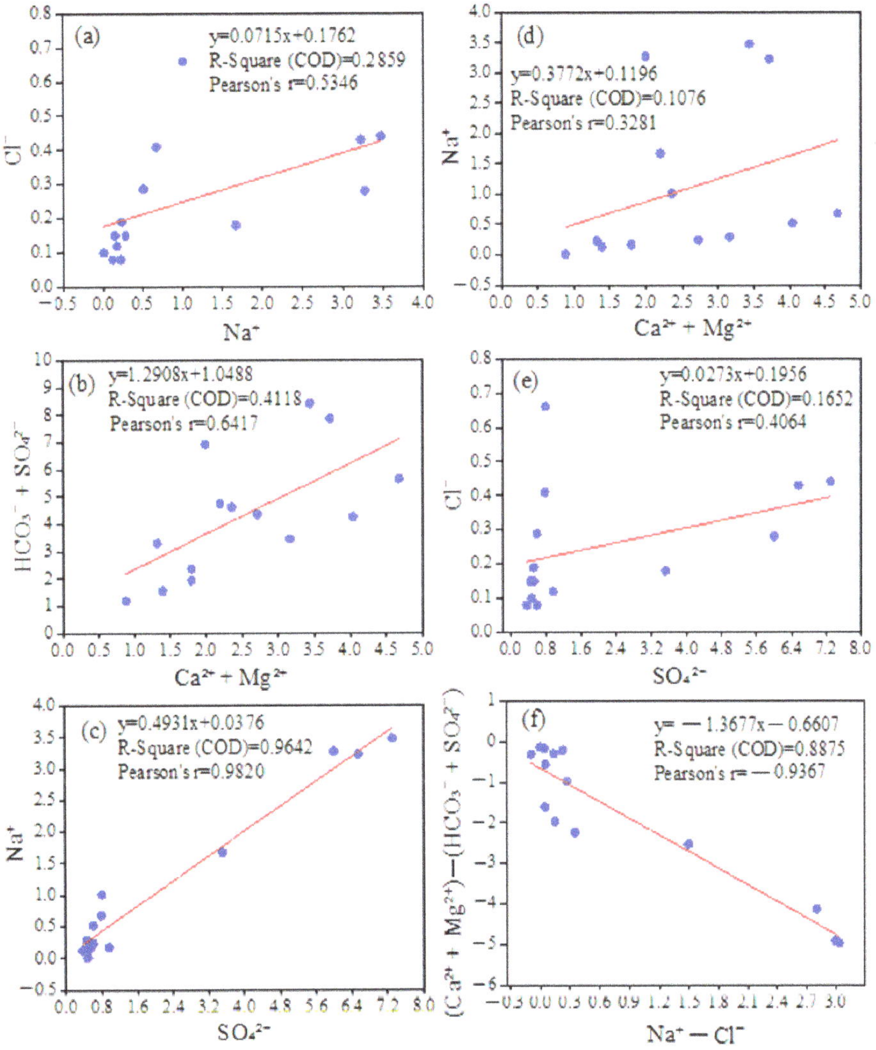

Figure 4. The relationships between major ions and their ionic ratios for the water samples acquired (meq/L): (**a**) Cl^- vs. Na^+, (**b**) HCO_3^- + SO_4^{2-} vs. Ca^{2+} + Mg^{2+}, (**c**) Na^+ vs. SO_4^{2-}, (**d**) Na^+ vs. Ca^{2+} + Mg^{2+}, (**e**) Cl^- vs. SO_4^{2-}, and (**f**) (Ca^{2+} + Mg^{2+}) − (HCO_3^- + SO_4^{2-}) vs. (Na^+ − Cl^-).

Figure 5 shows that most of the samples were located in the atmospherically dominated field of the rocks, indicating that the primary natural mechanism determining the water chemistry was the interaction between water and rocks. Additionally, the diagram suggests that evaporation influenced the waters. The $Cl^-/(Cl^- + HCO_3^-)$ ratios ranged from 0.02 meq/L to 0.69 meq/L in July, with an average of 0.28 meq/L, and from 0.14 meq/L to 0.64 meq/L in September, with an average of 0.35 meq/L, indicating strong anion exchange in the water system. The $Na^+/(Na^+ + Ca^{2+})$ ratios ranged from 0.06 meq/L to 0.23 meq/L in July, with an average of 0.12 meq/L, and from 0.03 meq/L to 0.29 meq/L in September, with an average of 0.13 meq/L (Figure 5).

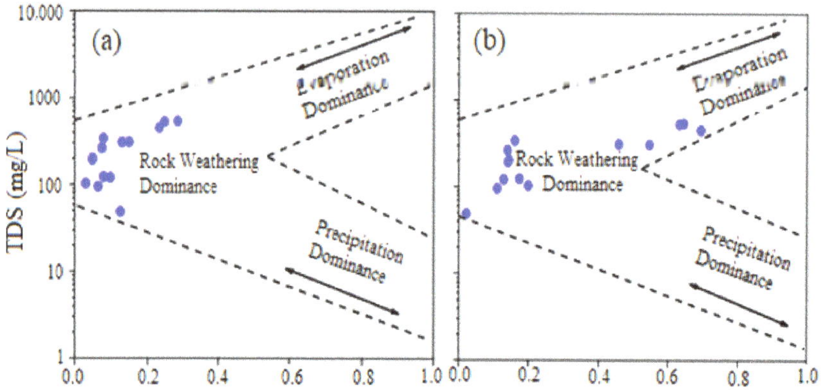

Figure 5. Chemical characteristics of surface waters in the Voghji River catchment basin based on Gibbs diagram: (**a**) $Na^+/(Na^+ + Ca^{2+})$, (**b**) $Cl^-/(Cl^- + HCO_3^-)$ [54].

4.3. Assessment of Irrigation Surface Water Quality

Figure 6 illustrates the distribution of several integrated water quality indices based on physicochemical characteristics, revealing a significant variation in the surface water quality within the Voghji River catchment basin for irrigation purposes. This variability was likely influenced by a combination of geological factors and anthropogenic impacts, which contributed to the differing water quality across the region.

SAR values ranged from 0.02 to 3.28 in July, with an average of 0.82, and from 0.21 to 2.66 in September, with an average of 1.02. As shown in Figure 4d, the surface waters were classified as excellent quality, indicating their suitability for irrigation without posing any alkalinity risk to crops. Furthermore, the SAR spatial distribution map demonstrates that the quality of the surface waters for irrigation remained consistent from upstream to downstream (Figure 6d). To classify irrigation water, the SAR was plotted against the EC, revealing that most surface water points were categorized within the C1-S1 and C2-S1 groups, characterized by medium salinity, low sodium content, and high suitability for irrigation (Figure 7). However, in September, the C3-S1 group included V-4 and A-5 sampling points. This indicates that an effective drainage system and adequate soil permeability are essential to optimize agricultural irrigation and mitigate salinity risks.

The Na% ranged from 1.01% to 65.53% in July, with an average of 21.81%, and from 9.75% to 54.94% in September, with an average of 26.71% (Figure 8). The surface water quality for crop irrigation varied from excellent to questionable (Figure 6c). Based on the Wilcox classification [48], the surface waters ranged from excellent to good and good to acceptable quality for irrigation, indicating that the waters were unlikely to cause soil alkalization and were suitable for irrigation (Figure 8).

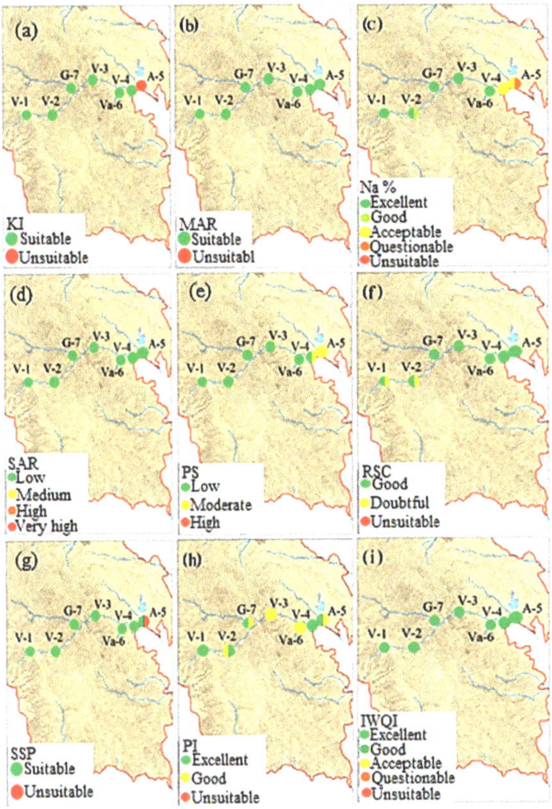

Figure 6. Maps showing the distribution of integrated water quality indices within the Voghji River catchment basin: (**a**) KI, (**b**) MAR, (**c**) Na%, (**d**) SAR, (**e**) PS, (**f**) RSC, (**g**) SSP, (**h**) PI, and (**i**) IWQI.

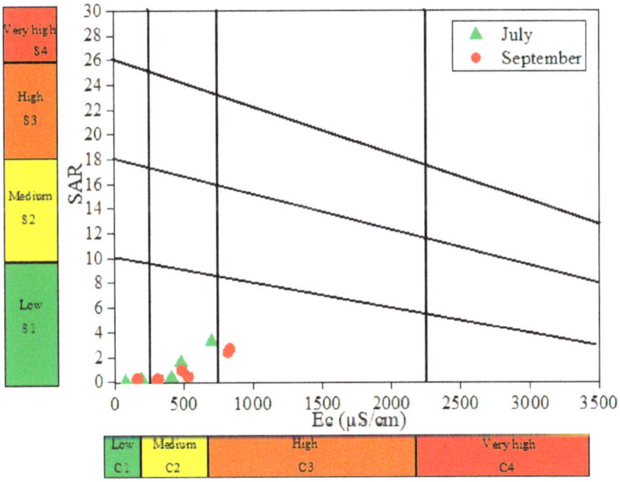

Figure 7. Irrigation water quality in the Voghji River catchment basin based on US Salinity Laboratory diagram [56].

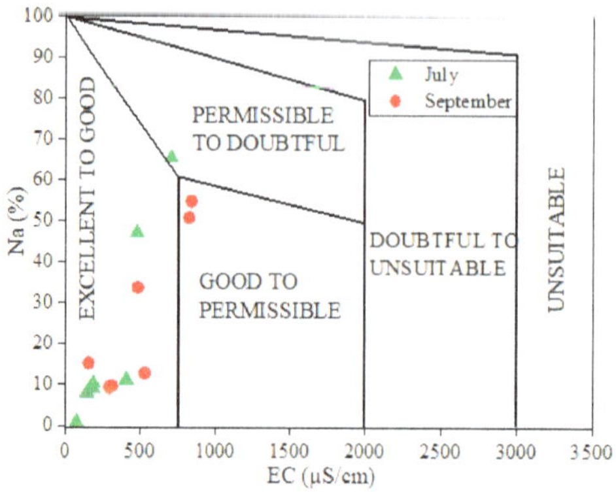

Figure 8. Irrigation water quality in the Voghji River catchment basin based on Wilcox diagram [57].

SSP values of the collected surface water samples ranged from 1.13% to 65.50% in July, with a mean of 21.82%, and from 9.46% to 54.98% in September, with a mean of 26.73%. Based on the SSP classification, the surface water samples with SSP levels below 60% were considered safe for irrigation (Figures 6g and 9). However, one exception was observed at the A-5 site in July, where the SSP value reached 65.5%, classifying the water as poor quality and unsuitable for irrigation. At the same site in September, the SSP value decreased to 54.98%, remaining high but within the acceptable limit (Figure 9).

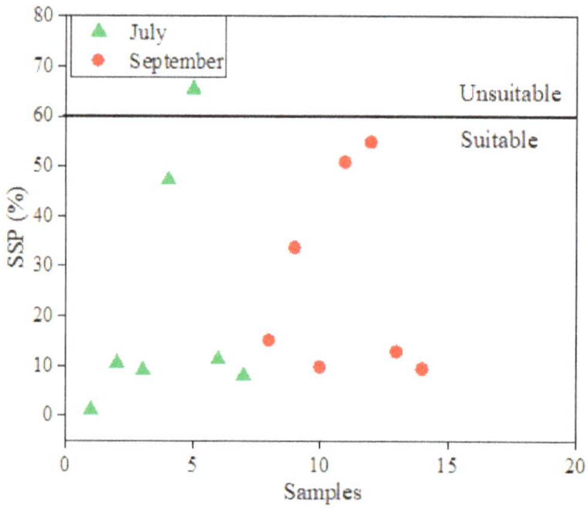

Figure 9. SSP values in the Voghji River catchment basin.

MAR values of the sampled water points ranged from 16.03% to 49.83% in July, with an average of 30.85%, and from 22.17% to 39.64% in September, with an average of 33.06% (Figure 10). These results indicate that the surface waters in all sampling points were suitable for irrigation (Figure 6b).

KI values of the sampled water points ranged from 0.01 to 1.64 in July, with an average of 0.40, and from 0.09 to 1.01 in September, with an average of 0.40 (Figure 11). Comparing the data of two months, KI values generally increased at most stations in September, with the largest rise at V-2 (+0.33) and V-1 (+0.16), while A-5 exhibited a notable decrease (−0.63), indicating possible seasonal or local influences. In July, the A-5 station had the highest KI value, making it unsuitable for irrigation, while all other stations had KI values below 1, indicating their waters were suitable for irrigation. In September, A-5 remained unsuitable with a KI value of 1.01, but the other stations continued to show suitability for irrigation, as their KI values remained below 1 (Figure 11). Overall, except for A-5, all other stations had waters suitable for irrigation (Figure 6a).

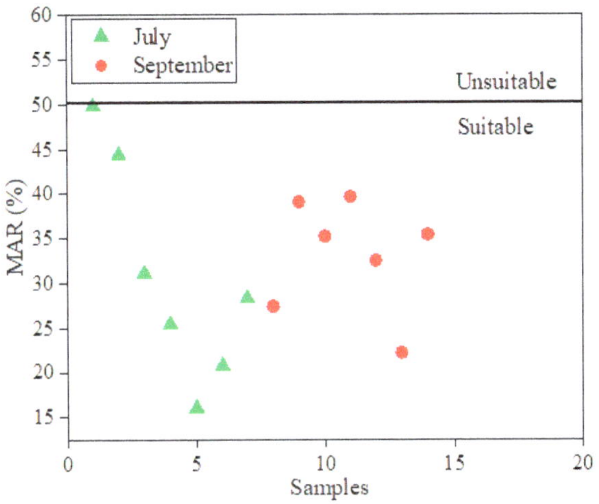

Figure 10. MAR values in the Voghji River catchment basin.

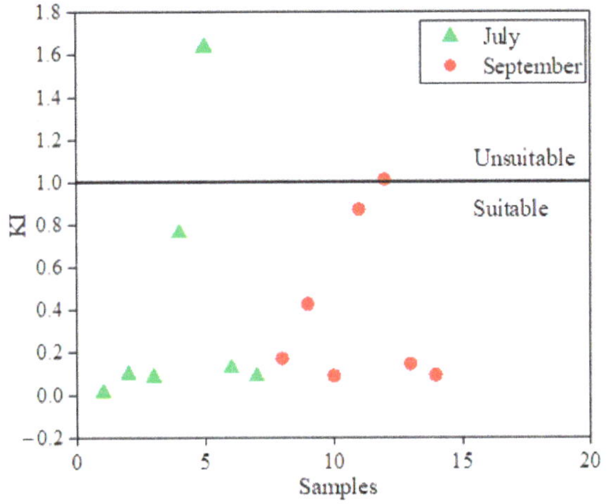

Figure 11. KI values in the Voghji River catchment basin.

PS values of the sampled water points ranged from 0.34 meq/L to 3.28 meq/L in July, with an average of 1.06 meq/L, and from 0.37 meq/L to 4.10 meq/L in September, with an average of 1.55 meq/L (Figure 12). In July, all stations except V-4 and A-5 remained in class I (low risk), indicating that the waters at V-1, V-2, V-3, Va-6, and G-7 had low salinity and posed minimal risk for crops. V-4 and A-5 were in class II (moderate risk), with PS values of 1.94 meq/L and 3.28 meq/L, respectively, suggesting a moderate risk that requires careful management to avoid salt buildup. In September, V-1, V-2, V-3, Va-6, and G-7 continued to be in class I, with PS values ranging from 0.26 meq/L to 0.80 meq/L. V-4 and A-5 showed an increase in PS, reaching 3.71 meq/L and 4.10 meq/L, respectively, placing them in class II (moderate risk), requiring ongoing monitoring and careful management to prevent salinity issues. No station fell into class III (high risk), meaning there was no immediate severe threat of excessive salinity accumulation. Overall, the waters from most stations were suitable for irrigation with minimal salinity concerns, though V-4 and A-5 need careful management due to their moderate salinity levels (Figure 6e).

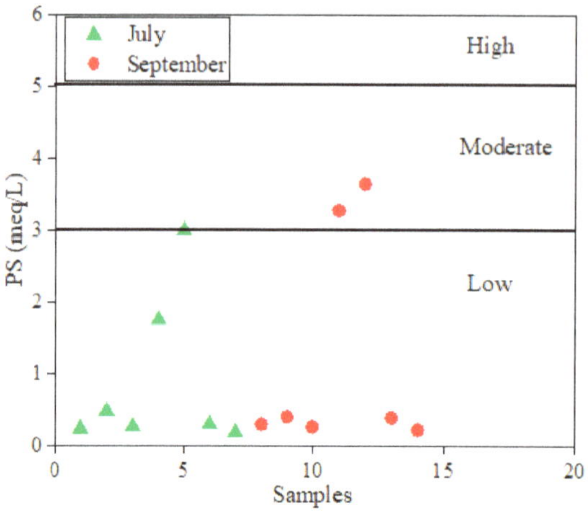

Figure 12. PS values in the Voghji River catchment basin.

RSC values of the sampled water points ranged from −1.08 meq/L to −0.1 meq/L in July, with an average of −0.50 meq/L, and from −2.42 meq/L to 1.44 meq/L in September, with an average of −0.20 meq/L (Figure 13). In July, all stations had negative RSC values, placing them in the safe category and indicating no risk of sodicity, making the waters suitable for irrigation. However, while negative RSC values suggest sufficient Ca^{2+} and Mg^{2+} to counteract Na^{+} accumulation, a slight risk of Na^{+} buildup remained. In contrast, positive RSC values indicate that HCO_3^{-} and CO_3^{2-} may have precipitated Ca^{2+} and Mg^{2+}, creating space for Na^{+} accumulation. In September, most stations remained in the safe category. However, V-1 and V-2 recorded RSC values of 1.37 meq/L and 1.44 meq/L, respectively, placing them in the borderline category and requiring close monitoring to prevent long-term soil degradation. The remaining stations maintained safe RSC levels. Notably, no site exceeded the 2.5 meq/L threshold, indicating no immediate risk of high Na^{+} accumulation. Overall, while the irrigation waters remained largely suitable, the increasing RSC values at V-1 and V-2 in September warrant attention and proactive irrigation management to mitigate potential long-term soil quality issues (Figure 6f).

PI values of the sampled water points ranged from 53.23% to 95.12% in July, with an average of 73.96%, and from 53.86% to 97.1% in September, with an average of 72.28% (Figure 14). In July, three stations (V-1, A-5, and G-7) exhibited excellent class, indicating minimal restrictions on irrigation. Four stations (V-2, V-3, V-4, and Va-6) fell into good class, indicating good but moderately suitable irrigation waters. In September, V-1 remained in excellent class, while V-2 improved from good to excellent class, reflecting the improvement in water quality at this location. However, A-5 and G-7 moved from excellent to good class, suggesting a slight decline in water quality. The other stations (V-3, V-4, and Va-6) remained in a good class, indicating stable but moderately suitable irrigation waters. Overall, the findings showed that all sites were classified as excellent and/or good (Figure 6h). Most of the water samples exhibited 75–100% of maximum permeability, while only two samples indicated 25% of maximum permeability (Figure 14).

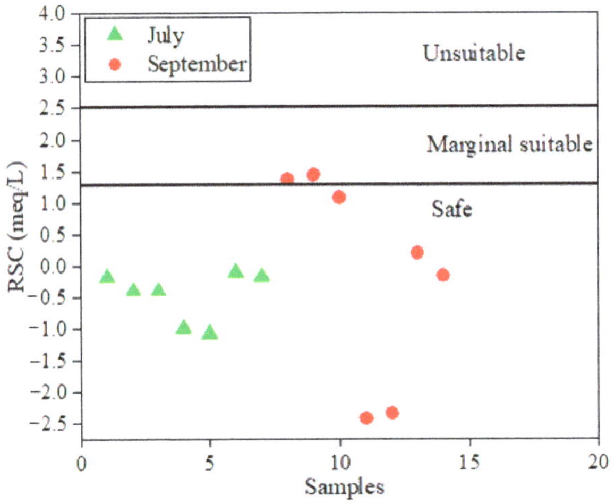

Figure 13. RSC values in the Voghji River catchment basin.

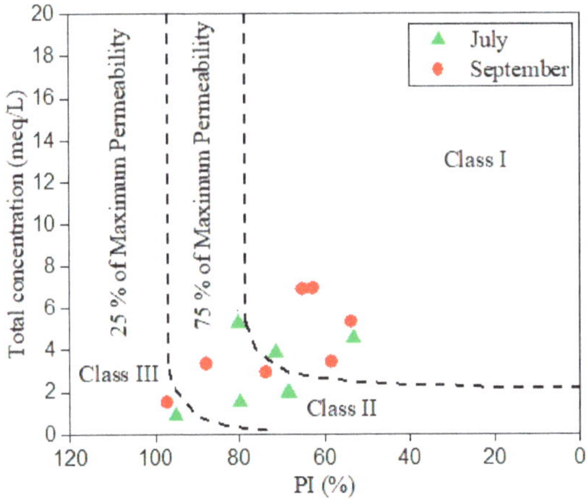

Figure 14. PI values in the Voghji River catchment basin based on the Doneen classification [51].

5. Limitations and Future Outlook

The study of irrigation water quality in the Voghji River catchment basin is limited by its spatial coverage, as the sampling sites may not fully represent the entire river basin, particularly the upper reaches. However, the selected sampling locations encompass key areas that are most likely to be influenced by both anthropogenic and natural contamination sources. It is important to note that the study may not comprehensively account for all human-made and natural impacts on water quality. Additionally, variations in water flow dynamics and contamination sources may further affect the results. Future research that includes groundwater sampling, alongside surface water, would provide a clearer understanding of the interactions between surface water and groundwater, and their combined impact on water quality [58,59]. Given that this study focused on the lower basin of the Voghji River catchment in Armenia, the findings may not be directly applicable to regions with different environmental and demographic characteristics. The upper basin, which may exhibit distinct characteristics, could be the focus of future studies to broaden the scope and applicability of the findings.

Additionally, the study focused on a specific set of physicochemical parameters related to water quality. To provide a more comprehensive assessment of water quality and its potential effects on both agriculture and human health, it would be beneficial to include additional factors, such as heavy metals, pesticides, and microbiological indicators. Expanding the scope to include these elements would enhance the overall understanding of water quality. Furthermore, integrating socioeconomic aspects, such as water demand, land use, and socio-cultural practices, into future studies could improve water quality management and inform decision-making processes.

6. Conclusions

This study evaluated the chemical characteristics of river waters in the Voghji River catchment basin for irrigation purposes, with samples taken from seven different locations. The findings indicated that Ca^{2+} and HCO_3^- were the dominant ions in the surface water samples. The hydrochemical analysis of the Voghji River catchment basin revealed that surface water facies were dominated by HCO_3^- in cations and exhibited a mixed Ca^{2+}-HCO_3^- and Ca^{2+}-K^+-SO_4^{2-}-Cl^- composition. Geochemical processes such as carbonate and gypsum dissolution, silicate weathering, and cation exchange played a key role in shaping the water chemistry. The Na^+-Cl^- relationship suggests contributions from both natural and anthropogenic sources, while ionic correlations highlight distinct geochemical pathways, including sodium enrichment and cation exchange mechanisms.

The study also assessed the suitability of waters from various sampling points for irrigation, considering key parameters such as Na^+, Ca^{2+}, Mg^{2+}, Cl^-, SO_4^{2-}, and others. The results revealed that the waters from most stations were suitable for irrigation, with low salinity and no immediate risk of soil alkalization or sodicity. However, the Achanan River near the mouth (A5) and the Voghji River downstream of Kapan Town (V4) exhibited moderate salinity levels, requiring careful management. The A-5 site, in particular, may have been impacted by potential effluent from the Artsvanik tailing dam, the largest tailings storage facility in Armenia, which could have subsequently influenced water quality at the V-4 station. Furthermore, water quality at the V-4 site may have been affected by domestic and mining discharges from Kapan Town and its surrounding areas. SAR and PI values indicated that most of the water sources had low salinity, high permeability, and minimal risk to soil quality. RSC values for the majority of stations were within the safe range.

The Gibbs diagram highlighted that the primary natural mechanisms influencing the water chemistry were rock–water interactions, with evaporation also playing a significant role. In conclusion, the study suggests that most of the analyzed water sources were generally suitable for irrigation, although A-5 and V-4 sites require careful management to avoid potential salinity and sodicity issues.

Author Contributions: Conceptualization, G.G. and A.V. (Ashok Vaseashta); methodology, G.G., G.K., A.V. (Anita Varagyan), and V.V.; software, G.K. and G.G.; validation, G.G., G.K., and A.V. (Ashok Vaseashta); formal analysis, G.K. and G.G.; investigation, G.G., A.V. (Anita Varagyan), V.V., and G.K.; resources, G.G.; data curation, G.G., G.K., A.V. (Anita Varagyan), and V.V.; writing—original draft preparation, G.K. and G.G.; writing—review and editing, G.G. and A.V. (Ashok Vaseashta); visualization, G.K., G.G., and A.V. (Ashok Vaseashta); supervision, G.G.; project administration, G.G.; funding acquisition, G.G. All authors have read and agreed to the published version of the manuscript.

Funding: This work was partly supported by the Higher Education and Science Committee of the Ministry of Education, Science, Culture, and Sports (MESCS) of the Republic of Armenia under research project No. 23LCG-1F005.

Data Availability Statement: All data are presented in the article. If additional information is needed, it is available via the corresponding author, upon reasonable request.

Conflicts of Interest: The authors declare no conflict of interest—financial or otherwise.

Abbreviations

The following abbreviations used in this manuscript may be helpful to readers.

EC	Electroconductivity
FAO	Food and Agriculture Organization
GIS	Geographic Information System
IWQI	Irrigation Water Quality Index
KI	Kelly Index
MAR	Magnesium Absorption Ratio
MESC	Ministry of Education, Science, Culture, and Sports
SAR	Sodium Adsorption Ratio
SSP	Soluble Sodium Percentage
PI	Permeability Index
RSC	Residual sodium carbonate
TDS	Total Dissolved Solids

References

1. Ingrao, C.; Strippoli, R.; Lagioia, G.; Huisingh, D. Water Scarcity in Agriculture: An Overview of Causes, Impacts and Approaches for Reducing the Risks. *Heliyon* **2023**, *9*, e18507. [CrossRef]
2. Feng, T.; Xiong, R.; Huan, P. Productive Use of Natural Resources in Agriculture: The Main Policy Lessons. *Resour. Policy* **2023**, *85*, 103793. [CrossRef]
3. Pawlak, K.; Kołodziejczak, M. The Role of Agriculture in Ensuring Food Security in Developing Countries: Considerations in the Context of the Problem of Sustainable Food Production. *Sustainability* **2020**, *12*, 5488. [CrossRef]
4. Vaseashta, A. Introduction to Water Safety, Security and Sustainability. In *Water Safety, Security and Sustainability*; Advanced Sciences and Technologies for Security Applications; Vaseashta, A., Maftei, C., Eds.; Springer International Publishing: Cham, Switzerland, 2021; pp. 3–22, ISBN 978-3-030-76007-6.
5. Mishra, B.; Kumar, P.; Saraswat, C.; Chakraborty, S.; Gautam, A. Water Security in a Changing Environment: Concept, Challenges and Solutions. *Water* **2021**, *13*, 490. [CrossRef]
6. Hanjra, M.A.; Qureshi, M.E. Global Water Crisis and Future Food Security in an Era of Climate Change. *Food Policy* **2010**, *35*, 365–377. [CrossRef]
7. Okello, C.; Githiora, Y.W.; Sithole, S.; Owuor, M.A. Nature-Based Solutions for Water Resource Management in Africa's Arid and Sem-Arid Lands (ASALs): A Systematic Review of Existing Interventions. *Nat.-Based Solut.* **2024**, *6*, 100172. [CrossRef]
8. Shemer, H.; Wald, S.; Semiat, R. Challenges and Solutions for Global Water Scarcity. *Membranes* **2023**, *13*, 612. [CrossRef]
9. Smith, P.; Ashmore, M.R.; Black, H.I.J.; Burgess, P.J.; Evans, C.D.; Quine, T.A.; Thomson, A.M.; Hicks, K.; Orr, H.G. REVIEW: The Role of Ecosystems and Their Management in Regulating Climate, and Soil, Water and Air Quality. *J. Appl. Ecol.* **2013**, *50*, 812–829. [CrossRef]
10. Musie, W.; Gonfa, G. Fresh Water Resource, Scarcity, Water Salinity Challenges and Possible Remedies: A Review. *Heliyon* **2023**, *9*, e18685. [CrossRef]
11. Baggio, G.; Qadir, M.; Smakhtin, V. Freshwater Availability Status across Countries for Human and Ecosystem Needs. *Sci. Total Environ.* **2021**, *792*, 148230. [CrossRef]
12. Cosgrove, W.J.; Loucks, D.P. Water Management: Current and Future Challenges and Research Directions. *Water Resour. Res.* **2015**, *51*, 4823–4839. [CrossRef]
13. Bashir, I.; Lone, F.A.; Bhat, R.A.; Mir, S.A.; Dar, Z.A.; Dar, S.A. Concerns and Threats of Contamination on Aquatic Ecosystems. In *Bioremediation and Biotechnology*; Hakeem, K.R., Bhat, R.A., Qadri, H., Eds.; Springer International Publishing: Cham, Switzerland, 2020; pp. 1–26, ISBN 978-3-030-35690-3.
14. Issaka, S.; Ashraf, M.A. Impact of Soil Erosion and Degradation on Water Quality: A Review. *Geol. Ecol. Landsc.* **2017**, *1*, 1–11. [CrossRef]
15. Anh, N.T.; Can, L.D.; Nhan, N.T.; Schmalz, B.; Luu, T.L. Influences of Key Factors on River Water Quality in Urban and Rural Areas: A Review. *Case Stud. Chem. Environ. Eng.* **2023**, *8*, 100424. [CrossRef]
16. Dinar, A. Challenges to Water Resource Management: The Role of Economic and Modeling Approaches. *Water* **2024**, *16*, 610. [CrossRef]
17. Levidow, L.; Zaccaria, D.; Maia, R.; Vivas, E.; Todorovic, M.; Scardigno, A. Improving Water-Efficient Irrigation: Prospects and Difficulties of Innovative Practices. *Agric. Water Manag.* **2014**, *146*, 84–94. [CrossRef]

18. Mishra, S.; Kumar, R.; Kumar, M. Use of Treated Sewage or Wastewater as an Irrigation Water for Agricultural Purposes-Environmental, Health, and Economic Impacts. *Total Environ. Res. Themes* **2023**, *6*, 100051. [CrossRef]
19. Stagnari, F.; Jan, S.; Angelica, G.; Michele, P. Sustainable Agricultural Practices for Water Quality Protection. In *Water Stress and Crop Plants*; Ahmad, P., Ed.; Wiley: Hoboken, NJ, USA, 2016; pp. 75–85, ISBN 978-1-119-05436-8.
20. World Bank. *Armenia—General Water Security Assessment*; Europe and Central Asia Region: Yerevan, Armenia, 2024; p. 48.
21. Varol, M.; Tokatli, C. Evaluation of the Water Quality of a Highly Polluted Stream with Water Quality Indices and Health Risk Assessment Methods. *Chemosphere* **2023**, *311*, 137096. [CrossRef]
22. Mohammed, M.A.A.; Szabó, N.P.; Szűcs, P. Multivariate Statistical and Hydrochemical Approaches for Evaluation of Groundwater Quality in North Bahri City-Sudan. *Heliyon* **2022**, *8*, e11308. [CrossRef]
23. Gevorgyan, G.; Mamyan, A.; Boshyan, T.; Vardanyan, T.; Vaseashta, A. Heavy Metal Contamination in an Industrially Affected River Catchment Basin: Assessment, Effects, and Mitigation. *Int. J. Environ. Res. Public Health* **2021**, *18*, 2881. [CrossRef]
24. Mammeri, A.; Tiri, A.; Belkhiri, L.; Salhi, H.; Brella, D.; Lakouas, E.; Tahraoui, H.; Amrane, A.; Mouni, L. Assessment of Surface Water Quality Using Water Quality Index and Discriminant Analysis Method. *Water* **2023**, *15*, 680. [CrossRef]
25. Chan Kujiok, D.; Sahile, Z.A. Water Quality Assessment of Elgo River in Ethiopia Using CCME, WQI and IWQI for Domestic and Agricultural Usage. *Heliyon* **2024**, *10*, e23234. [CrossRef]
26. Ushurhe, O.; Ozabor, F.; Onyeayana, W.V.; Adekunle, O.; Christabel, I.C.; Chike, D.F. Seasonal Sodium Percentage (%NA), Absorption Ratio (SAR) and Irrigation Water Quality Index (IWQI) Determination for Irrigation Purposes Along River Ethiope, Southern Nigeria. *J. Water Resour. Prot.* **2024**, *16*, 523–537. [CrossRef]
27. Satish Kumar, V.; Amarender, B.; Dhakate, R.; Sankaran, S.; Raj Kumar, K. Assessment of Groundwater Quality for Drinking and Irrigation Use in Shallow Hard Rock Aquifer of Pudunagaram, Palakkad District Kerala. *Appl. Water Sci.* **2016**, *6*, 149–167. [CrossRef]
28. Mokhtar, A.; Elbeltagi, A.; Gyasi-Agyei, Y.; Al-Ansari, N.; Abdel-Fattah, M.K. Prediction of Irrigation Water Quality Indices Based on Machine Learning and Regression Models. *Appl. Water Sci.* **2022**, *12*, 76. [CrossRef]
29. Eid, M.H.; Elbagory, M.; Tamma, A.A.; Gad, M.; Elsayed, S.; Hussein, H.; Moghanm, F.S.; Omara, A.E.-D.; Kovács, A.; Péter, S. Evaluation of Groundwater Quality for Irrigation in Deep Aquifers Using Multiple Graphical and Indexing Approaches Supported with Machine Learning Models and GIS Techniques, Souf Valley, Algeria. *Water* **2023**, *15*, 182. [CrossRef]
30. Gevorgyan, G.; Varagyan, V.; Varagyan, A. Agricultural Risk Assessment of the Heavy Metal and Mineral Salt Pollution of River Ecosystems in the Voghji River Catchment Area, Armenia. *Biol. J. Armen.* **2016**, *4*, 92–97.
31. Gad, M.; Saleh, A.H.; Hussein, H.; Elsayed, S.; Farouk, M. Water Quality Evaluation and Prediction Using Irrigation Indices, Artificial Neural Networks, and Partial Least Square Regression Models for the Nile River, Egypt. *Water* **2023**, *15*, 2244. [CrossRef]
32. Kramer, I.; Mau, Y. Review: Modeling the Effects of Salinity and Sodicity in Agricultural Systems. *Water Resour. Res.* **2023**, *59*, e2023WR034750. [CrossRef]
33. Varagyan, A.L.; Gevorgyan, G.A.; Varagyan, V.M. Mineral Salt Content in the Water of the Voghji River Catchment Basin (Armenia): Assessment of Potential Effects on Irrigated Soils and Crops. *Proc. YSU B Chem. Biol. Sci.* **2018**, *52*, 204–210. [CrossRef]
34. Viktorovich Larionov, M.; Haykaram Galstyan, M.; Garnik Ghukasyan, A.; Gagik Matevosyan, L.; Lendrush Hakobjanyan, I.; Arayik Gharibyan, P.; Yakhsibek Sayadyan, H.; Shahen Sargsyan, K. The Ecological and Sanitary-Hygienic Assessment of the River Systems Located in the Technogenic Polluted Zone of the Caucasus. *Egypt. J. Aquat. Res.* **2024**, *50*, 189–199. [CrossRef]
35. Mkrtchyan, S.S.; Gabrielyan, A.H.; Maghakyan, H.G.; Pafengolts, K.N.; Vardanyants, L.A. (Eds.) *Geology of the Armenian SSR, Volume I: Geomorphology*; Academy of Sciences Press: Yerevan, Armenia, 1962. (In Russian)
36. Fichtner Water & Transportation GmbH; Jrtuc LLC. *Armenian Small Municipalities Water Project: Environmental Impact Assessment—Kapan*; Project No. 610-1182; AWS CJSC: Yerevan, Armenia, 2013.
37. Gabrielyan, A.V.; Shahnazaryan, G.A.; Minasyan, S.H. Distribution and Identification of Sources of Heavy Metals in the Voghji River Basin Impacted by Mining Activities (Armenia). *J. Chem.* **2018**, *2018*, 7172426. [CrossRef]
38. Gevorgyan, G. Ecological Assessment of the Voghji and Meghriget Rivers and Their Catchment Basins. Ph.D Thesis, Yerevan State University, Yerevan, Armenia, 2011.
39. Rusydi, A.F. Correlation between Conductivity and Total Dissolved Solid in Various Type of Water: A Review. *IOP Conf. Ser. Earth Environ. Sci.* **2018**, *118*, 012019. [CrossRef]
40. *ISO 9964-3:1986*; Water Quality—Determination of Calcium and Magnesium—Atomic Absorption Spectrometric Method. International Organization for Standardization ISO: London, UK, 1986.
41. *ISO 9964-3:1993*; Water Quality—Determination of Sodium and Potassium—Part 3: Determination of Sodium and Potassium by Flame Emission Spectrometry. International Organization for Standardization ISO: London, UK, 1993.
42. *GOST 31957-2012*; Water—Methods for Determination of Alkalinity and Mass Concentration of Carbonates and Hydrocarbonates. Interstate Council for Standardization, Metrology, and Certification (ISC): Moscow, Russia, 2012.
43. *ISO 9297:2011*; Water Quality—Determination of Chloride—Silver Nitrate Titration with Chromate Indicator (Mohr's Method). International Organization for Standardization: London, UK, 2011.

44. *GOST 4389-72*; Drinking Water—Methods for Determination of Sulfate Content. Interstate Council for Standardization, Metrology, and Certification (ISC) Standard Publishing House: Moscow, Russia, 1972.
45. Ayers, R.; Westcott, D. *Water Quality for Agriculture—Food and Agricultural Organization of the United Nations*; Food and Agricultural Organization of the United Nations: Rome, Italy, 1994.
46. Kelly, W.P. Permissible Composition and Concentration of Irrigation Water. *Trans. Am. Soc. Civ. Eng.* **1941**, *106*, 849–855. [CrossRef]
47. Szabolcs, I.; Darab, C. The Influences of Irrigation Water of High Sodium Carbonate Contents on Soils. In Proceedings of the 8th International Congress Soil Science Sodic Soils, Bucharest, Romania, 31 August–9 September 1964; pp. 802–812.
48. Wilcox, L.V. *Classification and Use of Irrigation Waters*; Department of Agriculture: Washington, DC, USA, 1955.
49. Richards, L.A. Diagnosis and Improvement of Saline and Alkali Soils. *Soil Sci.* **1954**, *78*, 154. [CrossRef]
50. Doneen, L.D. *Notes on Water Quality in Agriculture*; University of California: Davis, CA, USA, 1964.
51. Doneen, L.D. The Influence of Crop and Soil on Percolating Waters. In Proceedings of the 1961 Biennial Conference Groundwater Recharge, Bekley, CA, USA, 26–27 June 1962; pp. 156–163.
52. Eaton, F.M. Significance of Carbonates in Irrigation Waters. *Soil Sci.* **1950**, *69*, 123–134. [CrossRef]
53. Todd, D.K. *Groundwater Hydrology*; Wiley: Hoboken, NJ, USA, 1980; ISBN 13: 9788126508365.
54. Gibbs, R.J. Mechanisms Controlling World Water Chemistry. *Science* **1970**, *170*, 1088–1090. [CrossRef]
55. Meireles, A.C.M.; Andrade, E.M.D.; Chaves, L.C.G.; Frischkorn, H.; Crisostomo, L.A. A New Proposal of the Classification of Irrigation Water. *Rev. Ciênc. Agron.* **2010**, *41*, 349–357. [CrossRef]
56. McGeorge, W.T. Diagnosis and Improvement of Saline and Alkaline Soils: By Staff of U. S. Salinity Laboratory, Agriculture Handbook No. 60 U. S. Dept. Agric., Supt. Documents, U.S. Government Printing Office Washington 25, D.C., 1954, 160 Pages, $2.00. *Soil Sci. Soc Amer. J.* **1954**, *18*, 348. [CrossRef]
57. Wilcox, L.V. *The Quality of Water for Irrigation Use*; United States Department of Agriculture (USDA): Washington, DC, USA, 1948. [CrossRef]
58. Vaseashta, A.; Gevorgyan, G.; Kavaz, D.; Ivanov, O.; Jawaid, M.; Vasović, D. Exposome, Biomonitoring, Assessment and Data Analytics to Quantify Universal Water Quality. In *Water Safety, Security and Sustainability*; Advanced Sciences and Technologies for Security Applications; Vaseashta, A., Maftei, C., Eds.; Springer International Publishing: Cham, Switzerland, 2021; pp. 67–114, ISBN 978-3-030-76007-6.
59. Vaseashta, A.; Duca, G.; Travin, S.O. *Handbook of Research on Water Sciences and Society*; Engineering Science Reference: Hershey, PA, Canada, 2022; ISBN 978-1-79987-356-3.

water

MDPI

Article

Variational Quantum Regression Application in Modeling Monthly River Discharge

Liu Zhen [1] and Alina Bărbulescu [2,*]

1 National Key Laboratory of Deep Oil and Gas, School of Geosciences, China University of Petroleum (East China), Qingdao 266580, China; lincoln110@foxmail.com

2 Department of Civil Engineering, Transilvania University of Brașov, 5, Turnului Street, 500152 Brașov, Romania

* Correspondence: alina.barbulescu@unitbv.ro

Abstract: In the framework of efficient water resources management, the hydrological forecast is the basis of the pertinent management of water resources. Therefore, this study applies the variational quantum regression (VQR), a novel machine learning approach inspired by quantum computing principles, to the series of water discharges from a river in Romania. The models were evaluated against the quantum neural network (QNN) and other classic artificial intelligence (AI) outputs on the same dataset. Performance was assessed based on the coefficient of determination (R^2), mean absolute error (MAE), and mean squared error (MSE). VQR outperformed classical neural networks and hybrid models with respect to MSE and MAE, demonstrating superior accuracy and generalization capability. Notably, the models exhibited exceptional skill in capturing monthly maxima—an area where other models often struggle, underscoring the potential of VQR as a powerful and reliable tool for hydrological forecasting, particularly in the context of nonlinear and high-variability data series.

Keywords: accuracy; forecast; VQR; river discharge

1. Introduction

Reliable hydrological models and the forecasts provided by them are essential to understanding the hydrological processes and efficiently manage water resources [1,2]. Accurately simulating hydrological processes across various spatial and temporal scales remains a central challenge, particularly in the face of increasing climatic variability and anthropogenic influences [3–5]. Traditionally, hydrological forecasting has relied on physics-based models (PBMs) and statistical techniques [6–8]. The integration of machine learning (ML) models in various domains [8–15] has brought transformative potential to hydrological modeling [16–21]. By leveraging vast datasets and learning from patterns within them, AI models can enhance forecasting accuracy, reduce computational demands, and provide real-time adaptability. ML models are more efficient in computational terms and need a reduced amount of data and characterization of the studied case when compared to PBMs.

Variational quantum regression (VQR), a variational quantum algorithm (VQA), is commonly used for regression problems. Its principle is based on the variational method, where a quantum state with variational parameters is constructed, and classical optimization algorithms iteratively adjust these parameters to minimize the difference between an observable quantity and the target value [22,23]. In VQR, a quantum state (parameterized by variational parameters) is first prepared, followed by measurements, to obtain the estimated value of a record. The variational parameters are further updated by a classical optimizer that considers the discrepancies between the simulated and target values. The process is repeated until an optimal solution is found.

The variational quantum circuits (VQCs) involved in quantum computation demonstrated performance comparable to classical neural networks on supervised and reinforcement learning tasks, utilizing significantly fewer trainable parameters [24–26]. Unlike classical models, which rely on kernel tricks to manage computational costs, quantum circuits directly leverage quantum resources to model functions intractable for classical systems [27]. Although VQCs currently involve longer training durations, their parameter efficiency highlights their potential as an alternative with high potential for specific machine learning applications [28].

Practical implementation of VQR faces challenges. As problem complexity increases, quantum circuit depth and complexity grow rapidly, exacerbating computational burden and error accumulation, which can diminish accuracy [29]. Furthermore, selecting appropriate variational forms and optimization algorithms is critical to VQR performance (yet no universal method exists), requiring problem-specific investigation and fine-tuning [22]. However, VQR has demonstrated potential in various scientific and industrial domains [30–39]. It offers a promising framework for learning nonlinear and complex mappings in data-rich environments by exploiting quantum superposition and entanglement, which allows for more effective modeling of complex patterns compared to classical methods [37].

The bibliographical search on quantum computing applications in *Environmental Sciences* and *Geosciences* returned only a few results [40–43]. Berger et al. [41] identified four critical areas where quantum technologies could have high-impact applications: simulating physical systems, combinatorial optimization, sensing, and energy efficiency. Their work emphasizes the need for interdisciplinary collaboration to explore these opportunities. Ahmad and Jas [42] introduced a quantum temporal convolutional network model targeting $PM_{2.5}$ levels in highly polluted regions. Their approach demonstrated improved predictive accuracy over traditional models. Grzesiak and Thakkar [43] explored the application of quantum machine learning in flood prediction, which integrates quantum algorithms with classical machine learning techniques. Their hybrid model achieved competitive training times and improved prediction accuracy for daily flood events along Germany's Wupper River.

Given the limited literature on hydrological modeling using quantum computing and the findings that VQR can provide compressed and nonlinear regression mappings in domains involving signal processing [44] (univariate time series might be considered nonlinear signals), we developed three VQR-based models for monthly river discharge data from 1955 to 2010 from the Buzău River (Romania). This study aimed to evaluate the suitability of VQR for hydrological forecasting by benchmarking the models against previous results obtained from classical approaches on the same dataset. The motivation stemmed from the inability of traditional AI and hybrid models to accurately predict monthly discharge values, especially in months with high water and floods. Our findings demonstrate that the quantum models outperformed classical neural networks and hybrid models, highlighting VQR's potential advantages in this domain.

2. Materials and Methods

2.1. Study Region and Data Series

The Buzău River (Romania) springs about 1800 m north of the Ciucaș Mountains. It has a direction from north to south and is 334.4 km long [45,46]. The river's hydrographic basin (h.b.) (Figure 1) is situated in the eastern part of Romania, within the Carpathian region, with an area of 5264 km^2. The Buzău River drains into the Siret, a main Danube tributary.

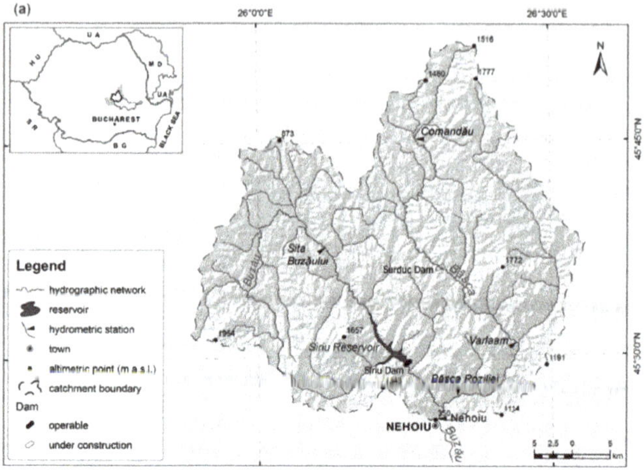

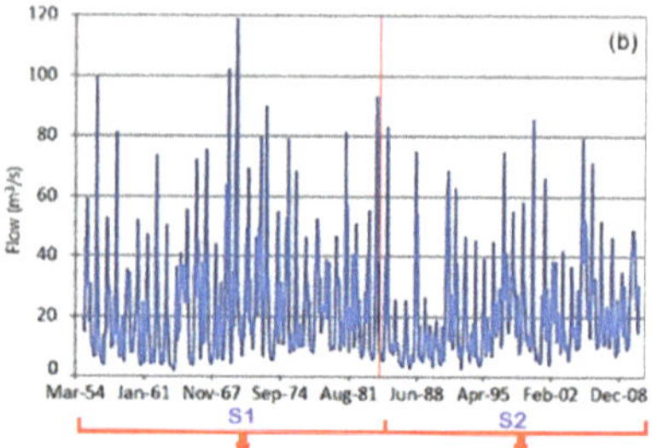

Anomalies of S1			
Apr-56	98.673	Jul-71	69.126
Apr-58	81.05	Aug-72	49.848
Apr-62	73.623	Sep-72	48.277
Apr-66	72.127	Oct-72	79.239
Apr-67	75.443	Apr-73	88.553
Jul-69	101.635	May-73	66.342
Mar-70	68.41	Jul-75	78.6
Apr-70	84.787	Mar-81	81.075
May-73	117.294	Aug-83	55.384

Anomalies of S2	
May-84	92.794
Jun-85	82.937
Apr-88	74.433
Jun-91	64.67
Jul-91	68.194
May-97	73.968
Apr-00	85.533
Jun-01	65.939
May-05	78.265
Aug-05	49.129
Apr-06	71.277

Figure 1. (**a**) The map of Romania, on which is indicated the catchment location, (**b**) the data series indicating the subseries S1 and S2, and the aberrant values of S1 and S2.

The river basin altitude varies from 114 m to 1915 m. Upstream Nehoiu, the catchment is characterized by steep slopes and a predominantly torrential hydrological regime. The average slope is about 11.7° in the upper and middle of h.b. Some parts of the catchment are covered by forest, with afforestation coefficients between 23% (Câlnău) and 84% (Bâsca Mare) [47]. Coniferous, broad-leaved, and mixed forests cover 41% of the study zone. Approximately 33% of the area is occupied by pastures, arable fields, and orchards [48], mainly in the southern part. Flysch formations, with heights between 900/1000 and 1700 m, are found in Carpathian regions. The Subcarpathians are made up of soft rock layers, mostly marl and clays, and rise to 900 m high. Clay-rich rocks from the Neogene period, containing minerals like montmorillonite and illite, are common at the surface. These soft materials cause the slopes' instability, so landslides often occur. The most frequent types are shallow slides, deeper rotational slides, and mudflows [49]. The region exhibits a humid continental climate with significant seasonal and interannual variability in precipitation, often leading to rapid runoff and frequent flash flood events [50,51].

These hydrological dynamics make catchment a critical area for flood risk assessment and hydrological modeling. The catchment is relatively well-instrumented, with several hydrometric stations providing long-term flow and rainfall data, making it a valuable case study for testing advanced forecasting and modeling techniques [52–54].

Due to its complex topography, heterogeneous land cover, and sensitivity to climate event variation, the Buzău Catchment has been the focus of numerous hydrological studies, particularly those related to runoff generation, flood forecasting, and the evaluation of climate change impacts mountainous water systems [55–57].

The Buzău River has experienced several major flood events over the past decades, with the most significant ones occurring in 1970, 1972, 1975, 1988, 1991, and 2005. The 1970s were particularly notable, with eight major floods and the highest average annual peak discharge for the decade (853 m^3/s). In contrast, the following decades—1980–1989 and 1990–2000—saw a marked reduction in flood frequency and intensity. However, this trend reversed in 2000–2010. The highest flood occurred in July 1975, when the discharge peaked at 2100 m^3/s. In comparison, during the July 2005 floods, the maximum discharge was significantly lower—925 m^3/s—primarily due to the attenuating effect of the Siriu and Cândești reservoir dams constructed upstream [57]. In January 1984, the dam on the Buzău

River at Siriu was partially commissioned, with the primary objectives of hydroelectric power generation and flood control.

The dataset analyzed consists of monthly river flow records spanning January 1955–December 2010, called S in the following, was divided into subseries S1 and S2, before and after January 1984. The records originated from the National Institute of Hydrology and Water Resources Management, where they were checked for accuracy. The series had no missing values.

Three models were developed: M was trained on data up to January 2005, M1 used series S1 for training, and M2 was trained on data from January 1984 to December 2005. In all cases, forecasting was performed on a common test set starting from January 2006. At the second study stage, other models, Mo, Mo1, and M2o, were built using the anomaly-filtered version of the series, called So, S1o, and S2o. The anomalies are presented in the Tables from Figure 1. Their detection was presented in detail in article [54]. The training and test sets covered the same period as in the case of models M, M1, and M2.

2.2. Basic Concept of Variational Quantum Regressor and VQR Structure

The qubit is the equivalent of a bit in classical computing. It is a 2D quantum system whose state can be written as:

$$|\phi\rangle = \alpha|0\rangle + \beta|1\rangle \tag{1}$$

where

$$|0\rangle = \begin{pmatrix} 1 \\ 0 \end{pmatrix}, \ \beta\,|1\rangle = \begin{pmatrix} 0 \\ 1 \end{pmatrix} \tag{2}$$

and the complex numbers α and β satisfy

$$\alpha^2 + \beta^2 = 1. \tag{3}$$

VQR is a quantum regression model based on a VQA. By leveraging the parameterized nature of variational quantum circuits (VQC), VQR achieves nonlinear mapping and optimization of input data, enabling regression tasks [23]. This approach combines quantum parallelism and classical optimization techniques, aiming to exploit quantum computing's potential advantages in high-dimensional feature mapping and computational efficiency, thereby enhancing regression modeling capabilities.

VQR relies on the following operations and elements:

- Feature encoding, which transforms the input data into quantum states;
- The variational quantum circuit (VQC) that uses parameterized quantum gates to perform quantum state transformations;
- Measurement and optimization, which extract output results via quantum measurement and optimize the circuit parameters using classical methods to achieve model convergence.

VQR performs regression tasks using a parameterized quantum circuit, with its core structure comprising [58]:

1. *Quantum feature encoding:* classical input data must be transformed into quantum states for quantum computation. This encoding process is typically performed using parameterized quantum gates (e.g., R_y, R_z, ZZ-interaction gates). For an input x, the encoding can be represented as the following:

$$|\,\psi(x)\rangle = R_y(x)|0\rangle \tag{4}$$

where ψ is a state. In multi-qubit systems, quantum entanglement (e.g., ZZ interaction) can be utilized to enhance feature representation.

2. *Quantum processing:* during this process, a VQC is applied to transform the quantum state. The VQC serves as the core of the VQR and consists of parameterized quantum gates and entangling gates akin to the hidden layers in classical neural networks. The VQC learns feature representations from the input and optimizes the trainable parameter θ by [58]:

$$U(\theta) = R_y(\theta_1)R_z(\theta_2)CNOT \tag{5}$$

   where $CNOT$ gates introduce quantum entanglement, and multiple circuit layers enable the modeling of complex nonlinear mappings.
3. *Measurement and regression computation*: since quantum computation results are stored in quantum states, quantum measurement is required to extract information. VQR applies Pauli-Z measurement to compute the expectation value of the quantum state by:

$$f(x,\theta) = \langle\psi(x,\theta)|Z|\psi(x,\theta)\rangle \tag{6}$$

   Due to the probabilistic nature of quantum measurements, shot-based sampling is typically employed to reduce measurement errors and improve result stability.
4. *Loss computation* using the standard loss functions:

$$\text{MSE} = \frac{1}{N}\sum_{i=1}^{N}(f(x_i,\theta) - y_i)^2 \tag{7}$$

$$\text{MAE} = \frac{1}{N}\sum_{i=1}^{N}|f(x_i,\theta) - y_i| \tag{8}$$

5. *Parameter optimization*: the process uses classical algorithms (e.g., L-BFGS-B [59,60], COBYLA [61,62], and Adam [63]) to optimize the parameters of quantum circuits using the equation:

$$\theta_{t+1} = \theta_t - \eta\nabla_\theta L(\theta) \tag{9}$$

   where η is the learning rate.
   The parameter θ is optimized through iterative training to minimize the regression error and finally obtain the optimal model.

2.3. Modeling Stages

The modeling stages of the study are as follows.

1. *Load data and preprocessing:*
   - Raw data was imported, checked for missing values and outliers, and normalized. No cleaning was necessary given the series origin and accuracy.
   - The normalized series were split into training and test sets.
2. *Quantum feature encoding and circuit design*
   - Training data are encoded via a single-qubit feature map.
   - A variational ansatz circuit is constructed.
3. *Optimization loop*
   - The L-BFGS-B optimizer iteratively updates circuit parameters.
   - A convergence check directs the loop until stopping criteria are met.
4. *Post-processing and evaluation*
   - Optimized outputs are de-normalized.
   - Standard goodness-of-fit metrics (MAE, MSE, RMSE, R^2) are calculated.
   - Results are displayed and saved for further analysis.

The study flowchart is shown in Figure 2. Table 1 contains the parameters settings.

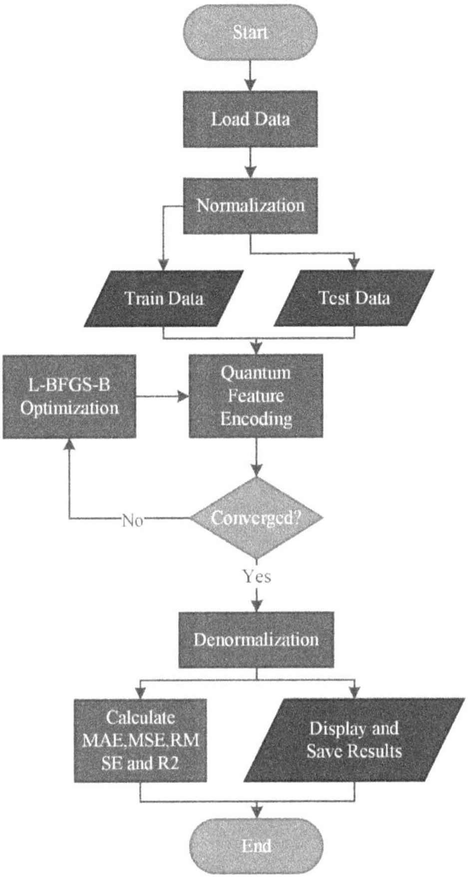

Figure 2. The study flowchart.

Table 1. Parameters settings.

Hyperparameter	**Description**
Number of qubits	1
Quantum register	1 qubit
Classical register	0 bits
Feature map	QuantumCircuit(1, name="fm"), with an ry(param_x) rotation on qubit 0
Variational circuit (ansatz)	QuantumCircuit(1, name="vf"), with an ry(param_y) rotation on qubit 0
Optimizer	L_BFGS_B(maxiter=50)
Callback	callback_graph, used to record the objective function value at each training iteration
Estimator	EstimatorQNN(circuit=qc, estimator=estimator): StatevectorEstimator
Loss function	MSE
Training call	vqr.fit(X_norm, y_norm), fitting the model on normalized inputs and targets

The experiments were conducted on a computer with an AMD RYZEN5 5500 processor (AMD, Santa Clara, CA, USA) and 24 GB RAM. Anaconda was used to manage the Python 3.12 environment and dependencies. Modeling was carried out using IBM's Qiskit framework. Additional libraries, including SciPy and NumPy, were employed to enhance computation performance and stability.

3. Results

The series chart is presented in Figure 1, and the basic statistics of the S, S1, and S2 are presented in Figure 3a,b. Figure 3c contains the violin plot.

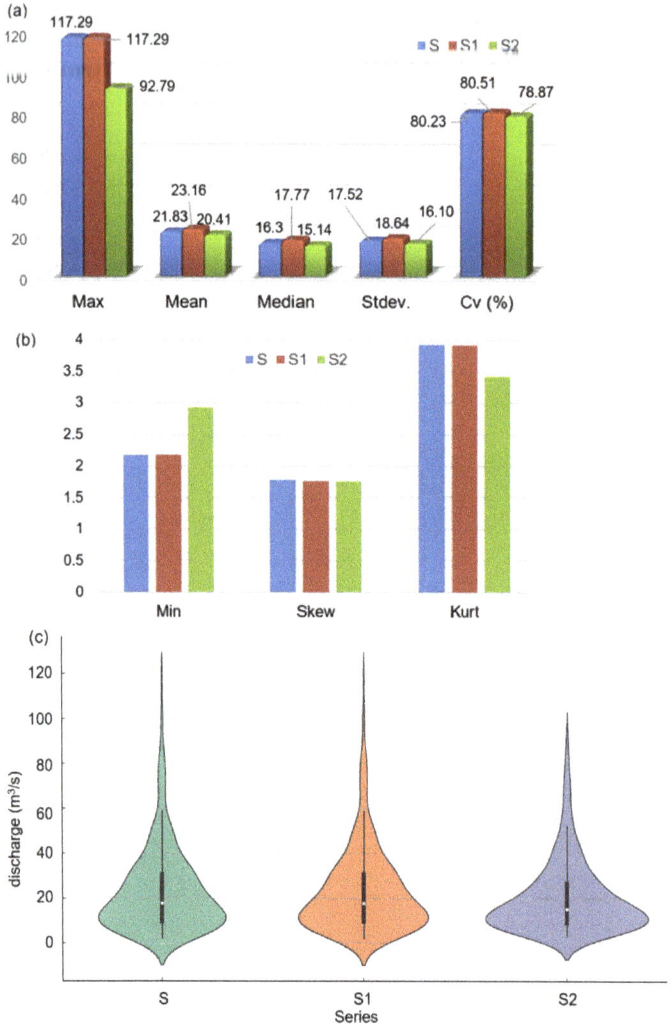

Figure 3. (**a**) Maximum (Max) [m^3/s], mean [m^3/s], median [m^3/s], standard deviation (Stdev) [m^3/s], and coefficient of variation (Cv%); (**b**) minimum (min) [m^3/s], skewness (Skew), and kurtosis coefficients (Kurt) of the series and its subseries; (**c**) violin plot of S, S1, and S2 series.

3.1. Models for Initial Datasets

Figures 4a, 5a and 6a display the recorded and computed values on the training sets of the models M, M1, and M2. A visual examination of these charts reveals no significant discrepancies between the recorded and estimated series, suggesting a good fit for the raw data. The computed and observed series exhibit similar overall patterns. Model M generally overestimates the recorded data, while M1 and M2 systematically underestimate it. Nonetheless, all models successfully capture the peak values.

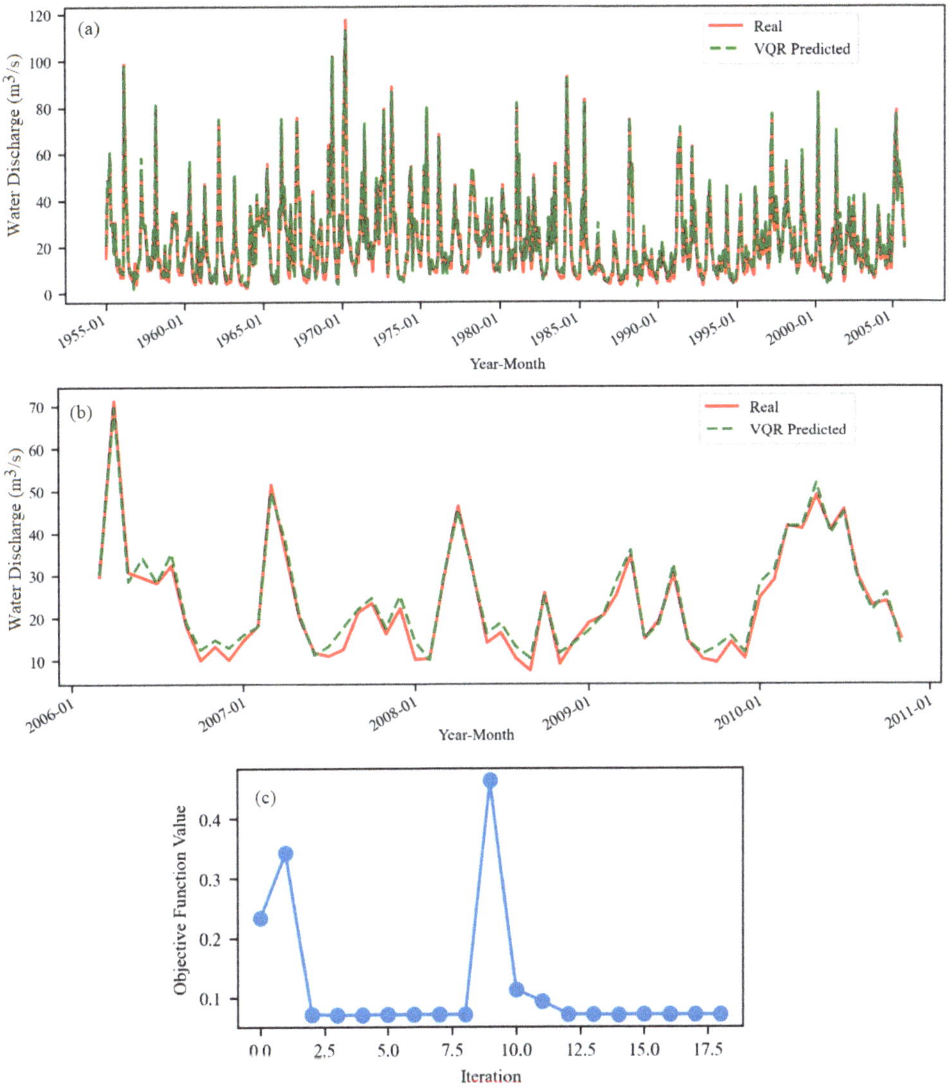

Figure 4. M model: registered and forecast on (**a**) training and (**b**) test sets; (**c**) objective function chart—variation in 18 epochs during training.

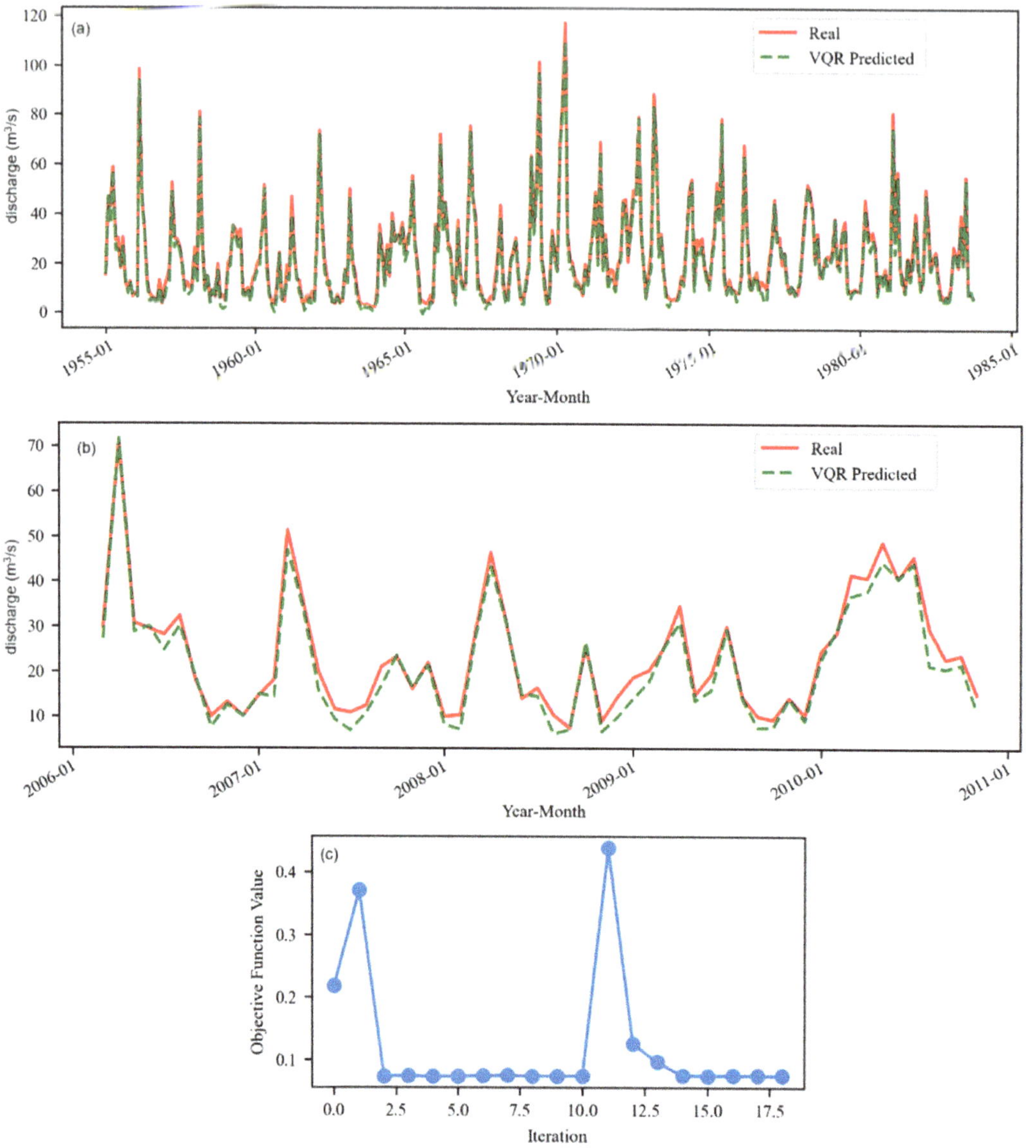

Figure 5. M1 model: record and forecast on (**a**) training and (**b**) test sets; (**c**) objective function chart—variation in 18 epochs during training.

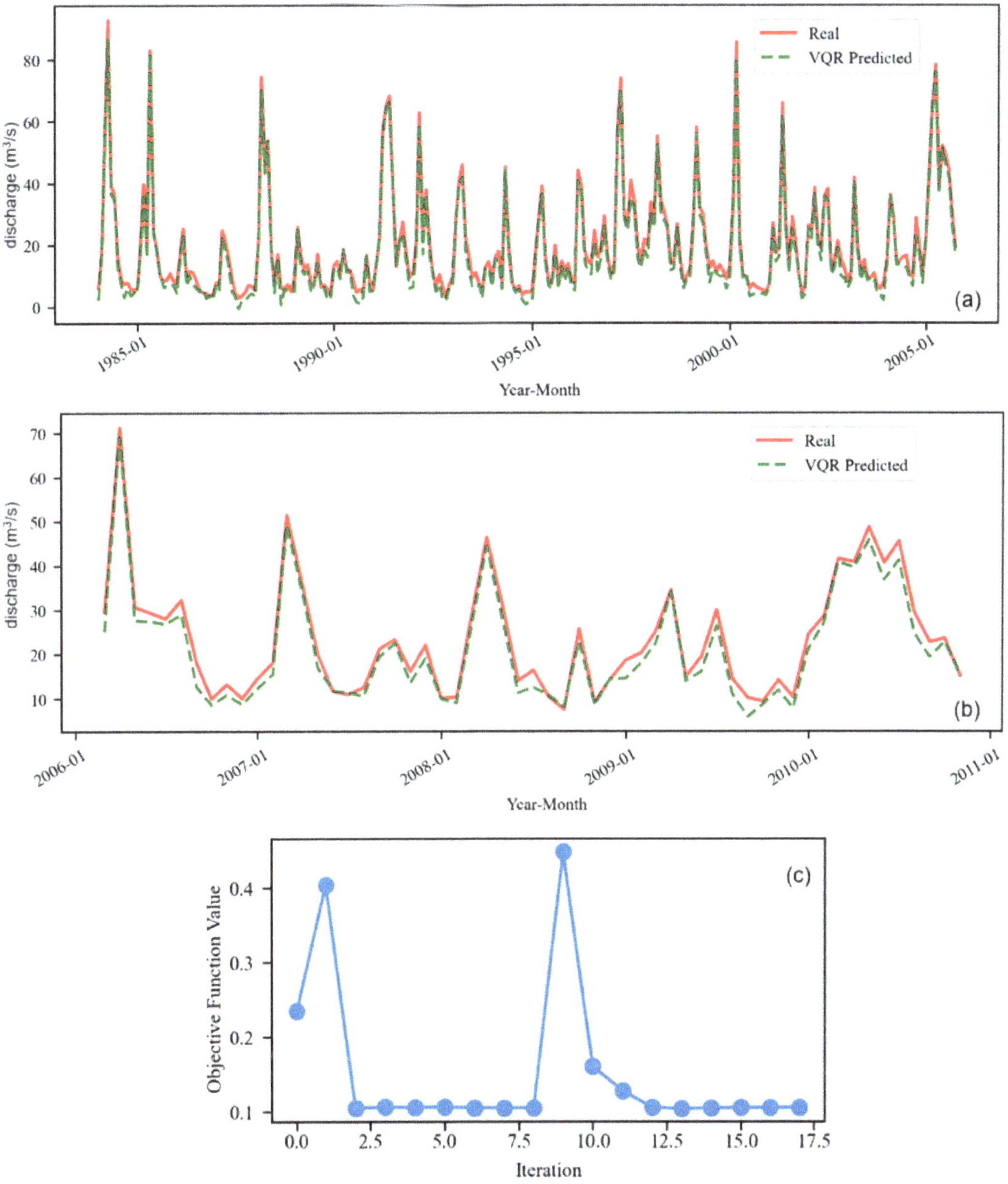

Figure 6. M2 model: record and forecast on (**a**) training and (**b**) test sets; (**c**) objective function chart—variation in 17 epochs during training.

Figure 4b present the recorded values from the test sets, alongside the estimated values produced by the M model using the VQR technique. Figure 4c displays the variation in the objective function over 18 training epochs. Some discrepancies between the initial and computed series trends are noticed on the test set of M model (Figure 4b) from March 2006 to May 2006, with a higher variation in the predicted than in the recoded values. During April–June 2007, the variability of the recorded series is lower than those of the registered data. In both cases, the consequence is an increase in the estimation error compared with the other segments in the test period.

The test set chart and corresponding forecasts for model M1 (Figure 5b) reveal the largest deviations between recorded and predicted values during the periods of November 2008 to March 2009 and after March 2010. Comparisons with Figure 4b, which presents the results for Model M, confirm these observations by indicating smaller deviations in the latter case. This finding is particularly relevant given that both models share the same test set, highlighting the importance of identifying the model that delivers the most accurate forecast for the test period.

The comparison of Figure 6a with Figures 4a and 5a reveals greater biases between the recorded and simulated values on the training set in M2, suggesting a less effective learning process in this model compared to M and M1. This may imply a potentially reduced prediction accuracy on the test set, where the model applies its learned patterns. However, Figure 6b highlights the highest alignment between the shape of the test set and the forecast in M2, indicating better performance in capturing the trend dynamics during January 2006 and December 2010 relative to M and M1.

In Figures 4c, 5c and 6c, we remark that after the first iteration, the values of the objective function increased to double, then leveled off below 0.1. At the ninth (eleventh and ninth, respectively) iteration, the objective function reaches its maximum, then decreases again to reach the minimum, under 0.1.

Visual assessments must be complemented by numerical evaluation to objectively measure model accuracy. Accordingly, the performance of the VQR models was assessed using three goodness-of-fit metrics: mean absolute error (MAE), mean squared error (MSE), and the coefficient of determination (R^2). The corresponding values for each model are reported in Table 2.

Table 2. Goodness-of-fit indicators in models M, M1 and M2.

Model	Set	MAE	MSE	R^2
M	Training	1.5158	3.6647	0.9886
	Test	1.7197	4.335	0.9742
M1	Training	2.3706	8.3552	0.9767
	Test	2.1569	7.3694	0.9561
M2	Training	2.4182	7.6084	0.9728
	Test	2.2678	6.8897	0.9589

Across all three indicators, M demonstrated the best overall performance, achieving the lowest MAE and MSE, as well as the highest R^2 on both the training and test sets. In terms of MAE, M1 ranked second, while considering MSE, M2 outperformed M1. With respect to R^2, M1 held the second-highest value, but only on the training set. Nevertheless, the R^2 values for M1 and M2 were relatively close, indicating comparable explanatory power.

3.2. *Models for the Series Without Aberrant Values*

In the second stage of the analysis, we excluded the aberrant values (indicated in Figure 1) and ran the VQR the algorithms on the adjusted series. This step aimed to evaluate the performance of the VQR on more homogeneous hydrological series. However, it is important to emphasize that the aberrant values may correspond to significant flood events with high return periods. Therefore, they must be retained when applying the models, as was performed in Section 3.1. The results of models Mo, M1o, and M2o are represented in Figures 7a, 8a and 9a for the training sets, and Figures 7b, 8b and 9b for the test sets.

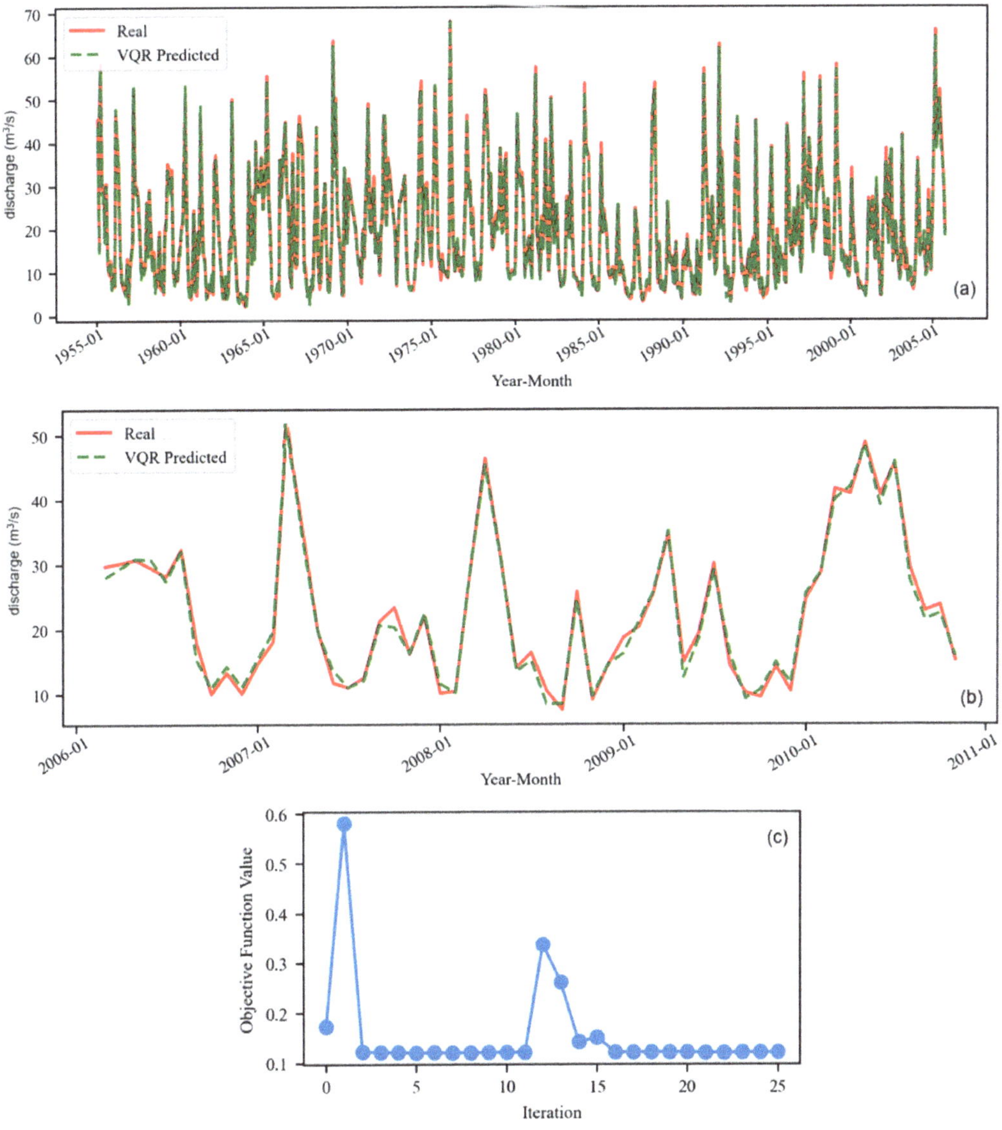

Figure 7. Mo model: record and forecast on (**a**) training and (**b**) test sets; (**c**) objective function chart—variation in 25 epochs during training.

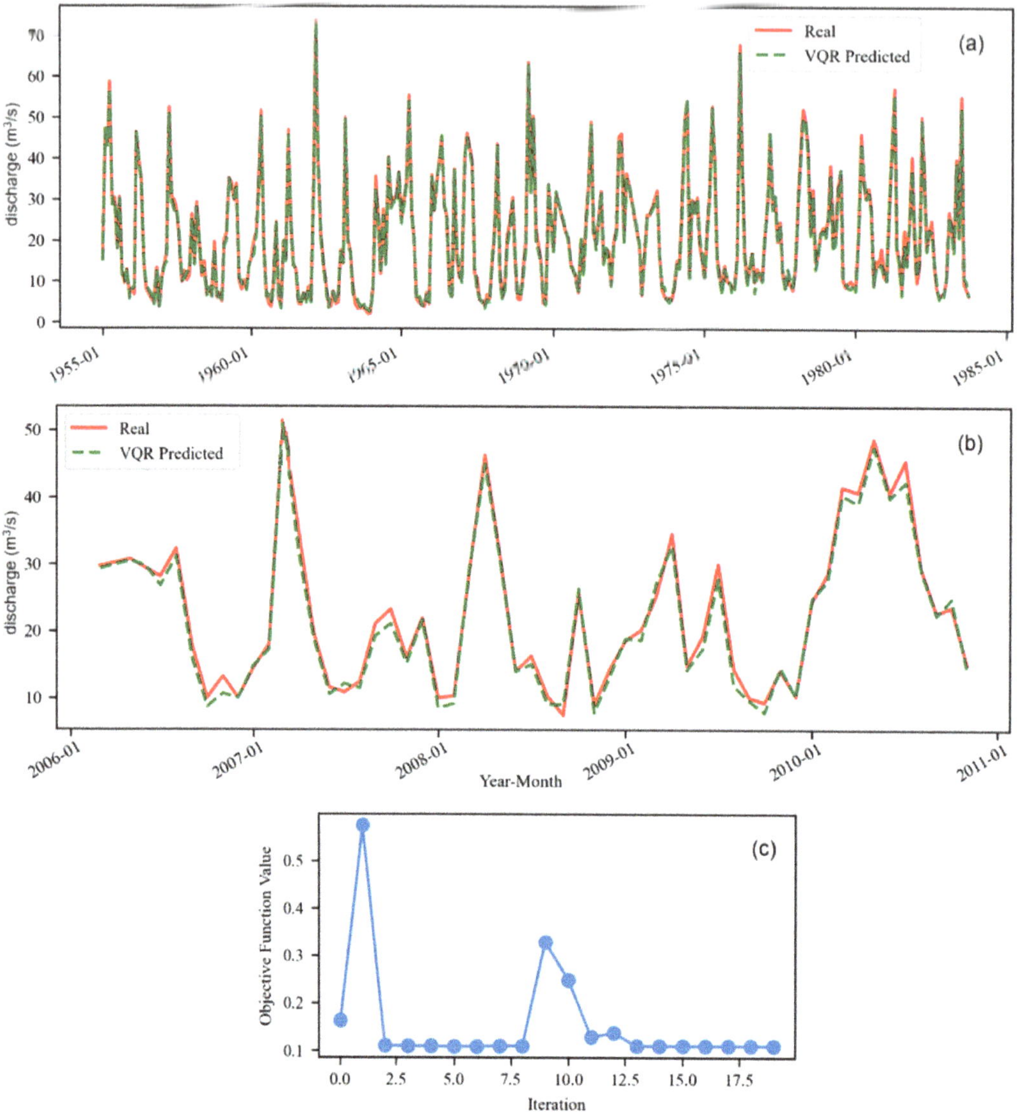

Figure 8. M1o model: record and forecast on (**a**) training and (**b**) test sets; (**c**) objective function chart-variation in 19 epochs during training.

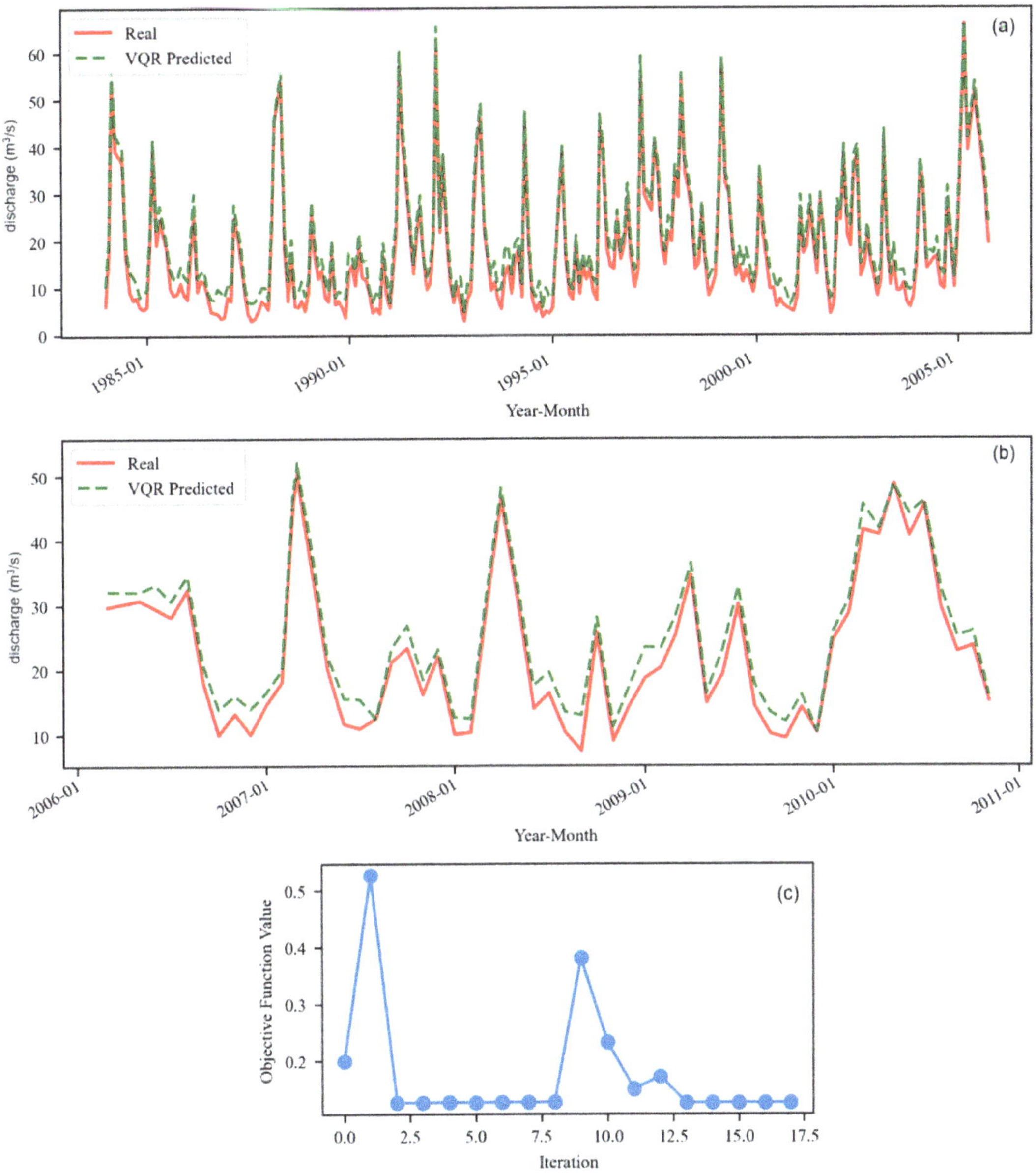

Figure 9. M2o model: record and forecast on (**a**) training and (**b**) test sets; (**c**) objective function chart.

Comparing the charts in Figures 4 and 7, a better fit of the recorded values—particularly on the test set—is observed in the Mo compared to the M. During the second iteration, the objective function rose to 0.58 before stabilizing around 0.12. This value remained relatively constant until the 12th iteration, at which point it abruptly rose to approximately 0.35. Following the 16th iteration, the objective function plateaued near 1.2. This pattern is like the behavior observed when the algorithm was applied to series S, with a similarly noticeable increase occurring around the 12th iteration.

An analysis of Figures 5 and 8 shows that the M1o offers a superior fit, especially in accurately capturing both the minimum and average discharge values. It also shows improved agreement between the observed and simulated trend patterns.

The objective function for the M1o declined after the second iteration and then stabilized around 0.1. Although this value remained constant, it is slightly higher than the minimum achieved in M1.

A comparison of Figures 6b and 9b reveals that the M2o exhibits significantly larger biases than the M2. The objective function values remained nearly constant at approximately 0.05 between the second and eighth iterations, and again from the thirteenth iteration onward, indicating a convergence pattern similar to that observed in the Mo and M1o. Although peak values were estimated with reasonable accuracy, M2o demonstrated the weakest overall performance, with a pronounced overestimation of the observed values.

Table 3 presents the goodness-of-fit indicators and their corresponding values for the **Mo, M1o,** and **M2o to enable a proper comparison.** Comparisons between the models trained on the raw series and those trained on the series with aberrant values removed indicate improved accuracy for both the Mo and M1o relative to M and M1, respectively. This is particularly evident in terms of MSE, which decreased by at least a factor of 2.5. In contrast, M2o demonstrated the poorest performance among all variants.

Table 3. Goodness-of-fit indicators in Mo, M1o, and M2o.

Model	Set	MAE	MSE	R^2
Mo	Training	0.9559	1.4420	0.9919
	Test	1.0103	1.5313	0.9881
M1o	Training	1.0585	1.8321	0.9906
	Test	1.1503	1.9007	0.9853
M2o	Training	2.8241	9.0615	0.9454
	Test	2.5055	7.5413	0.9416

4. Discussion

4.1. Discussions on Modeling Results

Some results from the VQR models were unexpected for the following reasons:

- The data series comprises discharge values recorded over two distinct periods: before and after the dam's construction. Prior to January 1984, the series features numerous high-peak floods that significantly elevated the monthly average discharge. Following January 1984, a marked reduction in both the frequency and intensity of floods was observed, leading to decreased variability in average discharge. Previous studies [51–53] have demonstrated that the subseries corresponding to the pre- and post-dam periods exhibit different statistical behaviors.
- In M2, both the training and test sets belong to the post-1984 period. As a result, the model was expected to generalize well, applying learned patterns effectively within the same hydrological regime.
- M1 was trained on data from the pre-1984 period and tested on data from the post-1984 period. Despite the distinct differences in flow patterns between the two sub-periods, it outperformed M2. This suggests that the richer variability and more dynamic patterns in the pre-1984 data may have enabled the model to learn more robust or generalizable features, even when applied to a different hydrological regime. This finding is contrary to the output of other kinds of neural networks and hybrid models [52–54].
- M was trained on a subseries that spans both the pre-1984 and 1984–2005 periods, allowing it to learn patterns from both the unregulated and regulated flow regimes.

It was then applied to a post-2005 subseries, which exclusively reflects the regulated flow conditions. It seems that M, benefiting from richer temporal coverage and greater variability in its training data, enabled stronger generalization to post-2005 conditions.

- All classical neural network models built on the same data series with the same training and test sets demonstrated better performance on more homogeneous time series. Therefore, it was expected that M2o would perform better than M2, but this was not the case.

The length of the series appears to influence the performance of the VQR output, with M (trained in the longest series) showing the best results, and the M2 (trained in the shortest series) performing the worst. However, this observation remains preliminary, and more extensive studies are required to rigorously validate the impact of series length on model performance.

The time complexity was in this study $O(\tau BpT_c')$, where

- τ = number of iterations, set by maxiter in the optimizer
- B = mini-batch size (in this case it was equal to the data size = n)
- P = circuit depth (the number of gate layers that must be executed sequentially)
- T_c' = time per circuit when using the StatevectorEstimator.

The method did not incur an $O(n^3)$ computational cost with respect to the total sample size n, instead maintaining a polynomial dependence on the number of circuit parameters and measurement efforts. It effectively leverages an exponentially large feature space without the need to explicitly construct or store an exponentially large kernel matrix. This characteristic highlights the potential of VQR for applications involving high-dimensional feature spaces, as is the case in this study.

Table 4 contains the runtime of the VQR algorithm, along with the number of epochs required for convergence during training. In general, longer or more complex time series tend to require increased computation time. However, this assumption does not entirely hold in our case. Although M and Mo exhibited the highest runtimes, the runtime for M1o exceeded that of M1, even though M1o was trained on a shorter time series.

Table 4. Runtime and number of epochs until convergence on the training sets.

Model	Time(s)	Epochs	Model	Time(s)	Epochs
M	21.8606	30	Mo	14.3117	25
M1	5.168	18	M1o	6.8727	19
M2	4.6312	17	M2o	4.3097	17

A similar pattern was observed in the number of epochs needed to reach the optimum of the objective function: 30, 18, and 17 epochs were required for the raw series in M, M1, and M2, respectively. In contrast, the number of epochs decreased for models Mo and M2o, while it slightly increased for M1o.

4.2. Assessment of Fitting Quality of Aberrant Values

An important direction for future research is the evaluation of network performance specifically on extreme values. Based on the results obtained of the models for the raw series, we determined the MAE, MSE, and R^2 corresponding only to the aberrant values. Note that only one such value belongs to the test set; therefore, we report the goodness-of -fit indicators from the training test only. They are as follows:

- In M: MAE = 1.6851, MSE = 2.1199, R^2 = 0.9814
- In M1: MAE = 4.0440, MSE = 4.2410, R^2 = 0.9917
- In M2: MAE = 3.7306, MSE = 4.083 R^2 = 0.9824.

It is worth noting that all models provided a good fit for the aberrant values. Based on MAE and MSE metrics, M performed best, followed by M2 and M1. However, when considering the R^2 coefficient, the best fit was achieved by M1, with M2 and M following in performance.

4.3. Comparisons of the VQR Models' Performance with Those of Classical Artificial Neural Networks and Quantum Neural Networks

In previous studies, we evaluated the effectiveness of artificial neural networks (ANNs) for modeling river discharge using the same dataset. The classical ANNs applied to the same series, and same training and test sets, included Backpropagation Neural Networks (BPNNs), Convolutional Neural Networks (CNNs), Echo State Networks (ESNs), Extreme Learning Machines (ELMs), Long Short-Term Memory (LSTM) networks, and Multilayer Perceptrons (MLPs). Additionally, we explored several hybrid ANN architectures, such as CNN-LSTMs, Particle Swarm Optimization with Extreme Learning Machines (PSO-ELMs), Sparrow Search Algorithm with Backpropagation Neural Networks (SSA-BPs), and Sparrow Search–Echo State Network (SSA-ESN). For details, the reader can see articles [51–53]. The best-performing models are summarized in Table 5.

Table 5. Goodness-of-fit indicators for the best classical AI and hybrid models.

Model	Set	MAE (Model)	MSE (Model)	R^2 (Model)
M	Training	5.7250 (SSA-BP)	80.5765 (ESN)	0.9976 (SSA-ESN)
	Test	4.2351 (CNN-LSTM)	32.4993 (SSA-BP)	0.9983 (LSTM)
M1	Training	6.5177 (CNN-LSTM)	102.9393 (ESN)	0.9899 (LSTM)
	Test	4.4784 (CNN-LSTM)	39.7982 (CNN-LSTM)	0.9917 (LSTM)
M2	Training	4.7433 (CNN-LSTM)	57.3421 (SSA-ESN)	0.9992 (LSTM)
	Test	3.5245 (CNN-LSTM)	29.8323 (CNN-LSTM)	0.9970 (LSTM)

The results in Tables 2 and 5 indicate that VQR outperformed other models in terms of MAE and MSE. In terms of the R^2 metric, the best-performing algorithm was SSA-ESN on the training set and LSTM on the test set of the S series, while LSTM achieved the highest R^2 for M1 and M2. The ESN model recorded the shortest runtime, completing in under 0.65 s. A comparative analysis between VQR and quantum neural networks (QNN) [64] on the raw series reveals that:

- When using QNN, the number of epochs to reach the objective function optimum was the lowest (9 epochs for S, 11 for Mo, 10 for M1 models, and 8 epochs for M2, M1o, and M2o models);
- When applying VQR, the smallest runtime was recorded on M1 (5.168 s) and M2o (4.3097 s) models, while running QNN resulted in shorter runtimes on M (10.8472 s), M2 (4.6218 s), Mo (12.1807 s), and M1o (5.5552 s) models.
- VQR achieved the best performance on the M model training set, while QNN produced the best results on the test set, with MAE = 1.1937, MSE = 2.3815, and R^2 = 0.9858;
- VQR showed superior performance compared to QNN on M1, M2, Mo, and M2o, whereas QNN outperformed VQR on M and M1o.
- QNN performed better than ESN and SSA-ESN on the series without aberrant values in terms of MSE and MAE. On So and M2o, ESN was the best with respect to R^2 on the test set. Moreover, the time necessary to run the algorithms was significantly lower for ESN and SSA-ESN.

To assess and compare the performance of previously studied AI models with those of VQR, we employed graphical representations. Figure 10 illustrates these comparisons:

the Taylor diagram, diagram on the left shows the results for the M model on the training set, using VQR as the reference, while the radar plot on the right offers a comprehensive overview of performance metrics.

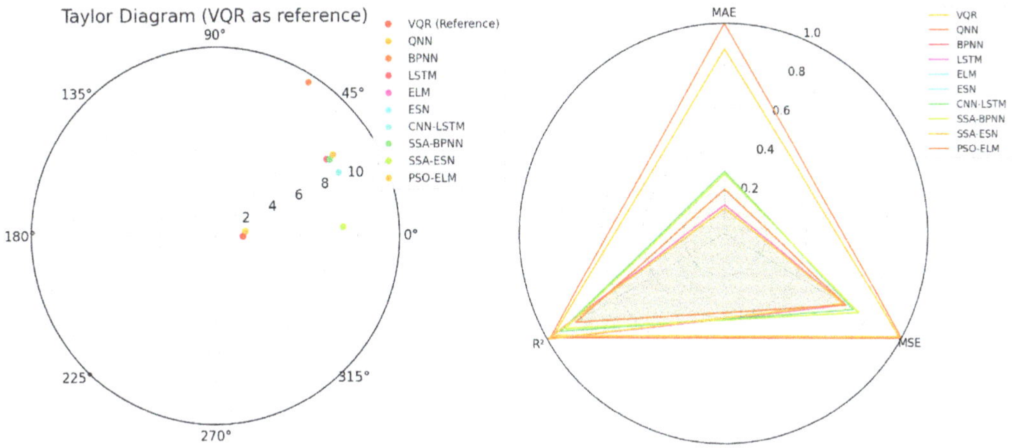

Figure 10. (**left**) Taylor diagram for training set of M model; (**right**) radar plot for test set of M model

From the Taylor diagram, the results show that:

- VQR, QNN, and SSA-ESN are the closest to the reference point, indicating the best performance.
- BPNN stands out with a high standard deviation and low correlation, showing its weak performance.
- ESN and SSA-ESN show high correlation and reasonable variance, performing very well.

From the radar plot it results in that

- QNN, VQR, and SSA-ESN show strong performance across all metrics.
- BPNN performs the worst, especially on MSE and R^2.
- LSTM and CNN-LSTM also perform well, particularly on R^2.

4.4. Limitations of the Actual Study

Despite the promising results of the VQR models, this study has several limitations, leaving a lot of space for improvement. First, all quantum experiments were conducted using classical simulations within the Qiskit framework, given current hardware constraints. Therefore, the results do not account for the noise and error rates inherent in real quantum devices, which could impact the models' performance.

Moreover, while the quantum model demonstrated superior predictive accuracy compared to classical and hybrid neural networks, this study does not provide a formal theoretical justification for the observed performance gains. Therefore, the conclusions are based on empirical comparisons and cannot be generalized for all hydrological modeling tasks.

A comprehensive study of VQR's optimization dynamics is also needed. Future efforts will employ loss landscape visualization, Hessian spectral analysis, and quantum Fisher information metrics to examine how initialization schemes, batch size, and measurement repetitions influence convergence speed and stability.

Deriving worst-case and average-case complexity lower bounds for VQR using quantum information complexity theory and quantifying encoding efficiency and generalization limits via quantum mutual information and quantum Fisher information will also highlight the extent of VQR performances for the problem at hand.

Beyond the L-BFGS-B optimizer and single-layer R_y circuit used here, alternative classical and quantum optimizers (e.g., Adam, SPSA, quantum natural gradient) should be evaluated. Systematic comparisons of variational ansatz designs—such as multi-layer circuits, parameter-sharing topologies, and domain-informed feature maps—will identify the architectures that best balance expressive power and trainability.

5. Conclusions

In this article, we evaluated the VQR's capability to forecast the water flow of the Buzău River. The findings indicate that VQR performed better than other NN and hybrid techniques in terms of MAE and MSE, effectively improving the evaluation of the maxima values of the water discharge. It should be noted that, in hydrology, keeping the maxima in the analysis is necessary given that they indicate exceptional phenomena. In this context, the results' accuracy is remarkable. However, the study was also performed after the aberrant values' removal to determine the models' performances when the series variability is lower. This process significantly enhanced series prediction accuracy, particularly in the M1 and M2, where the models exhibit notably improved generalization. Among the datasets, the highest prediction accuracy was achieved on Mo, followed by M1o.

Comparing the forecast provided by VQR with those obtained by other simple and hybrid models, except QNN, indicates the best accuracy provided by the former algorithm. Comparisons between VQR and QNN indicate that both algorithms exhibit high computational efficiency and perform better on longer datasets (e.g., S dataset), enabling fast convergence and relatively accurate predictions. For more complex datasets (e.g., S, S1o), QNN demonstrates superior predictive accuracy and robustness, particularly in handling outliers. Comparing the algorithms that gave the best results on the study series (LSTM, CNN-LSTM), one may notice that their run time was much lower than that for the quantum algorithms. Therefore, in practical applications regarding river water discharge, the choice of model should depend on the purpose of the study.

The findings of this study provide valuable insights into the applicability of quantum-based models in hydrological forecasting. They highlight the importance of dataset complexity and outlier management in predictive modeling.

Future studies should aim to test VQR models on actual quantum hardware, particularly as devices continue to improve in terms of coherence time, gate fidelity, and qubit count. Investigating alternative data encoding strategies—such as amplitude encoding or hybrid quantum–classical schemes—may improve the scalability and effectiveness of quantum models on larger and more complex hydrological datasets.

Furthermore, a theoretical analysis of VQR's representational power and generalization capacity in the context of nonlinear hydrological series would strengthen the foundation of quantum-based approaches for such modeling.

Applying VQR to multivariate hydrological time series and incorporating environmental predictors such as precipitation, snowmelt, and temperature could enhance model robustness. Moreover, comparative studies involving quantum-inspired algorithms and hybrid neural network architectures may offer further insight into the conditions under which quantum models provide real advantages in environmental modeling.

Author Contributions: Conceptualization, L.Z. and A.B.; methodology, L.Z.; software, L.Z.; validation, L.Z. and A.B.; formal analysis, A.B.; investigation, L.Z. and A.B.; resources, L.Z.; data curation, A.B.; writing—original draft preparation, L.Z. and A.B.; writing—review and editing, A.B.; visualization, L.Z.; supervision, A.B.; project administration, A.B.; funding acquisition, A.B. All authors have read and agreed to the published version of the manuscript.

Funding: The research received no external funding.

Data Availability Statement: Data will be available on request from the corresponding author.

Conflicts of Interest: The authors declare no conflicts of interest.

Abbreviations

The following abbreviations are used in this manuscript:

AI	Artificial Intelligence
ANN	Artificial Neural Networks
BPNN	Backpropagation Neural Networks
CNN	Convolutional Neural Network
CNN-LSTM	Convolutional Neural Network–Long Short-Term Memory
ELM	Extreme Learning Machine
ESN	Echo State Network and Sparow Search–Echo State Network
h. b.	Hydrographic basin
LSTM	Long-Short Term Memory
MAE	Mean absolute error
ML	Machine learning
MSE	Mean standard error
MLP	Multilayer Perceptron
NLS	Nonlinear system
NN	Neural networks
PBM	Physics-based model
PSO-ELM	Particle Swarm Optimization with Extreme Learning Machines
QNN	Quantum neural network
R^2	Coefficient of determination
SSA-ESN	Sparrow Search Algorithm–Echo State Network
VQA	Variational quantum algorithm
VQC	Variational quantum circuits
VQE	Variational quantum eigensolver
VQR	Variational quantum regression

References

1. Cristian, A.; Zuzeac, M.; Ciocan, G.; Iorga, G.; Antonescu, B. A thunderstorm climatology of Romania (1941–2022). *Rom. Rep. Phys.* **2024**, *76*, 710. [CrossRef]
2. Popescu, N.C.; Bărbulescu, A. On the flash flood susceptibility and accessibility in the Vărbilău catchment (Romania). *Rom. J. Phys.* **2022**, *67*, 811.
3. Birsan, M.-V.; Nita, I.-A.; Amihăesei, V.-A. Influence of large-scale atmospheric circulation on Romanian snowpack duration. *Rom. Rep. Phys.* **2024**, *76*, 708. [CrossRef]
4. Ene, A.; Moraru, D.I.; Pintilie, V.; Iticescu, C.; Georgescu, L.P. Metals and Natural Radioactivity Investigation of Danube River Water in the Lower Sector. *Rom. J. Phys.* **2024**, *69*, 802. [CrossRef]
5. Bărbulescu, A.; Băutu, E. Mathematical models of climate evolution in Dobrudja. *Theor. Appl. Clim.* **2010**, *100*, 29–44. [CrossRef]
6. Cui, H.; Ji, J.; Hürlimann, M.; Medina, V. Probabilistic and physically-based modelling of rainfall-induced landslide susceptibility using integrated GIS-FORM algorithm. *Landslides* **2024**, *21*, 1461–1481. [CrossRef]
7. Sannino, G.; Bordoni, M.; Bittelli, M.; Meisina, C.; Tomei, F.; Valentino, R. Deterministic Physically Based Distributed Models for Rainfall-Induced Shallow Landslides. *Geosciences* **2024**, *14*, 255. [CrossRef]
8. Saliba, Y.; Bărbulescu, A. A comparative evaluation of spatial interpolation techniques for maximum temperature series in the Montreal region, Canada. *Rom. Rep. Phys.* **2024**, *76*, 701.
9. Amisha, M.P.; Pathania, M.; Rathaur, V.K. Overview of artificial intelligence in medicine. *J. Family Med. Prim Care* **2019**, *8*, 2328–2331. [CrossRef]
10. Bărbulescu, A.; Dumitriu, C.S.; Dragomir, F.-L. Detecting Aberrant Values and Their Influence on the Time Series Forecast. In Proceedings of the 2021 International Conference on Electrical, Computer, Communications and Mechatronics Engineering (ICECCME), Mauritius, 7–8 October 2021. [CrossRef]

11. Bell, M. The Impact of AI and Generative Technologies on the Engineering Profession. Institution of Engineers Australia. 2025. Available online: https://www.engineersaustralia.org.au/sites/default/files/2025-01/impact-ai-generative-technologies-engineering-profession_0.pdf (accessed on 2 May 2025).
12. Berber, S.; Brückner, M.; Maurer, N.; Huwer, J. Artificial Intelligence in Chemistry Research—Implications for Teaching and Learning. *J. Chem. Educ.* **2025**, *102*, 1445–1456. [CrossRef]
13. Dragomir, F.-L. Thinking patterns in decision-making in information systems. *New Trends Psych.* **2025**, *7*, 89–98.
14. Dragomir, F.-L. Information System for Macroprudential Policies. *Acta Univ. Danubius Œconomica* **2025**, *21*, 48–57.
15. Ni, L.; Wang, D.; Singh, V.P.; Wu, J.; Wang, Y.; Tao, Y.; Zhang, J. Streamflow and rainfall forecasting by two long short-term memory-based models. *J. Hydrol.* **2019**, *583*, 124296. [CrossRef]
16. Xu, H.; Song, S.; Li, J.; Guo, T. Hybrid model for daily runoff interval predictions based on Bayesian inference. *Hydrol. Sci. J.* **2022**, *68*, 62–75. [CrossRef]
17. Bărbulescu, A.; Mohammed, N. Study of the river discharge alteration. *Water* **2024**, *16*, 808. [CrossRef]
18. Liang, W.; Chen, Y.; Fang, G.; Kaldybayev, A. Machine learning method is an alternative for the hydrological model in an alpine catchment in the Tianshan region, Central Asia. *J. Hydrol. Reg. Stud.* **2023**, *49*, 101492. [CrossRef]
19. Mosaffa, H.; Sadeghi, M.; Mallakpour, I.; Jahromi, M.N.; Pourghasemi, H.R. Chapter 43—Application of machine learning algorithms in hydrology. In *Computers in Earth and Environmental Sciences*; Pourghasemi, H.R., Ed.; Elsevier: Amsterdam, The Netherlands, 2022; pp. 585–591. [CrossRef]
20. Rozos, E. Assessing Hydrological Simulations with Machine Learning and Statistical Models. *Hydrology* **2023**, *10*, 49. [CrossRef]
21. Mushtaq, H.; Akhtar, T.; Hashmi, M.Z.u.R.; Masood, A. Hydrologic interpretation of machine learning models for 10-daily streamflow simulation in climate sensitive upper Indus catchments. *Theor. Appl. Climatol.* **2024**, *155*, 5525–5542. [CrossRef]
22. Carleo, G.; Cirac, J.I.; Gull, S.; Martin-Delgado, M.A.; Troyer, M. Solving the quantum many-body problem with artificial neural networks. *Science* **2017**, *355*, 602–606. [CrossRef]
23. McClean, J.R.; Romero, J.; Babbush, R.; Aspuru-Guzik, A. The theory of variational hybrid quantum—Classical algorithms. *New J. Phys.* **2016**, *18*, 023023. [CrossRef]
24. Kölle, M.; Feist, A.; Stein, J.; Wölckert, S.; Linnhoff-Popien, C. Evaluating Parameter-Based Training Performance of Neural Networks and Variational Quantum Circuits. *arXiv* **2025**. [CrossRef]
25. Schuld, M.; Killoran, N. Quantum machine learning in feature Hilbert spaces. *Phys. Rev. Lett.* **2019**, *122*, 040504. [CrossRef] [PubMed]
26. Schuld, M.; Bocharov, A.; Svore, K.M.; Wiebe, N. Circuit-centric quantum classifiers. *Phys. Rev. A* **2020**, *101*, 032308. [CrossRef]
27. Mitarai, K.; Negoro, M.; Kitagawa, M.; Fujii, K. Quantum circuit learning. *Phys. Rev. A* **2018**, *98*, 032309. [CrossRef]
28. Du, Y.; Hsieh, M.H.; Liu, T.; Tao, D. Expressive power of parametrized quantum circuits. *Phys. Rev. Res.* **2020**, *2*, 033125. [CrossRef]
29. Preskill, J. Quantum computing in the NISQ era and beyond. *Quantum* **2018**, *2*, 79. [CrossRef]
30. Zhou, J. Quantum Finance: Exploring the Implications of Quantum Computing on Financial Models. *Comput. Econ.* **2025**. [CrossRef]
31. Ullah, U.; Garcia-Zapirain, B. Quantum Machine Learning Revolution in Healthcare: A Systematic Review of Emerging Perspectives and Applications. *IEEE Access* **2024**, *12*, 11423–11450. [CrossRef]
32. Batra, K.; Zorn, K.M.; Foil, D.H.; Minerali, E.; Gawriljuk, V.O.; Lane, T.R.; Ekins, S. Quantum Machine Learning Algorithms for Drug Discovery Applications. *J. Chem. Inf. Model.* **2021**, *61*, 2641–2647. [CrossRef]
33. Ajagekar, A.; You, F. Molecular design with automated quantum computing-based deep learning and optimization. *npj Comput Mater* **2023**, *9*, 143. [CrossRef]
34. Tao, Y.; Zeng, X.; Fan, Y.; Liu, J.; Li, Z.; Yang, J. Exploring Accurate Potential Energy Surfaces via Integrating Variational Quantum Eigensolver with Machine Learning. *J. Phys. Chem. Lett.* **2022**, *13*, 6420–6426. [CrossRef] [PubMed]
35. Hailu, N.A. Predicting Materials' with Machine Learning. Available online: https://trepo.tuni.fi/bitstream/handle/10024/119253/HailuNahomAymere.pdf?sequence=2 (accessed on 1 May 2025).
36. Blanco, C.; Santos-Olmo, A.; Sánchez, L.E. QISS: Quantum-Enhanced Sustainable Security Incident Handling in the IoT. *Information* **2024**, *15*, 181. [CrossRef]
37. Mohamed, Y.; Elghadban, A.; Lei, H.I.; Shih, A.A.; Lee, P.H. Quantum machine learning regression optimisation for full-scale sewage sludge anaerobic digestion. *npj Clean Water* **2025**, *8*, 17. [CrossRef]
38. Yan, F.; Huang, H.; Pedrycz, W.; Hirota, K. Review of medical image processing using quantum-enabled algorithms. *Artif. Intell. Rev.* **2024**, *57*, 300. [CrossRef]
39. Faccio, D. The future of quantum technologies for brain imaging. *PLoS Biol.* **2024**, *22*, e3002824. [CrossRef]
40. Nammouchi, A.; Kassler, A.; Theocharis, A. Quantum Machine Learning in Climate Change and Sustainability: A Short Review. *Proc. AAAI Symp. Ser.* **2024**, *2*, 107–114. [CrossRef]

41. Berger, C.; Di Paolo, A.; Forrest, T.; Hadfield, S.; Sawaya, N.; Stęchły, M.; Thibault, K. Quantum Technologies for Climate Change: Preliminary Assessment. Available online: https://arxiv.org/pdf/2107.05362 (accessed on 2 May 2025).
42. Ahmad, N.; Jas, S. Quantum-inspired neural networks for time-series air pollution prediction and control of the most polluted region in the world. *Quantum Mach. Intell.* **2025**, *7*, 9. [CrossRef]
43. Grzesiak, M.; Thakkar, P. Flood Prediction using Classical and Quantum Machine Learning Models. *Int. J. Comp. Sci. Mob. Appl.* **2024**, *12*, 84–98.
44. Dutta, S.; Basarab, A.; Kouamé, D.; Georgeot, B. Quantum Algorithm for Signal Denoising. *IEEE Signal Proc. Lett.* **2024**, *31*, 31–156. [CrossRef]
45. Ujvari, I. *Geography of Romanian Waters*; Editura Științifică: București, Romania, 1972. (In Romanian)
46. Grecu, F.; Benabbas, C.; Teodor, M.; Yakhlefoune, M.; Săndulache, I.; Manchar, N.; Kharchi, T.E.; Vișan, G. Risk of Dynamics of the River Stream in Tectonic Areas. Case studies: Curvature Carpathian-Romania and Maghrebian Chain—Algeria. *Forum Geogr.* **2021**, *XX*, 5–22. [CrossRef]
47. Costache, R. The identification of suitable areas for afforestation in order to reduce the potential for surface runoff in the upper and middle sectors of Buzău catchment. *Cinq Cont.* **2015**, *5*, 93–103.
48. Popa, M.C.; Peptenatu, D.; Drăghici, C.C.; Diaconu, D.C. Flood Hazard Mapping Using the Flood and Flash-Flood Potential Index in the Buzău River Catchment, Romania. *Water* **2019**, *11*, 2116. [CrossRef]
49. Zumpano, V.; Hussin, H.; Reichenbach, P.; Bălteanu, D.; Micu, M.; Sterlacchini, S. A landslide susceptibility analysis for Buzau County, Romania. *Rev. Roum. Géogr./Rom. Journ. Geogr.* **2014**, *58*, 9–16.
50. Posea, G.; Popescu, N.; Ielenicz, M. *The Relief of Romania*; Editura Științifică: București, Romania, 1974. (In Romanian)
51. Posea, G. *Geomorphology of Romania*; Editura Fundaţiei România de Mâine: Bucureşti, Romania, 2005. (In Romanian)
52. Bărbulescu, A.; Zhen, L. Forecasting the River Water Discharge by Artificial Intelligence Methods. *Water* **2024**, *16*, 1248. [CrossRef]
53. Zhen, L.; Bărbulescu, A. Comparative Analysis of Convolutional Neural Network-Long Short-Term Memory, Sparrow Search Algorithm-Backpropagation Neural Network, and Particle Swarm Optimization-Extreme Learning Machine for the Water Discharge of the Buzău River, Romania. *Water* **2024**, *16*, 289. [CrossRef]
54. Zhen, L.; Bărbulescu, A. Echo State Network and Sparrow Search: Echo State Network for Modeling the Monthly River Discharge of the Biggest River in Buzău County, Romania. *Water* **2024**, *16*, 2916. [CrossRef]
55. Costache, R.; Pal, S.C.; Pande, C.B.; Islam, A.R.M.T.; Alshehri, F.; Abdo, H.G. Flood mapping based on novel ensemble modeling involving the deep learning, Harris Hawk optimization algorithm and stacking based machine learning. *Appl. Water Sci.* **2024**, *14*, 78. [CrossRef]
56. Gherghe, A.; Dobre, R.R.; Apotrosoaei, V.; Briceag, A.; Melinte-Dobrinescu, M. The Bâsca Rozilei river drainage model, Romanian Carpathian belt. *Geo-Eco-Marina* **2021**, *27*, 37–54. [CrossRef]
57. Grecu, F.; Zaharia, L.; Ioana-Toroimac, G.; Armas, I. Floods and Flash-Floods Related to River Channel Dynamics. In *Landform Dynamics and Evolution in Romania*; Radoane, M., Vespremeanu-Stroe, A., Eds.; Springer Geography; Springer: Cham, Switzerland, 2017; pp. 821–844. [CrossRef]
58. Gong, C.; Guan, W.; Gani, A.; Han, Q. Network attack detection scheme based on variational quantum neural network. *J. Supercomput.* **2022**, *78*, 16876–16897. [CrossRef]
59. Chen, H.; Wu, H.-C.; Chan, S.-C.; Lam, W.-H. A Stochastic Quasi-Newton Method for Large-Scale Nonconvex Optimization with Applications. *IEEE T. Neur. Net. Lear.* **2020**, *31*, 4776–4790. [CrossRef]
60. L_BFGS_B. Available online: https://qiskit-community.github.io/qiskit-algorithms/stubs/qiskit_algorithms.optimizers.L_BFGS_B.html (accessed on 3 May 2025).
61. Powell, M.J.D. A direct search optimization method that models the objective and constraint functions by linear interpolation. In *Advances in Optimization and Numerical Analysis*; Gomez, S.S., Hennart, J.-P., Eds.; Kluwer Academic: Dordrecht, The Netherlands, 1994; pp. 51–67.
62. COBYLA. Available online: https://qiskit-community.github.io/qiskit-algorithms/stubs/qiskit_algorithms.optimizers.COBYLA.html (accessed on 15 February 2025).
63. Kingma, D.P.; Ba, J.L. Adam: A method for stochastic optimization. *arXiv* **2017**. [CrossRef]
64. Liu, Z.; Bărbulescu, A. Quantum Neural Networks Approach for Water Discharge Forecast. *Appl. Sci.* **2025**, *15*, 4119. [CrossRef]

water

MDPI

Article

Seasonally Contrasting Sensitivity of Minimal River Runoff to Future Climate Change in Western Kazakhstan: A CMIP6 Scenario Analysis

Lyazzat Makhmudova [1], Sayat Alimkulov [1], Aisulu Tursunova [1], Lyazzat Birimbayeva [1,2,*], Elmira Talipova [1,2], Oirat Alzhanov [1], María Elena Rodrigo-Clavero [3] and Javier Rodrigo-Ilarri [3,*]

[1] Joint Stock Company «Institute of Geography and Water Security», Almaty 050000, Kazakhstan; mlk2002@mail.ru (L.M.), sayat.alimkulov@mail.ru (S.A.); ais.tursun@bk.ru (A.T.); elmira_280386@mail.ru (E.T.); oiratmuratovich@gmail.com (O.A.)

[2] Department of Meteorology and Hydrology, Al-Farabi Kazakh National University, Almaty 050040, Kazakhstan

[3] Instituto de Ingeniería del Agua y del Medio Ambiente (IIAMA), Universitat Politècnica de València, 46022 Valencia, Spain; marodcla@upv.es

* Correspondence: birimbayeva_l@mail.ru (L.B.); jrodrigo@upv.es (J.R.-I.)

Abstract

This study presents a scenario-based assessment of the future sensitivity of minimal low-water runoff to climate change in Western Kazakhstan. An ensemble of global climate models from the Coupled Model Intercomparison Project Phase 6 (CMIP6), combined with dynamically downscaled projections for Central Asia, was applied to estimate minimal monthly runoff during the summer–autumn and winter low-water periods for the rivers of the Zhaiyk–Caspian water management basin. The analysis covers three future time horizons: 2040 (2031–2050), 2060 (2051–2070), and 2080 (2071–2090), under two greenhouse gas concentration scenarios: SSP3-7.0 (moderately high emissions) and SSP5-8.5 (high emissions). The results reveal a pronounced seasonal contrast in the projected hydrological response. During the winter low-water period, a steady increase in minimal runoff is projected for all rivers, with the most significant changes observed for the Or, Zhem, Temir, and Shagan rivers. This increase is primarily driven by higher winter precipitation, increased thaw frequency, and enhanced infiltration recharge. Conversely, despite modest increases in summer–autumn precipitation, minimal runoff during the summer–autumn low-water period is projected to decline significantly, particularly in the southern basins, due to elevated evapotranspiration rates and soil moisture deficits associated with rising air temperatures. These findings emphasize the importance of developing seasonally differentiated, climate-resilient water management strategies to mitigate low-flow risks and ensure water security under future climate conditions in arid and semi-arid regions.

Keywords: minimal runoff; seasonal hydrological response; climate change; CMIP6; Western Kazakhstan; water resource management

1. Introduction

Uncertainty regarding the evolution of natural and climatic conditions remains one of the most critical challenges for environmental and water resource management today. The impacts of climate change on freshwater systems have been extensively addressed in international scientific forums and high-level reports, including those by the United Nations (UN), the World Meteorological Organization (WMO), and the Intergovernmental

Panel on Climate Change (IPCC). The latest IPCC assessment highlights the urgent need for adaptation measures, as the adverse effects of the ongoing climate crisis are already evident across multiple regions [1].

While the causes of the observed warming over recent decades result from both natural variability and anthropogenic greenhouse gas emissions, the hydrological implications are evident. Climate change directly or indirectly alters river regimes, affecting water availability and contributing to the increased frequency of extreme hydrological events, such as floods, droughts, and low-flow periods [2,3].

Modern hydrological and water management challenges driven by ongoing climate change require not only solid theoretical foundations but also the development of advanced methods and approaches to addressing these challenges under non-stationary hydrological conditions. The need to establish methodological principles for solving contemporary engineering hydrology problems highlights the importance of creating tools that enable quantitative assessments and reliable forecasts of river flow under changing climatic and environmental conditions.

Scenario-based forecasting of minimum river discharge using the sixth phase of the Coupled Model Intercomparison Project (CMIP6) models represents a key direction in hydrological research, particularly in the context of observed and projected changes in climatic variables. Improving our understanding of potential changes in minimum river flow within the study area is essential for developing effective water resource management strategies and strengthening adaptation to climate change.

In this context, the development of improved methodologies for characterizing and forecasting water resources under non-stationary climate conditions has become essential [4,5]. Accurate determination of river flow parameters, particularly during critical low-flow periods, is fundamental for water resource planning, infrastructure design, and risk management, especially in regions with uneven water distribution and high water demand [6].

The present study constitutes the second phase of a previously conducted research effort [7], which provided a contemporary assessment of observed changes in minimal river runoff during the summer–autumn and winter low-water seasons for the period 1974–2021. The objective of this study is to develop a long-term, scenario-based projection of minimal river runoff for the summer–autumn and winter low-water periods in the rivers of the Zhaiyk–Caspian water management basin, located in the western region of Kazakhstan, using CMIP6 scenarios that account for projected future climate change.

To achieve this objective, several specific tasks were undertaken. First, an analysis of climatic predictors—including air temperature and atmospheric precipitation—was conducted to assess their influence on minimal river runoff in the Zhaiyk–Caspian water management basin. A key component of the study also involved the selection and justification of CMIP6 models that most accurately capture potential future climate changes in Western Kazakhstan, considering regional climate sensitivity. Based on the selected models, scenario-based projections of minimal runoff for the basin's rivers were developed for the SSP3-7.0 (moderately high emissions) and SSP5-8.5 (high emissions) scenarios for the time horizons of 2040 (2031–2050), 2060 (2051–2070), and 2080 (2071–2090). These projections provide a scientific basis for assessing potential changes in water resources under future climate change conditions.

2. Materials and Methods

2.1. Study Area

Western Kazakhstan is one of the largest economic and geographical regions of the Republic of Kazakhstan by area and is located at the intersection of Eastern Europe and

Central Asia. It comprises four oblasts: Aktobe, West Kazakhstan, Atyrau, and Mangystau, which together encompass 36 administrative districts. The region shares borders with Russia to the north and Uzbekistan and Turkmenistan to the south; it opens onto the Caspian Sea to the west and is adjacent to other Kazakh oblasts to the east.

Extending from the eastern edge of the Volga River delta in the west to the Turan Lowland in the southeast, and from the southern foothills of the Ural Mountains and the Obshchiy Syrt to the Ustyurt Plateau, the region covers a total area of 736,241 km^2, with a north–south length of approximately 1200 km and an east–west extent of about 1300 km. The Mugaldzhar Ridge, which rises 200–300 m above the Zhaiyk-Zhem Plateau, serves as the watershed dividing Western and Central Kazakhstan and constitutes the source region for the area's main rivers: the Oiyl, Zhem, Elek, Or, and others. The relief is characterized by undulating hilly plains, sand dunes, solonchaks (salt flats), and endorheic depressions. Of particular geographical significance is the Pre-Caspian Lowland, which stretches along the northern coast of the Caspian Sea and is distinguished by its leveled surface, numerous lakes, and salt marshes [8–12] (Figure 1).

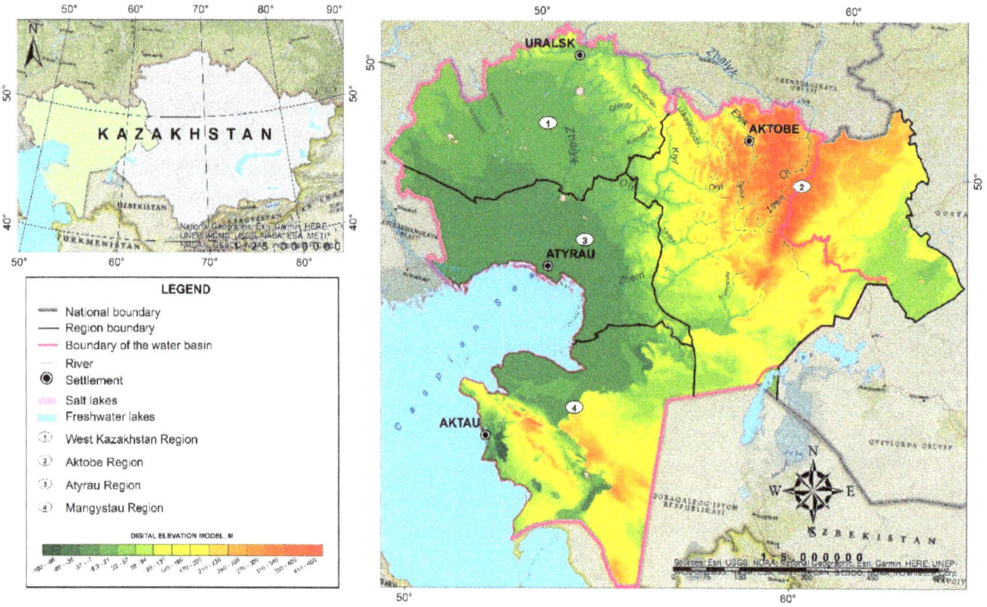

Figure 1. Geographical location of the Zhaiyk–Caspian water management basin in Western Kazakhstan.

The study area is not only the largest in terms of territory but also one of the most economically significant regions of the country due to its extensive oil industry.

The region contains over 1500 rivers and temporary watercourses longer than 10 km, with a total combined length exceeding 45,000 km. The main surface water sources of the region include the northeastern and eastern coasts of the Caspian Sea, as well as the rivers Zhaiyk, Oiyl, Zhem, Sagyz, Elek, Or, Kobda, Shagan, Karaozen, and Saryozen, together with the delta branches of the Volga-Kigach, Sharanovka, and other smaller rivers.

2.2. Source Data

In this study, official meteorological cadastral data from RGP "Kazhydromet" were used to analyze recent changes in climatic characteristics. The analysis utilized records from 40 meteorological stations, including daily air temperature and atmospheric precipitation, covering the long-term period from 1940 to 2021. In addition, data from stationary observations of river discharge for medium-sized rivers within the Zhaiyk–Caspian water management basin were used for both the winter and summer–autumn low-water periods. These discharge data, covering the entire observation period up to and including 2021, were obtained from nine key hydrological posts where runoff during low-water periods is systematically monitored. The hydrological data are available on the official website: https://www.kazhydromet.kz/ru/gidrologiya/ezhegodnye-dannye-o-rezhime-i-resursah-poverhnostnyh-vod-sushi-eds (last accessed on 6 July 2025). The list of meteorological stations and hydrological posts used for the calculations and analyses is provided in Tables 1 and 2.

Table 1. Meteorological stations used in the study.

Meteorological Station									
1	Novorossiyskoe	11	Uralsk	21	Temir	31	Akkuduk		
2	Aktobe	12	Kamenka	22	Karaulkeldy	32	Kyzan		
3	Martuk	13	Chapayevo	23	Uil	33	Tuschibek		
4	Rodnikovka	14	Taipak	24	Sagiz	34	Aktau		
5	Kos-Istek	15	Makhambet	25	Karabau	35	Fort Shevchenko		
6	Il'insky	16	Atyrau	26	Kulsary	36	Zhalpaktal		
7	Novoalekseevka	17	Dzhambeity	27	Shalkar	37	Dzhanybek		
8	Chingirlau	18	Karatobe	28	Ayakkum	38	Urda		
9	Aksai	19	Mugodzharskaya	29	Beineu	39	Novy Ushtogan		
10	Yanvartsevo	20	Emba	30	Sam	40	Ganyushkino		

Table 2. Hydrological posts used in the study.

N°	Hydrological Post	Watershed Area (km^2)	Average Height of the Basin, (m)
1	Or-Bogetsay	7480	350
2	Elek-Aktobe	11,000	340
3	Kargaly-Kargaly	5000	370
4	Ulken Kobda-Kobda	8110	240
5	Karakobda-Alpasay	2240	270
6	Shagan-Kamenny	4000	130
7	Zhem-Zharkamys	26,000	260
8	Temir-Leninski	5310	280
9	Oiyl-Taltogay	17,000	200

Based on geographic system analysis, which considers both zonal and azonal factors, as well as orography, river basins, and atmospheric circulation, the region is divided into the following hydrologically homogeneous areas (Figure 2):

- I—Left-bank tributaries of the Zhaiyk River: Or, Elek, Ulken Kobda, Shyngyrlau;
- II—Right-bank tributaries of the Zhaiyk River: Embulatovka, Rubezhka, Shagan, Derkul; rivers of the western part of Obschiy Syrt not flowing into the Zhaiyk River: Shyzhyn 1st, Shyzhyn 2nd, Karaozen, Saryozen;
- III—Zhaiyk River;
- IV—Rivers of the eastern part of the Caspian Lowland not flowing into the Zhaiyk River: Olenty, Kuperanakty, Buldurty, Shiderty;
- V—Southern rivers: Oiyl, Sagyz, Zhem;
- VI—Arheic area;
- VII—Arheic area.

Figure 2. Location of meteorological stations and hydrological posts within the Zhaiyk–Caspian water management basin in relation to hydrologically homogeneous areas.

The hydrological information used in this study included data on average monthly water discharges for the summer–autumn and winter seasons. The summer–autumn season was defined as the period from June to November, with the lowest monthly runoff within this period considered as the minimal average monthly runoff for the summer–autumn low-water period. The winter season was defined as the period from December to February, with the lowest monthly runoff during this period considered as the minimal average monthly runoff for the winter low-water period.

In addition to the observational dataset, forecasted values of climatic characteristics from CMIP6 models were used to develop long-term scenario-based projections of water resources under various future climatic and socioeconomic conditions. The forecasted climatic variables were obtained from the online resource: https://www.isimip.org/ (last accessed on 6 July 2025). Within the ISIMIP projection framework, forecasted climate data were applied for two socioeconomic development scenarios: SSP3-7.0 (moderately high emissions) and SSP5-8.5 (high emissions), extending up to the year 2100. This approach enables the assessment of a wide range of possible future conditions and facilitates the evaluation of climate risks while accounting for uncertainty.

2.3. Methods

To identify relationships between minimal runoff and the climatic factors influencing it, an analysis was conducted on the dynamics of streamflow-forming climatic characteristics within the basin. The study examined long-term series of precipitation sums for the summer–autumn (June–November), autumn (September–October), and winter (December–February) seasons; long-term series of average monthly air temperatures for the summer–autumn (June–November) and winter (December–February) seasons; long-term series of cumulative negative (November–March) and positive (December–March) air temperatures; and the number of days with above-zero air temperatures during the cold season.

The statistical significance of the linear trend was determined in accordance with the recommendations of Rozhdestvensky and Lobanova [13]. It is proposed to reduce the assessment of trend significance to the evaluation of the correlation coefficient (R) for the dependence Y = f(t), where Y is the studied characteristic and t is time. The trend is considered statistically significant if the following condition is met (Equation (1)):

$$\frac{R}{\sigma_R} \geq S \tag{1}$$

where R is the correlation coefficient, σ_R is the random root-mean-square error. At a 5% significance level or at a 95% confidence limit, $S = 2$.

Scenario-based forecasts of minimal runoff for the summer–autumn and winter low-water periods were developed based on established relationships between minimal runoff and meteorological characteristics for the baseline period, considering two scenarios: SSP3-7.0 (a scenario of a fragmented world with high regional risks) and SSP5-8.5 (a scenario of intensified economic growth with maximum emissions) for the years 2040, 2060, and 2080. The forecasts are based on global CMIP6 models from the ISIMIP projection. The ISIMIP project provides regional climate modelling results for Central Asia based on CMIP6 scenarios, using dynamic downscaling techniques [14]. The models were calibrated to account for regional characteristics, thereby improving the accuracy of climate risk assessments and supporting adaptive decision-making processes.

Dynamic downscaling of global climate models (GCMs) is a key approach to enhancing the accuracy of regional climate projections in areas with complex orography and high spatial heterogeneity. This method is particularly important for accurately reproducing extreme precipitation events, which are highly sensitive to terrain and atmospheric thermodynamics. Despite certain uncertainties, dynamic downscaling provides physically based and spatially detailed climate data essential for regional impact assessments.

In this study, a targeted selection of climate models from the CMIP6 project was carried out for forecasting purposes. Models developed by leading climate research centres were selected: GFDL (USA), MRI (Japan), MPI (Germany), IPSL (France), and UKESM

(United Kingdom). This approach ensures the formation of a representative ensemble that accounts for differences in the parameterization of atmospheric processes, circulation patterns, and interactions between components of the climate system, thereby covering a wide range of climate sensitivities: GFDL-ESM4, IPSL-CM6A-LR, and MPI-ESM1-2-HR, which are characterized by low climate sensitivity, and MRI-ESM2-0 and UKESM1-0-LL, which are models with high climate sensitivity [15]. The models were selected based on their structural independence, proven reliability in reproducing climate processes, and consistency with previous phases of the ISIMIP project. This limited yet representative set of models allows for a balance between projection accuracy and computational efficiency, avoiding the need to utilize the entire CMIP6 ensemble [16].

The criteria for forecast applicability were determined in accordance with [17]. The ratio $\overline{S}/\overline{\sigma}$ is used as an empirical criterion for both the applicability and quality of the forecasting method. The accuracy of the forecasting method is assessed using the root-mean-square error of the verification forecasts, calculated by the following formula:

$$S = \sqrt{\frac{\sum_{i=1}^{n}\left(y_i - y_i^{/}\right)^2}{n-m}} \quad (2)$$

where y_i and $y_i^{/}$ are the observed and predicted values, respectively;

n is the number of observations in the series;
m is the number of degrees of freedom, equal to the number of constants in the predictive equation.

The standard deviation of the predicted variable from its mean is calculated as:

$$\sigma = \sqrt{\frac{\sum_{i=1}^{n}(y_i - \overline{y})^2}{n-1}} \quad (3)$$

where y_i is the predicted value of the variable, y is its mean value, and n is the number of observations in the series.

Depending on the number of observations, the following conditions determine the applicability of the forecasting method (Table 3).

Table 3. Applicability criteria for forecasting methods.

Condition	Applicability Criteria
$n \leq 15$	$S/\sigma \leq 0.70$
$15 < n < 25$	$S/\sigma \leq 0.75$
$n \geq 25$	$S/\sigma \leq 0.80$

The strength of the relationship between the predicted phenomenon and its influencing factors is also characterized by the correlation ratio:

$$\rho = \sqrt{1 - (S/\sigma)^2} \quad (4)$$

For linear relationships, the correlation ratio numerically coincides with the correlation coefficient (r = ρ). A decrease in the $\overline{S}/\overline{\sigma}$ ratio or an increase in the ρ value indicates improved forecasting accuracy. The quality of the method is assessed based on the values of $\overline{S}/\overline{\sigma}$ or ρ (Table 4).

Table 4. Indicators of forecasting method quality for $n \geq 25$.

Quality Category of the Method	S/σ	ρ
Good	$\leq$0.50	$\geq$0.87
Satisfactory	0.51–0.80	0.86–0.60

3. Results

The magnitude and variability of low-water runoff are determined by multiple factors, with groundwater levels playing a key role. Groundwater levels, in turn, depend on both climatic conditions and the characteristics of the underlying surface [18]. To identify the climatic factors driving changes in minimal runoff, the first stage of the study involved an analysis of statistically significant changes in seasonal precipitation totals and air temperature within the basin over the representative period of 1940–2022.

Table 5 presents the linear trend coefficients for these climatic characteristics in the Zhaiyk–Caspian water management basin, with statistically significant trends highlighted.

Table 5. Coefficients of the linear trends of climatic characteristics for meteorological stations of the Zhaiyk–Caspian water management basin for 1940–2021.

Meteorological Station	ΣP for the Summer–Autumn Low-Water Period (VI-XI), mm/10 Years	T_{avg} for the Summer–Autumn Low-Water Period (VI-XI), °C/10 Years	ΣP for Autumn (IX-X), mm/10 Years	ΣP for Winter (XII-II), mm/10 Years	T_{avg} for Winter (XII-II), °C/10 Years	ΣT+ for Winter (XII-III), °C/10 Years	N° of Days with T+ in the Winter Season (XII-III), Days/10 Years	ΣT- for the Cold Period (XI-III), mm/10 Years
Novorossiyskoye	−3.5	**0.2**	−1.8	**6.7**	**0.4**	**1.0**	**0.4**	**55**
Aktobe	−0.2	**0.3**	−1.1	**7.6**	**0.5**	2.2	0.8	**67**
Martuk	−2.0	**0.2**	−1.1	**7.6**	**0.5**	1.7	0.7	**64**
Rodnikovka	**−8.6**	**0.2**	**−4.2**	**4.1**	**0.4**	**0.7**	**0.3**	**55**
Kos-Istek	−7.1	**0.2**	−3.1	**5.1**	0.3	1.6	0.6	38
Ilyinsky	**−7.3**	**0.3**	**−4.0**	−1.2	**0.5**	2.6	0.6	**54**
Novoalekseevka	3.6	**0.2**	**−2.9**	**7.1**	**0.5**	2.1	0.8	**56**
Chingirlau	−4.5	**0.3**	−2.1	**4.1**	**0.4**	**2.1**	**0.8**	**50**
Aksai	−1.4	**0.3**	−1.2	**5.6**	**0.5**	**3.1**	**1.1**	**70**
Yanvartsevo	−2.3	**0.3**	−0.4	**4.7**	**0.4**	3.7	1.3	**59**
Uralsk	0.3	**0.3**	0.0	**3.5**	**0.5**	**4.3**	**1.6**	**68**
Kamenka	−4.6	**0.3**	0.2	−1.2	**0.4**	**3.5**	**1.0**	**56**
Chapayevo	−2.5	**0.4**	−1.2	0.4	**0.5**	**5.3**	**1.8**	**68**
Taipak	−3.5	**0.3**	−1.5	**2.3**	**0.5**	**7.1**	**2.3**	**66**
Makhambet	2.1	**0.5**	0.8	2.0	**0.8**	**14.3**	**4.3**	**79**
Atyrau	−1.2	**0.3**	0.0	1.4	**0.5**	**11.5**	**3.0**	**58**
Dzhambeity	−3.0	**0.3**	−1.6	1.8	**0.5**	**3.3**	**1.2**	**64**
Karatobe	−2.4	**0.3**	−1.6	2.0	**0.5**	4.9	1.6	**65**
Mugodzharskaya	−3.3	**0.2**	**−2.8**	**3.8**	**0.4**	**2.6**	**2.6**	**51**
Emba	−1.6	**0.2**	**−1.9**	1.9	**0.4**	2.2	0.5	**46**
Temir	−3.1	**0.2**	**−2.7**	**4.8**	**0.5**	**3.1**	**0.9**	**64**
Karaulkeldy	−0.2	**0.3**	−1.2	**3.7**	**0.4**	**3.7**	**0.9**	**54**
Uil	−1.8	0.1	−1.4	**2.7**	0.2	1.1	0.1	22
Sagiz	−4.2	**0.2**	−2.3	0.2	**0.5**	5.9	1.8	**47**
Karabau	−1.7	**0.3**	−1.1	1.0	**0.6**	**8.0**	**2.5**	**69**
Kulsary	−3.7	**0.4**	**−3.3**	1.1	**0.7**	**16.9**	**4.3**	**75**
Shalkar	−1.0	**0.2**	−1.3	**1.4**	**0.3**	**6.6**	**1.4**	**46**
Ayakkum	−2.1	**0.3**	**−1.9**	−0.3	**0.4**	**9.5**	**1.7**	**49**
Beineu	**−4.5**	**0.3**	**−2.5**	−0.4	**0.4**	**17.4**	**3.4**	**44**
Sam	0.3	**0.3**	−1.0	**2.9**	0.3	12.9	2.1	27
Akkuduk	0.2	**0.3**	0.5	−0.3	**0.3**	20.5	3.3	**32**
Kyzan	**−5.2**	**0.4**	−1.6	0.0	**0.4**	**20.2**	**4.1**	**37**
Tuschibek	−1.8	**0.3**	−1.1	0.0	**0.2**	15.8	3.3	**25**
Aktau	3.4	**0.6**	−0.7	1.0	**0.4**	**29.2**	**3.7**	14
Fort Shevchenko	−2.3	**0.4**	−1.3	−0.2	**0.5**	**37.6**	**6.1**	**24**
Zhalpaktal	−2.5	**0.3**	0.9	**2.9**	**0.6**	**7.0**	**2.4**	**72**
Dzhanybek	0.1	**0.3**	1.9	**1.7**	**0.6**	**9.6**	**2.9**	**77**
Urda	−0.8	**0.3**	1.6	**5.4**	**0.5**	**10.9**	**3.1**	**60**
Novy Ushtogan	−0.6	**0.3**	0.1	1.4	**0.5**	**13.5**	**3.6**	**62**
Ganyushkino	−2.3	**0.2**	−1.2	−1.0	**0.5**	**12.7**	**3.2**	**46**

Note: P—atmospheric precipitation, T—air temperature; statistically significant trends are written in bold.

An analysis of changes in summer–autumn precipitation totals revealed a general decreasing trend across most of the study area, though these trends are statistically insignificant. The linear trend coefficient for precipitation at the Rodnikovka meteorological station (northwest) is −8.6 mm per decade; at the Mugodzhar meteorological station (Mugodzhar Mountains), it is −3.3 mm per decade; at the Zhalpaktal meteorological station (west of the basin), it is −2.5 mm per decade; and at the Kyzan meteorological station (south of the basin), it is −5.2 mm per decade. However, several meteorological stations exhibit only a weak positive trend, including Novoalekseevka, Uralsk, Makhambet, Aktau, Dzhanibek, and Akkuduk (Figure 3).

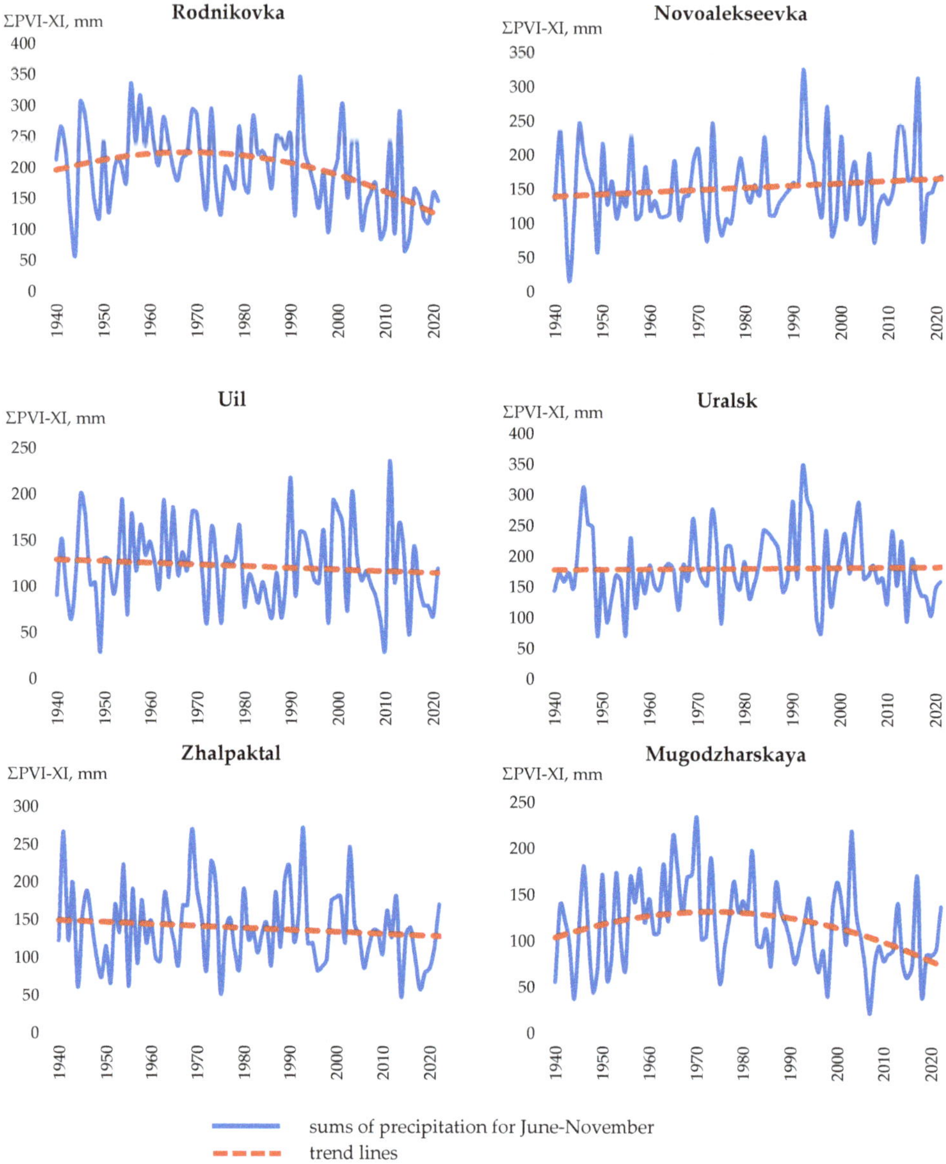

Figure 3. Long-term changes in sums of atmospheric precipitation during the summer–autumn season for meteorological stations of the Zhaiyk–Caspian water management basin.

Thus, summer–autumn precipitation totals alone cannot explain the observed changes in minimal summer–autumn runoff. In addition to precipitation, air temperature during this season should also be taken into account. An analysis of changes in air temperature during the summer–autumn low-water period revealed a significant and widespread increase in average air temperatures across the entire basin. The warming trend is statistically significant and intensifies from the north to the south.

The coefficients of the linear trend for average air temperature range from 0.2 to 0.3 °C per decade in the northern, northeastern, and western regions, as well as along the northern coast of the Caspian Sea. In the central part of the basin, this value increases to 0.5 °C per decade. The most pronounced warming is observed in the southern part of the basin and along the eastern coast of the Caspian Sea, where the trend reaches up to 0.6 °C per decade. The most pronounced increase in average air temperature across the basin has been observed since the mid-2000s. Approximately from the same period, a slight decrease in summer–autumn precipitation sums has also been recorded.

During the winter low-flow period, autumn precipitation is the primary source of groundwater recharge. Among the climatic factors influencing the formation of groundwater, the most significant role is attributed to atmospheric precipitation of the current or preceding month, season, year, or even longer period [19]. However, the analysis showed that the linear trend coefficients of autumn precipitation sums are statistically insignificant for most of the meteorological stations. In the majority of the study area, the trend is negative, with magnitudes not exceeding −4.2 mm per decade, as recorded at the Rodnikovka station. At several meteorological stations in the western part of the basin, small positive trends of up to 1.9 mm per decade are observed, such as at the Dzhanibek station.

This suggests that autumn precipitation alone cannot be considered the primary cause of the observed increase in minimal runoff of winter low flow; however, completely disregarding it would also be a mistake as it contributes to soil moisture recharge prior to the onset of the cold season.

In contrast to the autumn season, long-term winter precipitation sums indicate an overall increase across most of the basin, although significant spatial differences are observed. Statistically significant positive linear trends in winter precipitation, reaching up to 7.6 mm per decade, were recorded at meteorological stations in the northern part of the basin, including Aktobe and Martuk. An exception is the Ilyinsky station, which shows a slight negative trend of −1.2 mm per decade. In the eastern part of the basin, within the Mugodzhar Mountains, linear trend coefficients for winter precipitation reach 4.8 mm per decade, as observed at Temir station.

In contrast, weak positive trends, not exceeding 2.9 mm per decade, are recorded in the central part of the basin, along the Caspian Sea coast, and in endorheic areas, such as at the Sam station. Negative trends in winter precipitation were observed at the Ayakkum, Beineu, Akkuduk, Fort Shevchenko, and Ganyushkino stations. In the western part of the basin, winter precipitation increases range from 1.4 to 5.4 mm per decade, with statistically significant trends recorded at the Urda and Zhalpaktal meteorological stations (Figure 4).

Thaws can also play an important role in the formation of winter runoff, provided they occur with sufficient frequency and duration—a characteristic feature of the study region. To analyse changes in the thaw regime during the winter season, long-term air temperature data were evaluated.

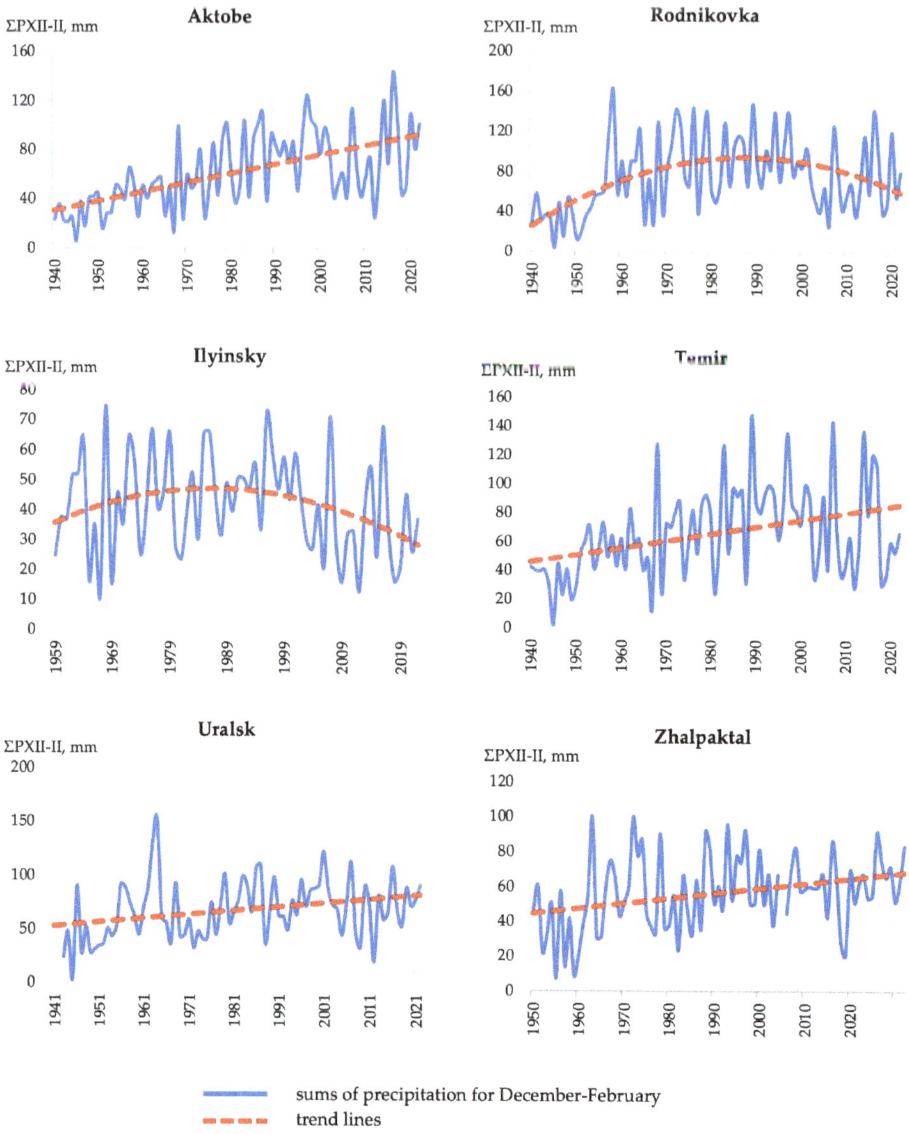

Figure 4. Long-term changes in sums of atmospheric precipitation during the winter season for meteorological stations of the Zhaiyk–Caspian water management basin.

In terms of winter air temperature, a statistically significant increase has been recorded at all meteorological stations across the basin. The intensity of warming increases from the northeast to the southwest, with the most pronounced warming observed along the northeastern coast of the Caspian Sea, reaching up to 0.8 °C per decade at the Makhambet station. The linear trend coefficients in the northeastern part of the basin range from 0.3 °C per decade at the Kosistek station to 0.5 °C per decade at stations such as Aktobe, Novoalekseevka, Temir, and others (Figure 5).

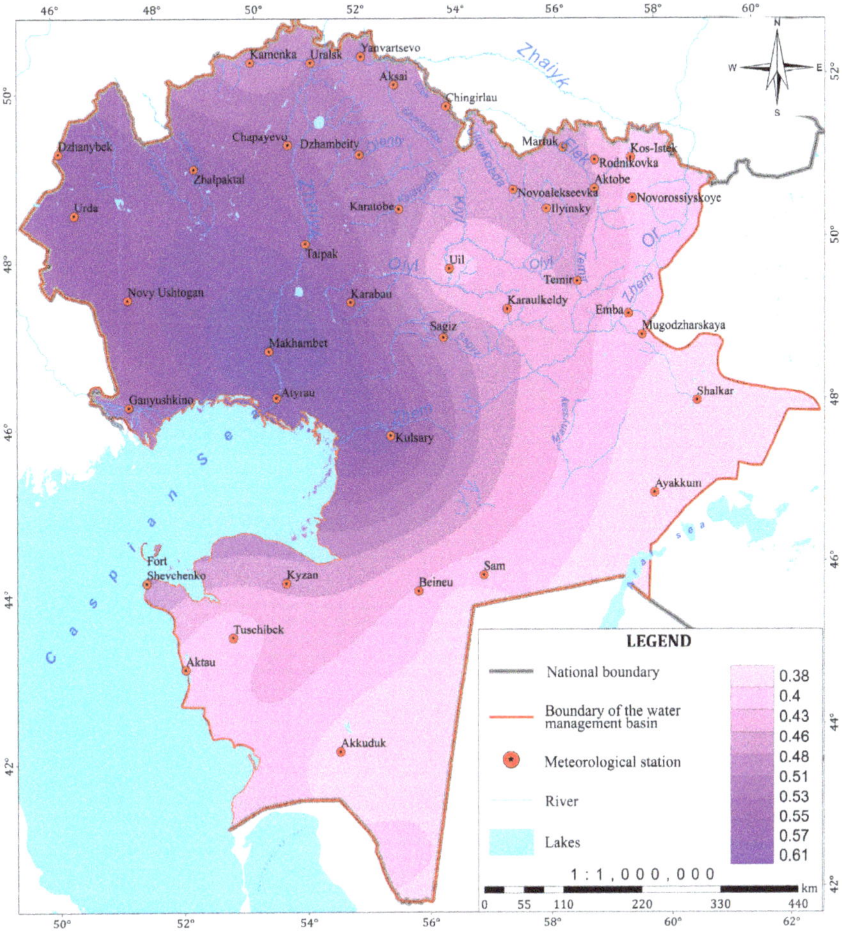

Figure 5. Coefficients of the linear trend of winter air temperature (December–February), in °C per 10 years for the period 1940 to 2021 in the territory of the Zhaiyk–Caspian water management basin.

The sums of positive temperatures and the number of days with positive average daily temperature during the winter period also show a pronounced increase across the entire territory of the studied basin (Figures 6 and 7). As a result, both the frequency and duration of thaw events are increasing. The observed trends are mostly statistically significant throughout the entire study area, with the maximum values recorded in the southern part of the basin (MS Akkuduk, Beineu, Kyzan, Tushchibek) and along the coast of the Caspian Sea (MS Fort Shevchenko, Aktau, Atyrau), where the sum of positive temperatures reaches up to 36.6 °C per decade and the duration of thaw periods increases by up to 6 days per decade.

In the northeastern direction, the intensity of temperature changes decreases, reaching 0.8 °C per decade in the central part of the basin and 0.3 °C per decade in the Mugodzhar Mountains area. A similar decreasing pattern is observed for the duration of positive temperature days, amounting to 2–4 days per decade in the central part of the basin, 0–1 day per decade in the northern part, and 0–3 days per decade in the northeastern region of the basin.

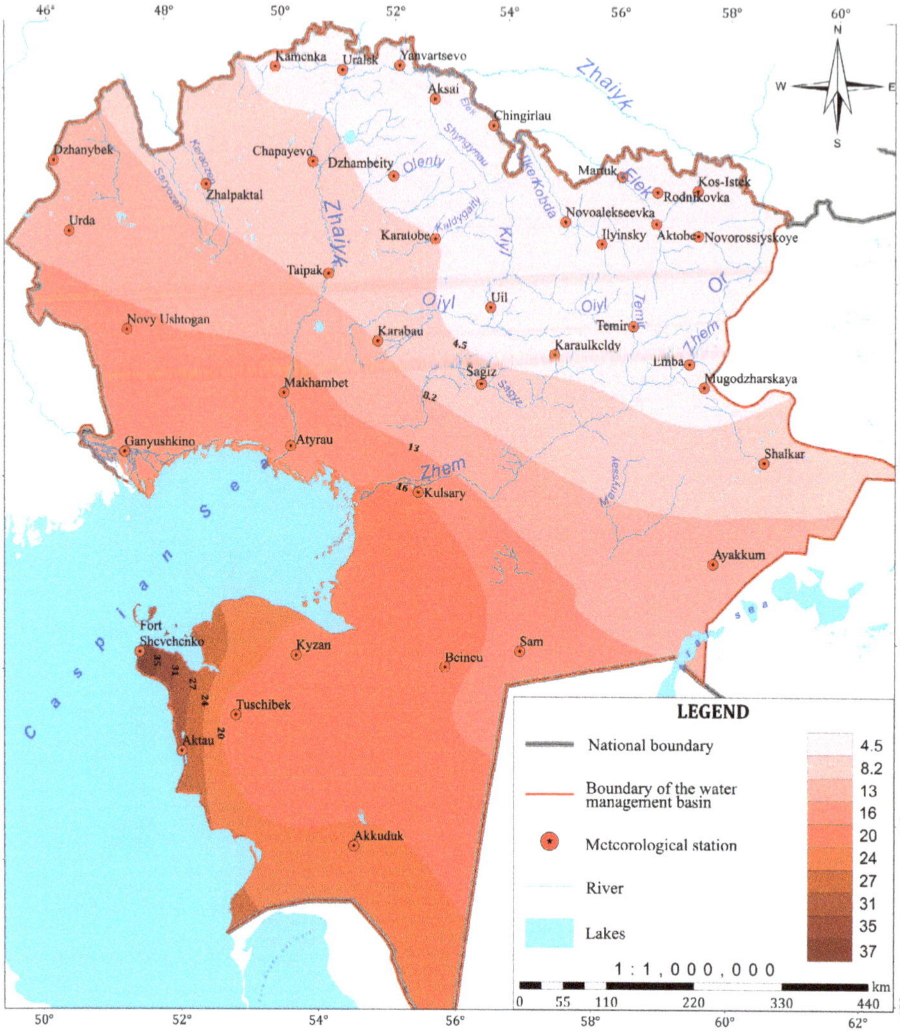

Figure 6. Coefficients of the linear trend of changes in sums of positive winter air temperatures (December–March), in °C per 10 years for the period 1940 to 2021 in the territory of the Zhaiyk–Caspian water management basin.

The analysis of climatic characteristics across the Zhaiyk–Caspian water management basin confirms the existence of statistically significant and persistent changes in air temperature and precipitation regimes, which exert a notable influence on the formation of minimal runoff in low-flow periods.

Atmospheric precipitation and air temperature during the summer–autumn period are identified as the key climatic factors governing the magnitude of minimal summer–autumn runoff.

The observed increase in minimal winter runoff is presumably associated with enhanced infiltration recharge of the soil layer, driven by the increased frequency and duration of thaw events, as well as a rise in winter precipitation.

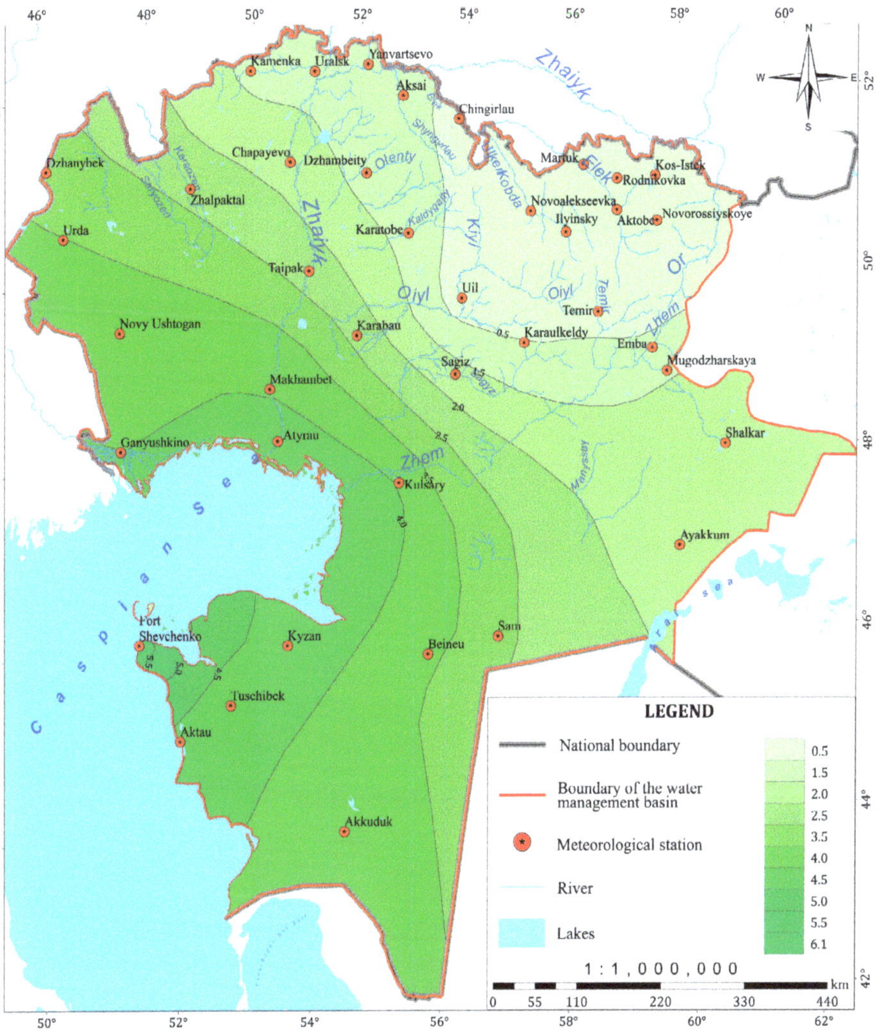

Figure 7. Coefficients of the linear trend of changes in the number of days with positive average daily air temperature during the winter season (December–March), days per 10 years for the period 1940 to 2021 in the territory of the Zhaiyk–Caspian water management basin.

3.1. Long-Term Forecast of Minimal Runoff

In the second phase of the study, based on the conducted analysis, seasonal meteorological factors with statistically significant trends were identified—factors that may exert both direct and indirect influence on the dynamics of minimal runoff.

To quantitatively assess the future state of minimal runoff under projected climate change, statistical relationships were established between minimal runoff and predictors (meteorological variables) that influence its magnitude. For the summer–autumn low-flow period, atmospheric precipitation sums and average air temperatures for the summer–autumn season were selected as predictors. For the winter season, multiple regression models were constructed using the sum of autumn–winter precipitation for the left- and right-bank tributaries of the Zhaiyk River, and the sum of winter precipitation for the southern rivers of the basin. The average air temperature during the winter months was used as a temperature predictor.

The dependencies were developed for the baseline period 1991–2020, following the new climatological standard normals recommended by the World Meteorological Organization. The calculation scheme for forecasting minimal runoff during the summer–autumn and winter low-water periods is presented in Table 6. The multiple regression correlation coefficients for minimal runoff during the summer–autumn low-water period ranged from R = 0.60 to 0.75, while for the winter low-water period, the coefficients ranged from R = 0.61 to 0.71.

Table 6. Calculation scheme for forecasting the minimal runoff of rivers in the Zhaiyk–Caspian water management basin.

No.	Hydrological Area	River Basin	Calculated Dependencies
		Summer–autumn low-water period	
1	Left-bank tributaries of the Zhaiyk River	Or, Elek, Kargaly, Ulken Kobda, Karakobda	$Q_{min} = f(\sum P_{VI-XI}; \overline{T}_{VI-XI})$
2	Right-bank tributaries of the Zhaiyk River	Shagan	$Q_{min} = f(\sum P_{VI-XI}; \overline{T}_{VI-XI})$
3	Southern rivers	Zhem, Temir, Oiyl	$Q_{min} = f(\sum P_{VI-XI}; \overline{T}_{VI-XI})$
		Winter low-water period	
1	Left-bank tributaries of the Zhaiyk River	Or	$Q_{min} = f(\sum P_{IX-II}; \overline{T}_{XII-II})$
		Elek, Kargaly	$Q_{min} = f(\sum P_{IX-II}; \overline{T}_{I-II})$
		Ulken Kobda	$Q_{min} = f(\sum P_{XII-II}; \overline{T}_{I-II})$
		Karakobda	$Q_{min} = f(\sum P_{X-II}; \overline{T}_{I-II})$
2	Right-bank tributaries of the Zhaiyk River	Shagan	$Q_{min} = f(\sum P_{X-II}; \overline{T}_{XII-II})$
3	Southern rivers	Zhem, Temir, Oiyl	$Q_{min} = f(\sum P_{XII-II}; \overline{T}_{XII-II})$

Note: P—atmospheric precipitation, T—air temperature.

3.2. *Projection of Minimal Runoff During the Summer–Autumn Low-Water Period*

The scenario-based forecasts of minimal runoff for the summer–autumn and winter low-water periods, developed based on the established statistical relationships, were prepared for two climate development scenarios: SSP3-7.0 (moderately high emissions) and SSP5-8.5 (high emissions) for the years 2040, 2060, and 2080, relative to the baseline period. The results are graphically presented in Figure 8. Seasonal changes in precipitation and air temperature for the Zhaiyk–Caspian water management basin under future climate conditions are provided in Tables 7 and 8.

According to the forecast data for both the SSP3-7.0 and SSP5-8.5 scenarios, a reduction in minimal runoff during the summer–autumn low-water period is expected for most of the left-bank tributaries of the Zhaiyk River, with the exception of the Or River.

For the Elek River and its tributary, the Kargaly River, a significant reduction in runoff is projected under both scenarios. Compared to the baseline period, minimal runoff is expected to decrease by 37–45% under SSP3-7.0 and by 41–47% under SSP5-8.5 by 2040; by 53–56% (SSP3-7.0) and 54–60% (SSP5-8.5) by 2060; and by 59–68% (SSP3-7.0) and 68–74% (SSP5-8.5) by 2080. Runoff in this part of the basin is projected to decline steadily, especially under the more extreme climate scenario, despite an expected increase in summer–autumn precipitation of up to 14% under SSP3-7.0 by 2040, up to 10% under SSP5-8.5 by 2060, and up to 12% under SSP3-7.0 by 2080.

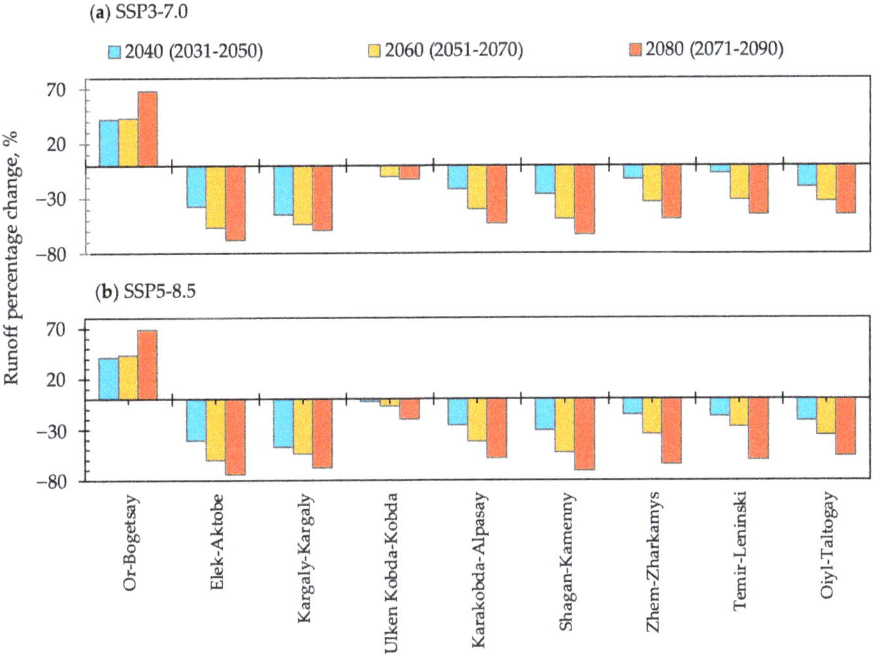

Figure 8. Projected percentage changes in the average monthly minimal runoff of the summer–autumn low-water period of rivers in the Zhaiyk–Caspian water management basin relative to the historical average (1991–2020).

Table 7. Projected changes in summer–autumn season precipitation in the territory of the Zhaiyk–Caspian water management basin.

Hydrological Area	River Basin	Predictor	Base Period (1991–2020)	Future Scenario	2040 (2031–2050)	2060 (2051–2070)	2080 (2071–2090)
			Actual Measured Value, mm		Rate of Change, %	Rate of Change, %	Rate of Change, %
Left-bank tributaries of the Zhaiyk River	Or River basin	$\sum P_{VI\text{-}XI}$	164	SSP3-7.0 SSP5-8.5	36 32	20 15	34 18
	Elek River basin	$\sum P_{VI\text{-}XI}$	161	SSP3-7.0 SSP5-8.5	14 9	7 10	12 5
	Ulken Kobda River basin	$\sum P_{VI\text{-}XI}$	152	SSP3-7.0 SSP5-8.5	−7 −11	−21 −13	−21 −19
Right-bank tributaries of the Zhaiyk River	Shagan River basin	$\sum P_{VI\text{-}XI}$	192	SSP3-7.0 SSP5-8.5	−2 −2	−4 −11	−10 −18
Southern rivers	Emba River basin	$\sum P_{VI\text{-}XI}$	116	SSP3-7.0 SSP5-8.5	14 11	3 22	10 7
	Oiyl River basin	$\sum P_{VI\text{-}XI}$	117	SSP3-7.0 SSP5-8.5	14 14	0 17	−5 −15

At the same time, air temperature is projected to rise significantly, particularly under SSP5-8.5, with increases of 2.6 °C (SSP3-7.0) and 2.8 °C (SSP5-8.5) by 2040; 3.7 °C (SSP3-7.0) and 4.1 °C (SSP5-8.5) by 2060; and 4.8 °C (SSP3-7.0) and 5.3 °C (SSP5-8.5) by 2080.

For the Ulken Kobda River, a slight decrease in minimal runoff during the summer–autumn low-water period is expected for all forecast periods under both scenarios. By 2040, minimal runoff is projected to decrease by 1% under SSP3-7.0 and by 2% under SSP5-8.5.

By 2060, the reduction is expected to reach 10% under SSP3-7.0 and 7% under SSP5-8.5, and by 2080, the decrease will amount to 13% under SSP3-7.0 and 20% under SSP5-8.5.

Table 8. Projected changes in summer–autumn season air temperature in the territory of the Zhaiyk–Caspian water management basin.

Hydrological Area	River Basin	Predictor	Base Period (1991–2020)	Future Scenario	2040 (2031–2050)	2060 (2051–2070)	2080 (2071–2090)
			Actual Measured Value, °C		Projected Increase, °C	Projected Increase, °C	Projected Increase, °C
Left-bank tributaries of the Zhaiyk River	Or River basin	$T_{avg.VI\text{-}XI}$	12.3	SSP3-7.0	2.6	3.6	4.7
				SSP5-8.5	2.7	4.0	5.7
	Elek River basin	$T_{avg.VI\text{-}XI}$	13.4	SSP3-7.0	2.6	3.7	4.8
				SSP5-8.5	2.8	4.1	5.3
	Ulken Kobda River basin	$T_{avg.VI\text{-}XI}$	14.2	SSP3-7.0	2.3	3.6	4.8
				SSP5-8.5	2.5	4.0	5.9
Right-bank tributaries of the Zhaiyk River	Shagan River basin	$T_{avg.VI\text{-}XI}$	14.1	SSP3-7.0	2.3	3.6	4.7
				SSP5-8.5	2.6	4.1	5.5
Southern rivers	Emba River basin	$T_{avg.VI\text{-}XI}$	14.6	SSP3-7.0	1.7	2.8	4.1
				SSP5-8.5	1.9	3.4	5.1
	Oiyl River basin	$T_{avg.VI\text{-}XI}$	15.3	SSP3-7.0	2.5	3.8	5.0
				SSP5-8.5	2.7	4.2	6.1

For the Karakobda River, a progressive and more pronounced decline in runoff is forecasted. By 2040, minimal runoff is expected to decrease by 22% under SSP3-7.0 and by 26% under SSP5-8.5. The decline will intensify, reaching 40% (SSP3-7.0) and 42% (SSP5-8.5) by 2060, and 53% (SSP3-7.0) and 59% (SSP5-8.5) by 2080.

Precipitation dynamics in this area are projected to be synchronous with runoff. According to forecasts, precipitation will also decline progressively: under SSP3-7.0, a decrease of 7% is expected by 2040, increasing to 21% by both 2060 and 2080; under SSP5-8.5, precipitation is projected to decrease by 11% by 2040 and by 19% by 2080. By the end of the century, average summer–autumn air temperatures are projected to rise by 4.8 °C under SSP3-7.0 and by 5.9 °C under SSP5-8.5.

Unlike these rivers, minimal runoff during the summer–autumn low-water period in the Or River is projected to increase significantly under both scenarios. Under SSP3-7.0, minimal runoff is expected to increase by 42% by 2040, 43% by 2060, and 68% by 2080. Under SSP5-8.5, the increase is projected to reach 41% by 2040, 44% by 2060, and 68% by 2080.

Precipitation in the Or River basin is expected to increase synchronously with runoff. Under SSP3-7.0, precipitation is projected to rise by 36% by 2040, 20% by 2060, and 34% by 2080; under SSP5-8.5, precipitation is expected to increase by 32% by 2040, 15% by 2060, and 18% by 2080. By the end of the century, summer–autumn season temperatures are projected to rise by 4.7 °C under SSP3-7.0 and to 5.7 °C under SSP5-8.5, based on observations at the Novorossiyskoye meteorological station.

In the right-bank area of the Zhaiyk River, projected trends of decreasing precipitation combined with a significant increase in air temperature will lead to a stable decline in the minimal summer–autumn low-water runoff of the Shagan River. According to the SSP3-7.0 scenario, minimal runoff is expected to decrease by 26% by 2040, 49% by 2060, and 63% by 2080. Under the SSP5-8.5 scenario, the corresponding reductions are projected at 31%, 53%, and 71%, respectively. Precipitation is forecasted to decrease by 2%, 4%, and 10% under SSP3-7.0 by 2040, 2060, and 2080, respectively, and by 2%, 11%, and 18% under SSP5-8.5 for the same periods. Temperature is expected to increase by 2.3 °C, 3.6 °C, and 4.7 °C under SSP3-7.0, and by 2.6 °C, 4.1 °C, and 5.5 °C under SSP5-8.5 across the three forecast periods.

The southern rivers of the basin demonstrate the most pronounced and systematic reductions in minimal runoff under both climate scenarios. By 2040, the reduction in minimal runoff is projected to range from 7% for the Temir River under SSP3-7.0 to 22% for the Oiyl River under SSP5-8.5. By 2060, the reduction intensifies to 32–33% for all rivers under SSP3-7.0, and to 35% for the Zhem and Oiyl rivers under SSP5-8.5. By 2080, significant decreases in minimal runoff are forecasted, with reductions ranging from 45% (Temir and Oiyl rivers) to 49% (Zhem River) under SSP3-7.0, and from 56% (Oiyl River) to 65% (Zhem River) under SSP5-8.5.

The divergent trends in forecasted precipitation for the southern river basins suggest that summer–autumn precipitation exerts a relatively minor influence on minimal runoff, while air temperature plays a dominant role. According to projections, summer–autumn temperatures in the region will gradually increase. By the end of the century, the maximum projected temperature increase reaches 4.1 °C for the Zhem River basin under SSP3-7.0 and 5.1 °C under SSP5-8.5. For the Oiyl River basin, temperature increases are projected at 5.0 °C (SSP3-7.0) and 6.1 °C (SSP5-8.5).

3.3. Projection of Minimal Runoff During the Winter Low-Water Period

Forecast results presented in Figure 8 indicate that under both climate scenarios, a steady increase in minimal monthly runoff during the winter low-water period is expected across all considered rivers for the forecast horizons of 2040, 2060, and 2080. The results are graphically presented in Figure 9. Seasonal changes in precipitation and air temperature for the Zhaiyk–Caspian water management basin under future climate conditions are provided in Tables 9 and 10.

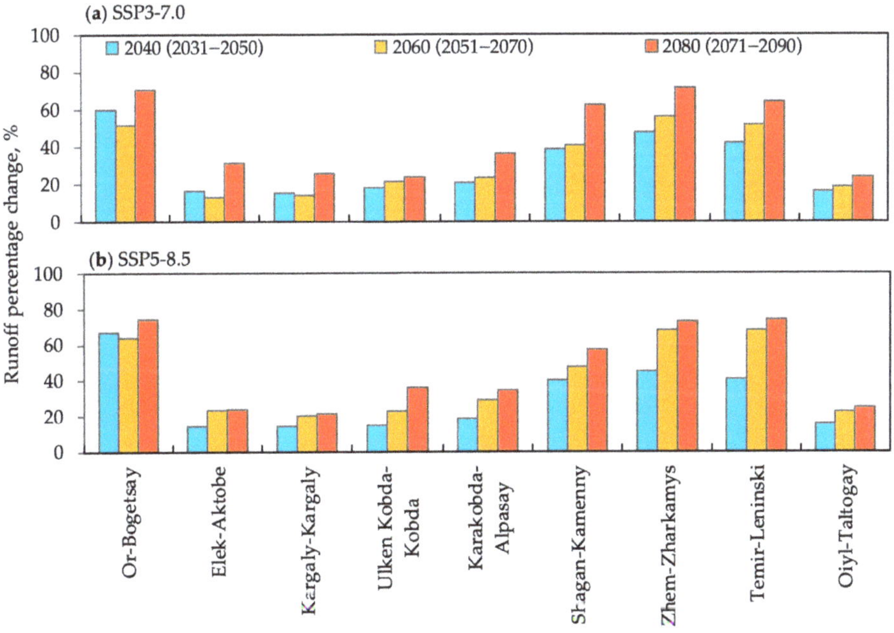

Figure 9. Projected percentage changes in the average monthly minimal runoff of the winter low-water period of rivers in the Zhaiyk–Caspian water management basin relative to the historical average (1991–2020).

Table 9. Projected changes in precipitation amounts during the winter season in the Zhaiyk–Caspian water management basin territory.

Hydrological Area	River Basin	Predictor	Base Period (1991–2020)	Future Scenario	2040 (2031–2050)	2060 (2051–2070)	2080 (2071–2090)
			Actual Measured Value, mm		Rate of Change, %	Rate of Change, %	Rate of Change, %
Left-bank tributaries of the Zhaiyk River	Or River basin	$\sum P_{IX-II}$	166	SSP3-7.0	43	36	61
				SSP5-8.5	42	48	46
	Elek River basin	$\sum P_{IX-II}$	160	SSP3-7.0	47	37	64
				SSP5-8.5	44	51	45
	Ulken Kobda River basin	$\sum P_{XII-II}$	86	SSP3-7.0	13	11	49
				SSP5-8.5	20	35	33
		$\sum P_{X-II}$	85	SSP3-7.0	47	41	74
				SSP5-8.5	42	55	47
Right-bank tributaries of the Zhaiyk River	Shagan River basin	$\sum P_{X-II}$	144	SSP3-7.0	26	23	42
				SSP5-8.5	27	31	28
Southern rivers	Emba and Oiyl river basins	$\sum P_{XII-II}$	61	SSP3-7.0	48	48	71
				SSP5-8.5	49	71	57

Table 10. Projected changes in air temperature during the winter season in the Zhaiyk–Caspian water management basin territory.

Hydrological Area	River Basin	Predictor	Base Period (1991–2020)	Future Scenario	2040 (2031–2050)	2060 (2051–2070)	2080 (2071–2090)
			Actual Measured Value, °C		Projected Increase, °C	Projected Increase, °C	Projected Increase, °C
Left-bank tributaries of the Zhaiyk River	Or River basin	$T_{avg.I-III}$	−12.7	SSP3-7.0	3.5	4.9	6.0
				SSP5-8.5	3.7	5.6	6.9
	Elek River basin	$T_{avg.I-II}$	−12.0	SSP3-7.0	1.7	2.9	4.3
				SSP5-8.5	1.9	3.7	5.6
	Ulken Kobda River basin	$T_{avg.I-II}$	−11.6	SSP3-7.0	2.2	3.6	5.0
				SSP5-8.5	2.5	4.2	6.1
Right-bank tributaries of the Zhaiyk River	Shagan River basin	$T_{avg.XII-II}$	−9.7	SSP3-7.0	2.0	3.3	4.3
				SSP5-8.5	2.3	3.9	6.0
Southern rivers	Emba River basin	$T_{avg.XII-II}$	−11.1	SSP3-7.0	2.1	3.4	4.7
				SSP5-8.5	2.4	4.0	5.7
	Oiyl River basin	$T_{avg.I-II}$	−11.1	SSP3-7.0	3.8	4.9	6.1
				SSP5-8.5	4.0	5.5	7.2

Among the left-bank tributaries of the Zhaiyk River, the Or River shows the most pronounced changes, with minimal runoff projected to increase by 60% (SSP3-7.0) and 67% (SSP5-8.5) by 2040; by 52% and 64%, respectively, by 2060; and by 70% (SSP3-7.0) and 74% (SSP5-8.5) by 2080.

The Elek River and its tributary, the Kargaly River, are also projected to exhibit positive dynamics. By 2040, minimal runoff is expected to increase by 16–17% under SSP3-7.0 and by 14% under SSP5-8.5. By 2060, the increase is projected to range from 13–14% (SSP3-7.0) to 20–23% (SSP5-8.5). By 2080, minimal runoff is expected to continue increasing, reaching 26–31% under SSP3-7.0 and 21–24% under SSP5-8.5.

For the Ulken Kobda and Karakobda rivers, a stable and gradual increase in minimal runoff is projected throughout the forecast period. By 2040, runoff is expected to increase by 18–21% under SSP3-7.0 and by 15–19% under SSP5-8.5. By 2060, increases are projected to reach 22–24% (SSP3-7.0) and 23–29% (SSP5-8.5). By 2080, minimal runoff is projected to increase by 24–36% under SSP3-7.0 and by 35–36% under SSP5-8.5.

The territory of the left-bank tributaries of the Zhaiyk River is characterized by synchronous dynamics between river flow and precipitation during the autumn season and the cold period, as evidenced by both historical observations and future projections. Winter precipitation in this area is expected to increase under the SSP3-7.0 scenario from 13% by 2040 to 49% by 2080, and under the SSP5-8.5 scenario from 20% by 2040 to 35% by 2060. The total precipitation for the autumn and winter seasons under SSP3-7.0 is projected to increase, on average, by 43–47% by 2040, 36–41% by 2060, and 61–74% by 2080. Under SSP5-8.5, precipitation is expected to increase by 42–47% by 2040, 48–55% by 2060, and 45–47% by 2080.

Forecast data also indicate a steady increase in average air temperatures for January and February across the area, continuing through the end of the century. The projected increase in average air temperature is 2.0 °C, 3.2 °C, and 4.6 °C under the SSP3-7.0 scenario, and 2.2 °C, 3.9 °C, and 5.8 °C under the SSP5-8.5 scenario for the forecast periods.

A pronounced increase in river flow, synchronous with future precipitation trends, is forecasted for the Shagan River. According to the SSP3-7.0 scenario, minimal runoff is expected to increase by 39% by 2040, 41% by 2060, and 62% by 2080. Under the SSP5-8.5 scenario, the projected increases are 40%, 48%, and 57%, respectively. During the first two forecast periods, precipitation in the region is projected to increase by up to 31% by 2060 under the SSP5-8.5 scenario. By the end of the century, precipitation is expected to increase by up to 42% under SSP3-7.0 and 28% under SSP5-8.5.

In addition, air temperature in the area of the right-bank tributaries of the Zhaiyk River is projected to increase by 4.3 °C under SSP3-7.0 and by 6.0 °C under SSP5-8.5 by the end of the century.

Among the southern rivers, significant increases in minimal winter runoff are projected for the Zhem River and its tributary, the Temir River. According to the SSP3-7.0 scenario, minimal runoff is expected to increase by 42–48% by 2040, 56–62% by 2060, and 64–71% by 2080. Under the SSP5-8.5 scenario, the projected increases are 41–45% by 2040, 68% by 2060, and 73–74% by 2080.

Scenario-based forecasts for the Oiyl River also indicate positive changes, though to a lesser extent. By 2040, minimal runoff is expected to increase by 16% under both scenarios, by 18–22% by 2060 (SSP3-7.0 and SSP5-8.5, respectively), and by 24–25% by 2080 under both scenarios.

Precipitation in this area is projected to increase under the SSP3-7.0 scenario by 48% in 2040 and 2060, reaching 71% by 2080. Under the SSP5-8.5 scenario, precipitation is expected to increase by 49% in 2040, 71% in 2060, and 57% in 2080.

The average air temperature for the December–February period by the end of the 21st century is projected to increase by 4.7 °C under SSP3-7.0 and by 5.7 °C under SSP5-8.5, relative to the baseline period. For the January–February period specifically, temperatures are projected to rise by 6.1 °C under SSP3-7.0 and by 7.2 °C under SSP5-8.5 by 2080.

The quality of the methodology was evaluated according to the criteria proposed by [17], taking into account the correlation coefficient (r) between observed and predicted values.

For verification periods longer than 25 years, the methodology demonstrated a satisfactory forecast performance for both the summer–autumn and winter minimal runoff, as interpreted in Table 11.

Table 11. Evaluation of the forecast quality for minimal runoff.

Low-Flow Period	S/σ	ρ	Methodological Quality Rating
Summer-Autumn	0.69–0.81	0.60–0.75	Satisfactory
Winter	0.77–0.80	0.61–0.67	Satisfactory

The accuracy indicators fall within the established standards, confirming the applicability of the selected approach for climate-sensitive assessment of minimal runoff.

4. Discussion

Long-term hydrological monitoring in various catchments of Kazakhstan plains reveals both intra-annual runoff redistribution and emerging trends in total annual runoff, highlighting the need for new approaches to characterizing low-flow conditions and supporting water resource management decisions [18,20]. Among the available methods, scenario-based hydrological projections that incorporate observed and projected climatic and hydrological changes are increasingly applied in research and practice. This trend has been facilitated by recent advances in global and regional climate modelling [21–23].

Within the framework of the World Climate Research Programme (WCRP), CMIP6 provides a new generation of climate scenarios, known as Shared Socioeconomic Pathways (SSPs). These scenarios combine trajectories of socioeconomic development with pathways of anthropogenic changes in atmospheric greenhouse gas concentrations, expanding upon the Representative Concentration Pathways (RCPs) framework introduced in the previous CMIP5 project. The RCPs—specifically RCP2.6, RCP4.5, RCP6.0, and RCP8.5—formed the basis for earlier climate change projections and are incorporated within the broader SSP framework of CMIP6 [24]. CMIP6 simulations offer a valuable basis for assessing regional climate change impacts, including those affecting water resources, by providing a wide range of plausible future climate conditions consistent with both socioeconomic and emission-related factors.

Most previous research has focused on scenario-based projections of annual river runoff. For instance, the authors of [25] highlighted considerable uncertainty in projections of annual runoff changes derived from an ensemble of nine global climate models participating in the CMIP3 project. In that study, substantial calculation errors and a wide inter-model spread were reported for the Volga River basin, with projected anomalies in total annual runoff ranging from −14% to +27%. In a more recent study by [26], multi-year average differences between precipitation and evaporation, calculated from 26 CMIP5 models, were used to characterize annual runoff across Russia. Positive anomalies were identified for almost all rivers, except those located in southern Russia. The largest positive anomalies by the end of the 21st century were projected under the unfavourable RCP8.5 scenario, reaching 34% for the Lena River, 27% for the Yenisei River, and 42% for the total runoff of rivers in northeastern Siberia.

The application of CMIP6 climate models is essential for assessing future water resources for several reasons. First, they provide a broad range of plausible future climate projections that can be applied to scenario-based assessments across different regions and time horizons. Second, CMIP6 includes key climate variables, such as air temperature and atmospheric precipitation, that directly affect the availability and distribution of water resources. Third, these scenario-based forecasts represent important tools for developing water resource management strategies, including water supply planning, irrigation management, and flood and hydrological drought risk assessment. Finally, improving knowledge of future climate conditions enables the implementation of adaptation measures to enhance the resilience of water systems under changing climatic conditions.

From a scientific and methodological perspective, CMIP6 offers notable advantages over the previous generation of models (CMIP5), providing improved accuracy and enhanced information for hydrological research, including scenario-based forecasting of low-water runoff. However, previous studies have demonstrated that assessments of water regime characteristics for individual river basins based on climate models often carry significant uncertainties, primarily due to inaccuracies in estimating atmospheric precipitation at the regional scale and substantial simplifications in the representation of hydrological processes [27,28]. Nevertheless, recent advancements in CMIP6 models, including increased computational efficiency and improved parameterizations of land surface hydrological processes, have partially mitigated these limitations [29].

The results of the present study confirm consistent changes in air temperature and atmospheric precipitation patterns within the Zhaiyk–Caspian water management basin, which have a direct influence on the formation of minimum low-water runoff. The primary factors affecting the quantitative characteristics of minimum river discharge during the summer–autumn low-water period are atmospheric precipitation and air temperature during the same season. Scenario-based projections derived from CMIP6 models indicate an expected increase in summer–autumn precipitation across much of the basin. Nevertheless, a consistent reduction in minimum runoff is projected for most rivers, particularly in the southern part of the basin, across all future periods and under both greenhouse gas scenarios. These results are consistent with established climatic trends for the northern mid-latitudes [30]. The apparent contradiction between increasing precipitation and decreasing minimum flow can be explained by the compensating effects of increased evapotranspiration and reduced soil moisture reserves due to rising air temperatures, ultimately contributing to reductions in both surface and groundwater contributions to river flow [31].

Conversely, during the winter low-water period, an increase in minimum river discharge is projected for all watercourses in the study region of Western Kazakhstan. This is likely associated with enhanced infiltration recharge resulting from more frequent winter thaws and increased winter precipitation. While autumn precipitation is traditionally considered the main source of groundwater recharge during the winter period, it alone cannot explain the observed and projected increases in minimum winter runoff, given the statistically insignificant trends in autumn precipitation observed at most meteorological stations in the basin.

Long-term dynamics of winter precipitation, in contrast, exhibit significant positive trends across most of the basin, albeit with substantial spatial variability. Similar conclusions were reached by other researchers [6], who demonstrated that autumn precipitation alone is insufficient to sustain winter runoff recharge in the Volga River basin. Therefore, when assessing the formation of minimum winter runoff under ongoing climate change, it is advisable to consider the combined influence of both autumn and winter precipitation.

Changes in groundwater levels must also be considered a contributing factor. Under significant winter warming, snowmelt generated by more frequent thaws, combined with increased winter precipitation, can provide an additional source of river recharge and contribute to higher groundwater levels. A substantial portion of the water generated during thaws is retained in the active soil layer, enhancing soil moisture and promoting infiltration recharge and groundwater replenishment. These processes are consistent with the findings of several other studies [6,32,33]. Furthermore, recent observations confirm a rise in the frequency and duration of winter thaws in the region [34]. Scenario-based projections for the basin indicate a stable increase in minimum river discharge during the winter low-water period across all forecast horizons and under both climate scenarios. These results demonstrate that rising winter temperatures are accompanied by increased precipitation, consistent with the findings of [35]. Together, these changes contribute to greater snowmelt volumes, ultimately leading to higher base flow during the winter period.

It is important to emphasize that the projections of minimum river discharge presented in this study are subject to inherent uncertainties associated with the use of global and regional climate models, as well as the applied methodological approach. The range of projected changes in minimum runoff across different rivers and time periods, as illustrated in the tables and figures, reflects both the variability among the selected CMIP6 models and the influence of different greenhouse gas concentration scenarios. Although the ensemble of models employed in this study provides a representative spectrum of possible future climate conditions, significant uncertainty persists regarding both the magnitude and spatial distribution of projected changes, particularly for the summer–autumn low-water period, where the combined effects of precipitation and temperature changes on runoff formation are complex and nonlinear.

In addition to model-related uncertainty, it is essential to recognize methodological limitations. The statistical relationships established between minimum runoff and meteorological predictors are based on historical observations and assume their stability under future climate conditions. However, potential non-stationarity in hydrological processes, changes in land use, and groundwater dynamics—factors not explicitly accounted for in the applied models—may alter these relationships, especially under extreme climate scenarios.

Despite these limitations, the consistent spatial and seasonal patterns identified in the projections, together with the physical plausibility of the underlying processes, increase confidence in the qualitative trends described. Nevertheless, future research should incorporate more comprehensive uncertainty assessments, including probabilistic approaches, multi-model ensembles, and improved representation of groundwater processes, to further enhance the robustness of runoff projections and support the development of climate-resilient water management strategies in Western Kazakhstan.

5. Conclusions

This study provides a scenario-based assessment of future changes in minimum river discharge in the Zhaiyk–Caspian water management basin of Western Kazakhstan under projected climate change conditions. Using an ensemble of dynamically downscaled CMIP6 climate models for Central Asia, low-flow conditions were estimated for the summer–autumn and winter low-water periods for three future time horizons (2040, 2060, and 2080) under two greenhouse gas concentration scenarios (SSP3-7.0 and SSP5-8.5).

The results reveal a pronounced seasonal contrast in the projected hydrological response to climate change. During the summer–autumn low-water period, a widespread reduction in minimum river discharge is expected, particularly for rivers in the southern part of the basin, including the Zhem, Oiyl, and Temir rivers. This projected decline, which may reach up to 64% by the end of the century, is primarily attributed to increased evapotranspiration and soil moisture deficits driven by rising air temperatures, which offset the potential positive effects of increased precipitation.

In contrast, during the winter low-water period, a steady increase in minimum discharge is projected for all analysed rivers, with the Or, Zhem, Temir, and Shagan rivers showing the most substantial changes. This increase is associated with higher winter precipitation and air temperatures, which collectively enhance the frequency of thaws, promote infiltration processes, and contribute to groundwater recharge, ultimately leading to higher winter baseflow.

The analysis also indicates that changes in winter discharge cannot be explained by autumn precipitation alone, emphasizing the need to account for the combined effects of seasonal precipitation patterns, snowmelt dynamics, and groundwater contributions when assessing future low-water conditions.

It should be acknowledged that these projections are subject to uncertainties arising from the limitations of climate models, regional downscaling techniques, and the simplified representation of hydrological processes. Nevertheless, the consistency of the spatial and seasonal patterns identified, supported by physically plausible mechanisms, provides a reliable basis for informing water resource management and climate adaptation planning in the region.

These findings underscore the importance of developing seasonally differentiated, climate-resilient water management strategies. Priority should be given to mitigating summer–autumn low-flow risks in the southern basins while also considering opportunities to manage increased winter runoff.

Further research is needed to improve the representation of groundwater processes in hydrological models, to integrate socioeconomic factors such as land use and water management changes, and to apply high-resolution regional modelling to reduce uncertainties in low-flow projections. In particular, detailed field investigations on groundwater–surface water interactions and the role of snowpack dynamics are essential for refining future assessments of minimum river discharge under evolving climate conditions.

Author Contributions: Conceptualization, L.M. and S.A.; methodology, L.M. and A.T.; validation, L.B. and E.T.; formal analysis, L.B. and E.T.; investigation, L.M. and L.B.; data curation, M.E.R.-C. and O.A.; writing original draft preparation, L.B., J.R.-I. and M.E.R.-C.; writing review and editing, J.R.-I. and M.E.R.-C.; visualization, J.R.-I. and O.A.; project administration, L.M., S.A. and A.T. All authors have read and agreed to the published version of the manuscript.

Funding: This research was funded by the Committee of Science of the Ministry of Science and Higher Education of the Republic of Kazakhstan (Grant No. BR21882122) "Sustainable Development of Natural-Industrial and Socioeconomic Systems of the West Kazakhstan Region in the Context of Green Growth: A Comprehensive Analysis, Concept, Forecast Estimates and Scenarios."

Data Availability Statement: Data are contained within the article.

Conflicts of Interest: Authors Lyazzat Makhmudova, Sayat Alimkulov, Aisulu Tursunova, Lyazzat Birimbayeva, Elmira Talipova, and Oirat Alzhanov were employed by the company JSC "Institute of Geography and Water Security". The remaining authors declare that the research was conducted in the absence of any commercial or financial relationships that could be construed as a potential conflict of interest. The authors declare no conflicts of interest.

References

1. Marshall, A.; Grubert, E.; Warix, S. Temperature overshoot would have lasting impacts on hydrology and water resources. *Water Resour. Res.* **2025**, *61*, e2024WR037950. [CrossRef]
2. Abu-Zeid, M.A.; Abu-Zeid, K.M.; Halim, M.K. Impact of climate change on water resources. In *Climate Changes Impacts on Aquatic Environment: Assessment, Adaptation, Mitigation, and Road Map for Sustainable Development*; Khalil, M.T., Emam, M., Negm, A., Eds.; Springer Nature: Cham, Switzerland, 2025; pp. 121–157. [CrossRef]
3. Birimbayeva, L.; Makhmudova, L.; Alimkulov, S.; Tursunova, A.; Mussina, A.; Tigkas, D.; Beksultanova, Z.; Rodrigo-Clavero, M.-E.; Rodrigo-Ilarri, J. Analysis of the Spatiotemporal Variability of Hydrological Drought Regimes in the Lowland Rivers of Kazakhstan. *Water* **2024**, *16*, 2316. [CrossRef]
4. Gaidukova, E.V.; Litvinova, N.V. Otsenka dolgosrochnykh izmenenii mnogoletnego godovogo stoka pod vliyaniem klimata v Arkticheskom regione RF [Assessment of long-term changes in annual runoff under climate influence in the Arctic region of the Russian Federation]. In Proceedings of the Third International Scientific and Practical Conference, Kazan, Russia, 4 November 2017; In four parts. Volume Part 4. Aktual'nye voprosy v nauke i praktike. pp. 251–262. (In Russian).
5. Loboda, N.S.; Vorobyeva, L.V.; Ilchenko, E.A. Metodicheskie podkhody k otsenke vliyaniya global'nogo potepleniya na sostoyanie vodnykh resursov gornykh rek Krymskogo poluostrova [Methodological approaches to assessing the impact of global warming on the water resources of mountain rivers of the Crimean Peninsula]. *Vestn. Odess. Derzhavnogo Ekoljgicheskogo Univ.* **2006**, *2*, 208–214. Available online: http://bulletin.odeku.edu.ua/ru/metodicheskie-podhody-k-oczenke-vliyaniya-globalnogo-potepleniya-na-sostoyanie-vodnyh-resursov-gornyh-rek-krymskogo-poluostrova/ (accessed on 10 August 2025). (In Russian).
6. Bolgov, M.V.; Trubetskova, M.D.; Filimonova, M.K.; Filippova, I.A. Sovremennye izmeneniya klimaticheskikh kharakteristik i veroyatnostnaya otsenka izmeneniy minimal'nogo stoka v basseine reki Volgi [Curent changes in climate characteristics and probabilistic assessment of minimum flow changes in the Volga River basin]. *Vodn. Khozyaistvo Ross.* **2014**, *3*, 83–99. (In Russian) [CrossRef]
7. Birimbayeva, L.; Makhmudova, L.; Alimkulov, S.; Tursunova, A.; Tigkas, D.; Abayev, N.; Dostayeva, A.; Birimbayev, Z.; Alzhanov, O.; Rodrigo-Clavero, M.-E.; et al. Reaction of Minimal Streamflow to Natural Factors in the Context of Climate Uncertainty. *Water Resour Manag.* **2025**. [CrossRef]

8. Tokbergenova, A.; Skorintseva, I.; Ryskeldiyeva, A.; Kaliyeva, D.; Salmurzauly, R.; Mussagaliyeva, A. Assessment of anthropogenic disturbances of landscapes: West Kazakhstan Region. *Sustainability* **2025**, *17*, 573. [CrossRef]
9. Sarsekov, A.S.; Visloguzova, A.V.; Vladimirov, N.M.; Gus'kova, A.I.; Medeu, A.R.; Nurmambetov, E.I.; Potapova, G.M. *Rel'ef Kazakhstana (Poiasnitel'naya Zapiska k Geomorfologicheskoy Karte Kazakhstana Masshtaba 1:1,500,000) [Relief of Kazakhstan: Explanatory Note to the Geomorphological Map of Kazakhstan at a Scale of 1:1,500,000]*, 2nd ed.; Научная библиотека: Almaty, Kazakhstan, 2017; Available online: https://library.qyzpu.edu.kz/lib/document/EK/8161433F-8D0C-49AA-B19A-35C81344D600/ (accessed on 10 August 2025). (In Russian)
10. Akiyanova, F.Z.; Abitbayeva, A.D.; Yegemberdiyeva, K.B.; Temirbayeva, R.K. Problems of desertification of the territory of Kazakhstan: Status and forecast. *Life Sci. J.* **2014**, *11* (Suppl. S10), 341–345. [CrossRef]
11. Akiyanova, F.Z.; Semenov, O.E.; Khalykov, E.E. Podvizhnye peski poluostrova Buzachi (Bozashy) [Mobile sands of the Buzachi (Bozashy) Peninsula]. *Gidrometeorol. I Ekol.* **2012**, *1*, 58–63. (In Russian)
12. Respublika Kazakhstan. *Tom 1: Prirodnye Usloviya i Resursy [Republic of Kazakhstan. Natural Conditions and Resources]*; Iskakov, N.A., Medeu, A.R., Eds.; Ministry for Environmental Protection: Almaty, Kazakhstan, 2006; Volume 1. (In Russian)
13. Rozhdestvenskiy, A.V.; Lobanova, A.G. *Metodicheskie Rekomendatsii po Otsenke Odnorodnosti Gidrologicheskikh Kharakteristik i Opredeleniyu Ikh Raschetnykh Znacheniy po Neodnorodnym Dannym [Methodological Recommendations for Assessing the Homogeneity of Hydrological Characteristics and Determining Their Design Values Based on Heterogeneous Data]*; Nestor Istoriya: St. Petersburg, Russia, 2010; ISBN 978 598187 550 2. (In Russian)
14. Fallah, B.; Rostami, M.; Russo, E.; Harder, P.; Menz, C.; Hoffmann, P.; Didovets, I.; Hattermann, F.F. Climate model downscaling in Central Asia: A dynamical and a neural network approach. *Geosci. Model Dev.* **2025**, *18*, 161–180. [CrossRef]
15. Lange, S. Trend-preserving bias adjustment and statistical downscaling with ISIMIP3BASD (v1.0). *Geosci. Model. Dev.* **2019**, *12*, 3055–3070. [CrossRef]
16. Frieler, K.; Lange, S.; Piontek, F.; Reyer, C.P.O.; Schewe, J.; Warszawski, L.; Zhao, F.; Chini, L.; Denvil, S.; Emanuel, K.; et al. Assessing the impacts of 1.5 °C global warming—simulation protocol of the Inter-Sectoral Impact Model Intercomparison Project (ISIMIP2b). *Geosci. Model. Dev.* **2017**, *10*, 4321–4345. [CrossRef]
17. Georgievsky, Y.M.; Shanoshkin, S.V. *Gidrologicheskie Prognozy [Hydrological Forecasts]*; RSHU Publishing: St. Petersburg, Russia, 2017; p. 436. (In Russian)
18. Smagulov, Z.; Makhmudova, L.; Alimkulov, S.; Talipova, E.; Zagidullina, A.; Birimbayeva, L.; Sailaubek, A. Transformation of seasonal distribution of river flow in the Zhaiyk-Caspian water basin under changing climate conditions. *J. Water Clim. Change* **2025**, *16*, 400–429. [CrossRef]
19. Vladimirov, A.M. Faktory formirovaniya ekstremal'nogo stoka v malovodnyy sezon [Factors affecting the formation of extreme runoff during the low-water season]. *Uchenye Zap. Ross. Gos. Gidrometeorol. Univ.* **2008**, *7*, 13–22. Available online: http://elib.rshu.ru/files_books/pdf/7-2.pdf (accessed on 10 August 2025). (In Russian).
20. Moldakhmetov, M.; Makhmudova, L.; Mussina, A.; Abdullayeva, A. Studying the non-stationarity of river flow during long-term fluctuations in plain rivers of Kazakhstan. *Geogr. Water Resour.* **2024**, *2*, 64–82. [CrossRef]
21. Patel, G.; Das, S.; Das, R.A. Comparative Approach to Understand the Performance of CMIP6 Models for Maximum Temperature near Tropic of Cancer Using Multiple Machine Learning Ensembles. *Water Resour. Manag.* **2025**, *39*, 3921–3936. [CrossRef]
22. Mentzafou, A.; Papadopoulos, A.; Dimitriou, E. Assessing the impacts of climatic and water management scenarios in a small mountainous Greek river. *Hydrology* **2025**, *12*, 13. [CrossRef]
23. Dioha, E.C.; Chung, E.S.; Ayugi, B.O.; Babaousmail, H.; Sian, K.T.C.L.K. Quantifying the added value in the NEX-GDDP-CMIP6 models as compared to native CMIP6 in simulating Africa's diverse precipitation climatology. *Earth Syst. Environ.* **2024**, *8*, 417–436. [CrossRef]
24. Voskresenskaya, E.N.; Marchukova, O.V.; Afanasyeva, V.V. Otsenka vozmozhnykh izmenenii povtoryaemosti sobytii El-Niño i LaNiña k koncuXXIveka po modelyam proekta CMIP6 [Assessment of possible changes in the repeatability of ElNiño and LaNiña events by the end of the twenty-first century using CMIP6 models]. *Monit. Syst. Environ.* **2021**, *46*, 14–21. [CrossRef]
25. Semenov, S.M. (Ed.) *Metody Otsenki Posledstvii Izmeneniya Klimata Dlya Fizicheskikh i Biologicheskikh System [Metody Ocenki Posledstvij Izmeneniya Klimata Dlya Fizicheskikh i Biologicheskikh Sistem]*; Roshydromet: Moscow, Russia, 2012; Chapter 2; pp. 53–86. Available online: http://downloads.igce.ru/publications/metodi_ocenki/02.pdf (accessed on 10 August 2025). (In Russian)
26. Shkolnik, I.M.; Meleshko, V.P.; Karol', I.L.; Kiselev, A.A.; Nadezhina, E.D.; Govorkova, V.A.; Pavlova, T.V. Ozhidaemye izmeneniya klimata na territorii Rossijskoj Federacii v XXI veke [Expected climate changes on the territory of the Russian Federation in the 21st century]. In *Second Roshydromet Assessment Report on the Climate Change and Its Consequences in the Russian Federation*; Roshydromet: Moscow, Russia, 2014; pp. 65–118, (In Russian). [CrossRef]
27. Didovets, I.; Krysanova, V.; Nurbatsina, A.; Fallah, B.; Krylova, V.; Saparova, A.; Niyazov, J.; Kalashnikova, O.; Hattermann, F.F. Attribution of current trends in streamflow to climate change for 12 Central Asian catchments. *Clim. Change* **2024**, *177*, 16. [CrossRef]

28. Kundzewicz, Z.W.; Krysanova, V.; Benestad, R.E.; Hov, Ø.; Piniewski, M.; Otto, I.M. Uncertainty in climate change impacts on water resources. *Environ. Sci. Policy* **2018**, *79*, 1–8. [CrossRef]
29. Gelfan, A.N.; Gusev, E.M.; Kalugin, A.S.; Krylenko, I.N.; Motovilov, Y.G.; Nasonova, O.N.; Frolova, N.L. Stok rek Rossii pri proiskhodyashchikh i prognoziruemykh izmeneniyakh klimata: Obzor publikatsiy. 2. Vliyanie izmeneniya klimata na vodnyy rezhim rek Rossii v XXI veke [Runoff of Russian rivers under current and projected climate change: A review of publications. Part 2. Climate change impact on the water regime of Russian rivers in the twenty-first century]. *Vodn. Resur.* **2022**, *49*, 270–285. (In Russian) [CrossRef]
30. Wang, A.; Miao, Y.; Kong, X.; Wu, H. Future changes in global runoff and runoff coefficient from CMIP6 multi-model simulation under SSP1-2.6 and SSP5-8.5 scenarios. *Earth's Future* **2022**, *10*, e2022EF002910. [CrossRef]
31. Zhao, T.; Dai, A. The magnitude and causes of global drought changes in the twenty-first century under a low–moderate emissions scenario. *J. Clim.* **2015**, *28*, 4490–4512. [CrossRef]
32. Danilovich, I.S.; Loginov, V.F.; Groisman, P.Y. Changes of Hydrological Extremes in the Center of Eastern Europe and Their Plausible Causes. *Water* **2023**, *15*, 2992. [CrossRef]
33. Georgiadi, A.G.; Groisman, P.Y. Extreme Low Flow during Long-Lasting Phases of River Runoff in the Central Part of the East European Plain. *Water* **2023**, *15*, 2146. [CrossRef]
34. Makhmudova, L.; Talipova, E.; Myrzakhmetov, A.; Birimbayeva, L.; Tursynbai, A.; Alipbek, A. Climatic characteristics of thaw in the territory of the Zhaiyk-Caspian water basin. *J. Geogr. Environ. Manag.* **2024**, *75*, 80–91. [CrossRef]
35. Bolgov, M.V.; Korobkina, E.A.; Filippova, I.A. Bayesian forecast of minimum runoff under nonstationary conditions considering possible climate change. *Meteorol. Hydrol.* **2016**, *7*, 7–16. [CrossRef]

water

MDPI

Article

Research on the Zoning of Watershed Aquatic Ecological Functions Based on a Distributed Hydrological Model

Bingqing Lin [1,*], Ying Chen [2], Musheng Lin [3], Lizao Ye [4] and Mingjiang Cai [1,*]

1 School of Resource and Environment Sciences, Quanzhou Normal University, Quanzhou 362000, China
2 School of Geographical Science, Fujian Normal University, Fuzhou 350117, China; chenying_nju@163.com
3 College of Geography and Oceanography, Minjiang University, Fuzhou 350007, China; lms081@126.com
4 School of Computer and Information Engineering, Chuzhou University, Chuzhou 239000, China; lizaoye@chzu.edu.cn
* Correspondence: bqlin2011@163.com (B.L.); mingjiangc@foxmail.com (M.C.); Tel.: +86-15960071950 (B.L.)

Abstract

This study aims to enhance the aquatic eco-functional zoning by incorporating the spatial variability of hydrological processes during the zoning process. We propose a method for watershed eco-functional zoning based on distributed hydrological modeling. Using the Jinjiang Basin in Southeast China as a case study, we applied the Soil Water and Assessment Tool (SWAT) model to delineate the basic zoning units and simulate their hydrological processes. We integrated natural environmental indicators—specifically topography, vegetation, meteorology, and hydrology—with land use as a measure of human activity, while accounting for their spatial variability. This approach enabled us to conduct both first-level and second-level eco-functional zoning of the watershed. The results indicated that (1) the Jinjiang Basin can be categorized into three main groups consisting of six first-level aquatic ecological zones, which reflect the spatial variability of terrestrial natural environmental factors and their influence on aquatic ecosystems; (2) building on this categorization, the first-level aquatic ecological regions were further divided into five categories comprising 18 second-level aquatic ecological functional zones, emphasizing the impact of human activities on aquatic ecosystems and their associated first-level ecological service functions; and (3) the application of hydrological simulation techniques allows for a comprehensive assessment of the spatial variability of hydrological processes, thereby enhancing the validity of the ecological function zoning results and providing robust technical support for watershed ecological function zoning.

Keywords: aquatic eco-functional zoning; hydrological processes; hydrological model; southeast China

1. Introduction

The concept of watershed aquatic ecological zoning is predicated on the spatial distribution patterns of watershed ecosystems, with the objective of enhancing the management of water resources, aquatic environments, and ecological systems [1]. Since the introduction of the ecoregion concept in 1967 [2], the delineation of aquatic ecosystems has become a critical foundation for the management and conservation of water environments. In 1987, the United States Environmental Protection Agency (EPA) proposed a framework for aquatic ecological zoning that was based on land use, soil type, vegetation, and hydrology, aimed at guiding water resource management and protection [3]. Biggs et al. (1990) [4] classified New Zealand into five aquatic ecological zones using indicators such as flow, water quality, benthic invertebrates, periphyton, and salmon, which were subsequently refined into 25 aquatic ecological zones by the country's environmental protection agency [5]. In 2000, the European Commission introduced the Water Framework Directive (WFD), which categorized surface water bodies into four types of aquatic ecosystems: rivers, lakes, transitional waters, and coastal waters [6]. Austria delineated four categories of aquatic ecological zones nationwide based on river landscape characteristics, further subdividing them into 15 aquatic biogeographic regions according to benthic community structure [7]. Lorenz et al. [8] and Tison et al. [9] classified rivers in countries such as the UK, Germany, and France based on the spatial differences in macroinvertebrate, benthic algae, and aquatic plant communities. Rivers act as essential conduits for the transfer of materials, energy flow, and species information, providing significant ecological and social services [10,11]. Different aquatic ecosystems not only exhibit variability in ecological factors but also in their functional services. Therefore, it is crucial to account for differences in service functions in studies of water ecological zoning, although relevant research remains relatively sparse.

Currently, some researchers have begun to investigate the functional zoning of aquatic ecosystems [12–14]. The characteristics of river ecosystems and their service functions are closely linked to the physical conditions of the surrounding environment; thus, aquatic functional zoning should incorporate both aquatic ecological factors and the spatial variability of terrestrial environmental factors [15,16]. Among various terrestrial environmental factors, hydrological processes serve as a critical link between aquatic ecosystems and terrestrial environments, facilitating significant energy and information exchange, which determines the spatial distribution patterns of aquatic ecosystems [16,17]. Researchers have identified hydrological indicators for watershed zoning, including river network density, river curvature, channel width, and lake–reservoir ratio [14,18–20]. While these indicators provide some indirect insight into watershed hydrological processes, they are insufficient for objectively reflecting the characteristics of hydrological processes within watersheds. Therefore, how to fully account for the spatial heterogeneity of hydrological processes in the zoning of watershed aquatic ecological functions requires further investigation.

In the context of aquatic functional zoning, sub-watersheds are typically employed as the fundamental spatial units [16]. However, the sparse distribution of hydrological monitoring stations presents a significant challenge in acquiring hydrological process data for individual sub-watersheds. The Soil and Water Assessment Tool (SWAT), which utilizes sub-watersheds as its primary simulation units and effectively captures the spatial variability of water yield, sediment transport, and water quality throughout the watershed. Its simulation outputs for each sub-watershed have established the SWAT as a valuable tool for watershed management [21–24]. Consequently, additional research can be pursued regarding the application of the SWAT model in the functional zoning of watershed aquatic ecological functions.

The Jinjiang watershed, located in Quanzhou City on the southeastern coast of China, is one of the regions in Fujian Province most significantly affected by human activities, which have substantially altered the hydrological characteristics of its rivers [25,26]. This study utilizes the Jinjiang watershed as a case study to propose a watershed-scale ecological function zoning scheme based on hydrological modeling. The objective is to provide a scientific foundation for decision-making in the ecological management of the watershed while also offering methodological guidance for similar studies in other watersheds. The specific components of the research include the following: first, the development of an indicator system suitable for ecological function zoning of the watershed; second, the establishment of a SWAT model for daily runoff simulation to derive the basic units and runoff indicators necessary for ecological function zoning; and finally, the delineation of first-level and secondary ecological function zones within the watershed.

2. Study Area Overview and Data

2.1. Study Area Overview

The Jinjiang Basin, located in the southeastern coast of China, is the third largest river in Fujian Province, with a basin area of 5629 km^2. The Jinjiang River is fed by two primary tributaries: the eastern tributary, which flows through the Shanmei Reservoir, and the western tributary, which passes by the Anxi Station. These tributaries converge 2.5 km upstream of the Shilong Station. The drainage area located upstream of Shilong encompasses 5042 km^2, which constitutes our study area (see Figure 1). The Shanmei Reservoir, the largest reservoir in Quanzhou City, has a total storage capacity of 655 million m^3. It plays a crucial role in agricultural irrigation, domestic water supply, power generation, and flood control for the Quanzhou region. The study area is characterized by subtropical monsoon climate, with an average annual temperature and precipitation of 20 °C and 1686 mm, respectively. Due to its location in Quanzhou City, the towns within the basin have a developed private economy and a high level of urbanization, with relatively active human activities that have significantly altered the underlying surface conditions of the basin [25] and have had a significant impact on the basin's water resources, water environment, and aquatic ecology.

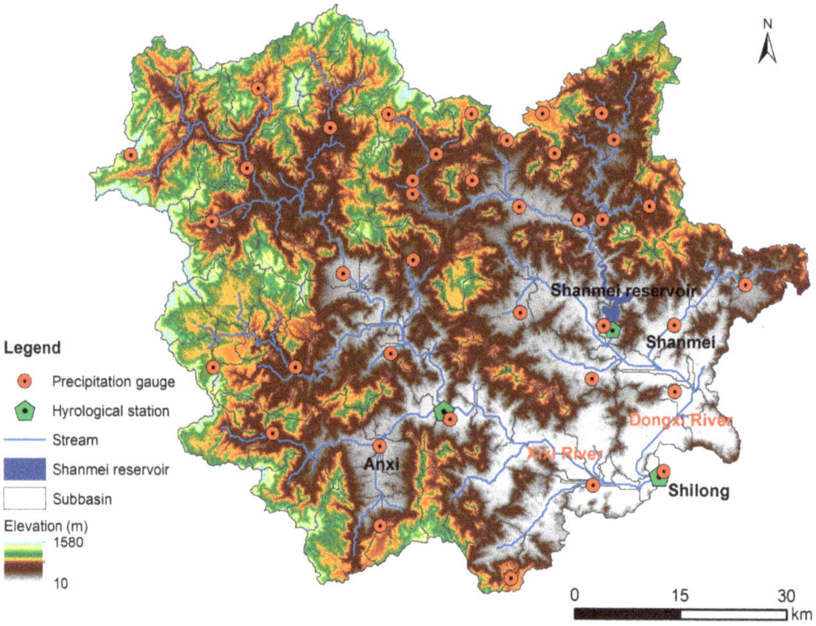

Figure 1. The overview of the Jinjiang Basin.

2.2. Data Sources

Spatial Data: This includes include a Digital Elevation Model (DEM), Normalized Difference Vegetation Index (NDVI, in 2005), soil types, and land use (in 2010). Considering the relatively short duration of the study, we assume that human activities have had minimal impact on land use changes; therefore, we treated the data as static. For the Normalized Difference Vegetation Index (NDVI), we selected the midpoint of the study period, specifically the year 2005, to represent the average conditions throughout the entire study duration. Both the DEM and NDVI data were obtained from the Resource and Environment Data Cloud Platform (https://www.resdc.cn/). Among them, the spatial resolution of the DEM is 30 × 30 m, and the NDVI is 1 × 1 km. The map of soil types was obtained by digitizing the soil type map (1:500,000) generated by the Fertilization Experiment Station of Fujian Province. The land use data were derived from the interpretation of the TM remote sensing images in 2010.

Hydrometeorological Data: Meteorological data (2001–2010) for driving the SWAT model, including 38 daily rainfall stations and two detailed weather elements gauges, were obtained from Fujian Provincial Meteorological Bureau. For hydrological model calibration, records of daily runoff (2001–2010) in Anxi, Shilong, were provided by Fujian Provincial Hydrological and Water Resources Survey Bureau. Additionally, daily inflow and outflow data from 2001 to 2010 for the Shanmei Reservoir were obtained from the Quanzhou Shanmei Reservoir Administration Bureau. Daily inflow data were utilized for the calibration of parameters in the runoff simulation for the catchment area of the Shanmei Reservoir. In contrast, daily outflow data were directly incorporated into the reservoir module of the model to simulate the outflow runoff from the reservoir.

3. Research Methods

3.1. Subdivision and Hydrological Simulation Based on the SWAT Model

The SWAT model is a distributed hydrological model with a strong physical mechanism [27], which has been widely applied to simulate the hydrological response in changing environment. Firstly, the model divides the basin into multiple sub-basins based on the DEM, and then each sub-basin further was divided into different hydrological response units (HRUs), with each HRU comprising the same land use and soil. Based on the principle of water balance, the hillslope water generation is simulated in each HRU; then the results of all the HRUs are summarized as the hillslope water generation within the sub-basin.

Previous research has shown that simulation outputs of model were seriously affected by watershed subdivision levels [28,29]. In order to ensure that the hydrological simulation indicators could reflect the spatial differences in the hydrological process as truly as possible, this paper refers to our existing research results and divides the Jinjiang Basin into 127 sub-basins [30]. In this study, the delineation of hydrological response units (HRUs) is based on four slope categories: 3.5%, 10.5%, 26.8%, and 46.6%. The thresholds for land use, soil, and slope are set at 5%, 10%, and 10%, respectively. Furthermore, to enhance the accuracy of spatial hydrological process simulations, a multi-site calibration approach is utilized [31]. This involves calibrating and validating daily runoff data at three hydrological stations: Anxi Station, Shanmei Station, and Shilong Station. The year 2001 was used as the model warm-up period, 2002–2006 as the calibration period, and 2007–2010 as the validation period. Three evaluation indicators, Nash–Sutcliffe efficiency coefficient (NSE), coefficient of determination (R^2), and relative error (Re), were used to determine the goodness-of-fit performance of the model's simulation [32].

3.2. Construction of the Regional Indicator System

In order to better reflect the spatial difference in natural environmental factors and human activities, as well as their impacts on the spatial differentiation of aquatic ecological system, this paper constructed an index system of regionalization with two different spatial scale levels. The first-level indexes mainly reflect the impact of large-scale natural environmental factors on aquatic ecological system, featuring relatively similar topographic, climatic, vegetation, and hydrological characteristics. Therefore, the specific indexes include the elevation, rainfall, NDVI. and runoff production. The secondary level indexes focus on the impact of human activities on the river ecosystem at a smaller spatial scale, which is a further subdivision of the first-level area. This study selects land use as the secondary level index reflecting the characteristics of human activities. The aquatic ecological significance of each indicator is shown in Table 1.

Table 1. The indices and their meaning of aquatic eco-function regionalization.

Classification Level	Indicator Type	Indicator	Ecological Significance
First level	Topography	Elevation	Reflect the morphological characteristics of rivers
	Climate	Precipitation	Reflect the status of water resources in rivers
	Vegetation	NDVI	Reflect the capacity for material production and water conservation
	Hydrology	Water yield	Reflect hydrological regulation capacity, and the ability of rivers to transport materials, energy, and sediment
Secondary level	Human Activity	Land Use	Reflect the impact of human activities on river ecosystems

3.3. Aquatic Eco-Functional Zoning

The 127 sub-basins delineated by the SWAT model form the foundational units for the eco-functional zoning of the Jinjiang Basin. Utilizing the developed indicator system (Table 1), we calculated the first-level zoning indicator values for each sub-basin. Initially, we derived annual averages of hydrological parameters—total water yield, evapotranspiration, surface runoff, lateral flow, and groundwater flow—from the simulation results of the SWAT model for the years 2002 to 2010. Total water yield was selected as the first-level zoning indicator, as it effectively reflects the spatial distribution characteristics of hydrology within the basin. The other hydrological parameters were utilized to validate the rationality of the zoning.

Subsequently, a spatial overlay analysis was conducted using measured data on elevation, precipitation, and vegetation, in conjunction with the sub-basin layers. This analysis provided the average elevation, annual average precipitation (2001–2010), and average Normalized Difference Vegetation Index (NDVI) for 2005 for each sub-basin. These metrics served as zoning indicators that reflect the spatial distribution characteristics of topography, climate, and vegetation within the basin.

Building on this foundation, we employed a bottom-up agglomerative clustering method to classify the first-level aquatic ecological zones. In this approach, each classification sample was treated as an independent class, and hierarchical merging was performed based on the similarity between samples until all samples were consolidated into a single class. In this study, each sub-basin was treated as an independent classification sample. The classification variables included the previously mentioned statistical indicators: elevation, precipitation, NDVI, and total water yield. Prior to conducting the cluster analysis, we standardized the data to scale them within the range of −1 to 1. Subsequently, the Euclidean distance method was employed to calculate the distance between each pair of sub-basins based on their zoning indicator values, thereby quantifying their similarity. The between-groups linkage method was applied to merge the two classes with the smallest average distance into a single class. This iterative process facilitated the hierarchical clustering of the sub-basins within the Jinjiang Basin. In the final classification results, if any ecological zone exhibited spatial discontinuity, manual fine-tuning were made in accordance with the principle of regional contiguity to ensure spatial continuity within the same ecological functional zone. In the process of manual fine-tuning, full consideration should be given to the physical connections between sub-watersheds, such as river connections. The first-level ecological zones represent spatial variations in natural environmental factors and their effects on aquatic ecosystems, which are inherently associated with topography. Consequently, the nomenclature for this level adheres to the format: "Topography + Aquatic Ecological Zone".

The secondary aquatic eco-functional zones primarily reflect the impact of human activities on aquatic ecosystems at a smaller spatial scale. Based on the secondary zoning indicators (Table 1), this study proposes a zoning method based on the dominant land use types. First, the area proportions of each land use type are calculated at the watershed scale, representing the average land use conditions. Second, the area proportions of each land use type are computed at the sub-watershed scale and compared with the average land use conditions. If only one type of land use exceeds the average land use conditions, that type is designated as the dominant land use type within the sub-watershed. In cases where two or more land use types exceed the average conditions, a comparative analysis of their proportions is required to identify the type that most significantly surpasses the average state of the entire watershed, which will then be classified as the dominant land use type for the sub-watershed. Subsequently, sub-watersheds exhibiting the same dominant land use type within the primary water ecological zone are combined to form secondary water ecological functional zones that reflect similar impacts of human activities. The secondary aquatic eco-functional zones primarily represent the dominant functions of river ecosystems; therefore, their nomenclature follows the format: "Dominant Function of Aquatic Ecosystem Service + Functional Zone".

4. Results

4.1. Daily Runoff Simulation Results

Seven sensitive model parameters (RCHRG_DP, SOL_AWC, CN2, GWQMIN, SOL_K, GW_DELAY, CH_K2, GW_REVAP) were manually calibrated, and the optimal values of the parameters were listed in Table 2. The results of the daily runoff simulation for the Jinjiang Basin are illustrated in Figure 2. The simulated daily runoff values for both the calibration and validation periods at the three hydrological stations closely match the observed data (Figure 2). An evaluation of the simulation performance metrics indicates that, with the exception of the Anxi station during the validation period ($NSE = 0.76$), the NSE values for the three stations during both calibration and validation periods reached 0.8, while the R^2 values also approached 0.8. The absolute Re remained within 10%, satisfying the accuracy requirements for model simulations. This suggests that the model effectively captures the spatial distribution of hydrological processes in the Jinjiang Basin, making the simulation results suitable for studies on watershed water ecological functional zoning.

Using the model simulation results, the multi-year average water yield (2002–2010) for each sub-watershed was calculated, as depicted in Figure 3. The statistical analysis reveals significant spatial variability in the hydrological characteristics of the Jinjiang Basin. The headwaters of the basin are characterized by abundant water resources, with an average annual water yield ranging from 850 to 1400 mm during the period from 2002 to 2010. In contrast, the upper and middle valley regions of Xixi, the downstream plain area, and the region below the Shanmei Reservoir in Dongxi are predominantly classified as low runoff areas, exhibiting a multi-year average water yield of only 600 to 800 mm. This highlights the importance of considering the spatial variability of hydrological processes when evaluating the impacts on aquatic ecosystems during the functional zoning process.

Table 2. Optimal values of the sensitive parameters for Jinjiang Basin.

Parameters	Description	Sensitivity (t Values)	Range	Optimal Value
RCHRG_DP	Deep aquifer percolation fraction	−5.75	0–1	0.25
SOL_AWC	Available water capacity of the soil layer	5.46	0–1	0.11–0.34
CN2	Initial SCS curve number for moisture	3.96	35–98	45–92
GWQMIN	Threshold depth of water in the shallow aquifer acquired for return flow to occur, mm	−1.85	0–500	26
SOL_K	Saturated hydraulic conductivity, mm/h	1.66	0–2000	0.2–79.2
GW_DELAY	Delay time for aquifer recharge, days	−1.58	0–500	31
CH_K2	Effective hydraulic conductivity of channel, mm/h	−1.04	−0.01–500	3
GW_REVAP	Revap coefficient	0.73	0.02–0.2	0.18

Note: the t-value reflects parameter sensitivity, with a higher absolute value indicating greater sensitivity.

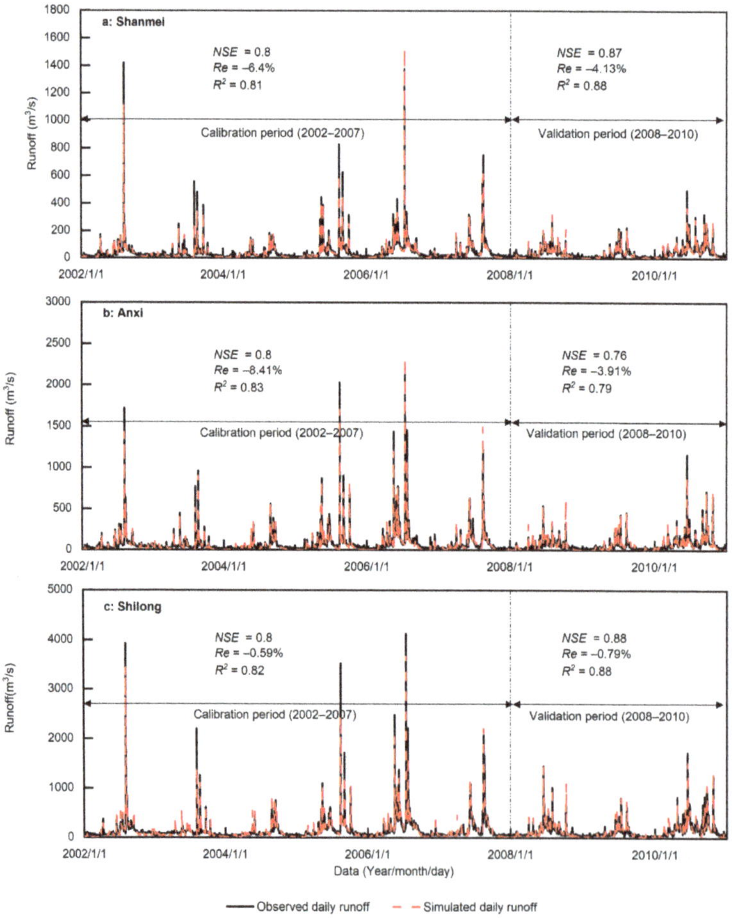

Figure 2. Performances in daily outlet runoff simulation in Jinjiang Basin.

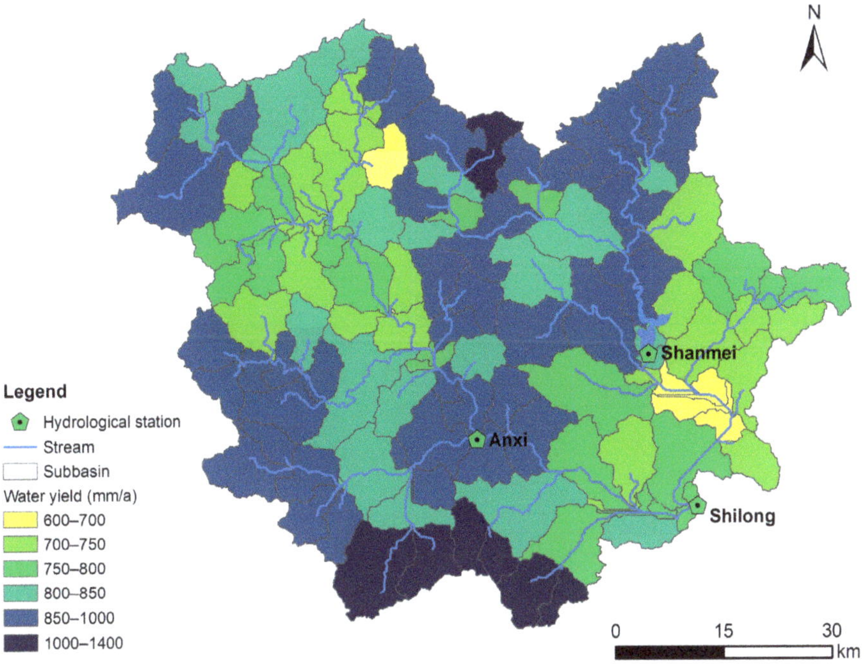

Figure 3. Average annual water yield (2002–2010) of each sub-basin in Jinjiang Basin.

4.2. First-Level Aquatic Zones and Their Characteristics

Utilizing statistical data on the average elevation, average Normalized Difference Vegetation Index (NDVI), annual water yield, and precipitation from 2002 to 2010 for each sub-basin, we conducted a hierarchical clustering analysis of the Jinjiang Basin. This analysis led to the classification of the Jinjiang Basin into three categories of first-level aquatic ecological zones: the Mountainous River Aquatic Ecological Zone, the Hilly River Aquatic Ecological Zone, and the Plain River Ecological Zone (Figure 4). Additionally, considering the independence of the Dongxi and Xixi tributaries, we further subdivided the same type of aquatic ecological zone that spans both tributaries into two distinct aquatic ecological zones. Consequently, we delineated a total of six first-level aquatic ecological zones for the entire watershed (Figure 4). The statistical characteristics of the natural environment for each first-level zone are summarized in Figure 5.

The first category is the Mountainous River Ecological Zone, which includes the I. Upper Xixi—Mountainous River—Aquatic Ecological Zone and (Figure 4) II. Upper Dongxi—Mountainous River—Aquatic Ecological Zone. This category is situated in high-altitude regions (average elevation of 616 m, Figure 5a) and is characterized by high vegetation coverage (average NDVI of 0.76, Figure 5b). Furthermore, this type of aquatic ecological zone also has the most abundant precipitation, with an annual rainfall of up to 1684 mm (Figure 5c).

The second category is the Hilly River Ecological Zone, which comprises the III. Midstream Xixi—Hilly River—Aquatic Ecological Zone and (Figure 4) IV. Midstream Dongxi—Hilly River—Aquatic Ecological Zone. Compared with the Mountainous River Ecological Zone, this type of aquatic ecological zone has a lower altitude and less vegetation coverage and precipitation, with average values of 303.5 m, 0.73, and 1646.7 mm, respectively (Figure 5).

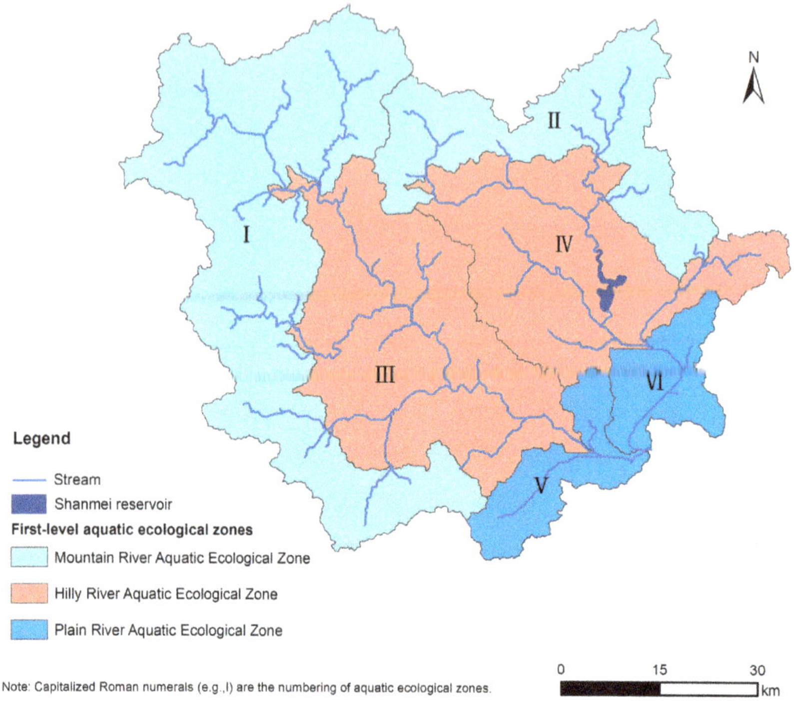

Figure 4. The first-level aquatic ecological regions in Jinjiang Basin.

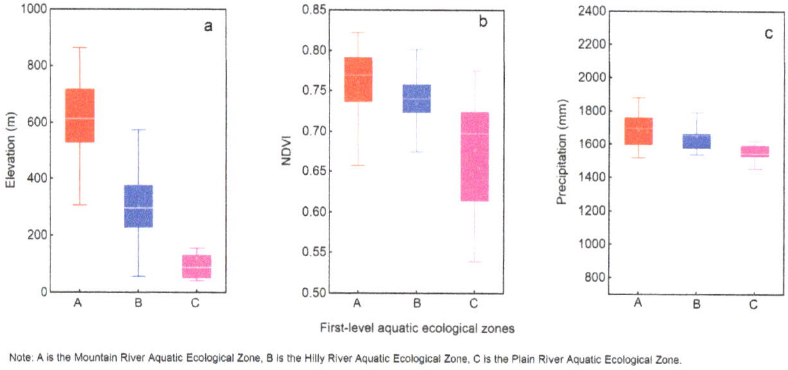

Figure 5. Characteristics of elevation (**a**), NDVI (**b**) and precipitation (**c**) in three types of first-level aquatic ecological zones.

The third category is the Plain River Ecological Zone, which includes the V. Downstream Xixi—Piedmont Plain River—Aquatic Ecological Zone and VI. Downstream Dongxi—Piedmont Plain River—Aquatic Ecological Zone (Figure 4). This category is located in downstream plain areas and is characterized by the least vegetation coverage and precipitation (Figure 5).

Figure 6 presents boxplots illustrating the hydrological factors, including evapotranspiration, soil moisture content, surface flow, groundwater flow, soil lateral flow, and total water yield, across the different aquatic ecological zones. An ANOVA test conducted to assess the differences in these hydrological elements among the three aquatic ecological zones revealed that the majority of p-values were less than 0.05, indicating significant differences in the hydrological elements across the various aquatic ecological zones. As illustrated in Figure 6a, there is minimal variation in evapotranspiration from the upstream Mountainous River Ecological Zone to the downstream Plain River Ecological Zone, with a slight increasing trend observed. In contrast, surface runoff exhibits a marked increase, with an average multi-year surface runoff of 268.4 mm in the upstream mountainous zone, escalating to 412.5 mm in the downstream zone (Figure 6c). Conversely, soil moisture content (Figure 6b), soil lateral flow (Figure 6e) and total water yield (Figure 6f) are substantially greater in the upstream mountainous river ecological zone (278.6 mm, 475.6 mm, 863.3 mm) compared to the midstream hilly ecological zones (262.4 mm, 396 mm, 817.5 mm) and downstream piedmont plain ecological zones (216.3 mm, 277 mm, 823.5 mm). From the upstream to the downstream, only the groundwater flow has no obvious pattern of change (Figure 6d).

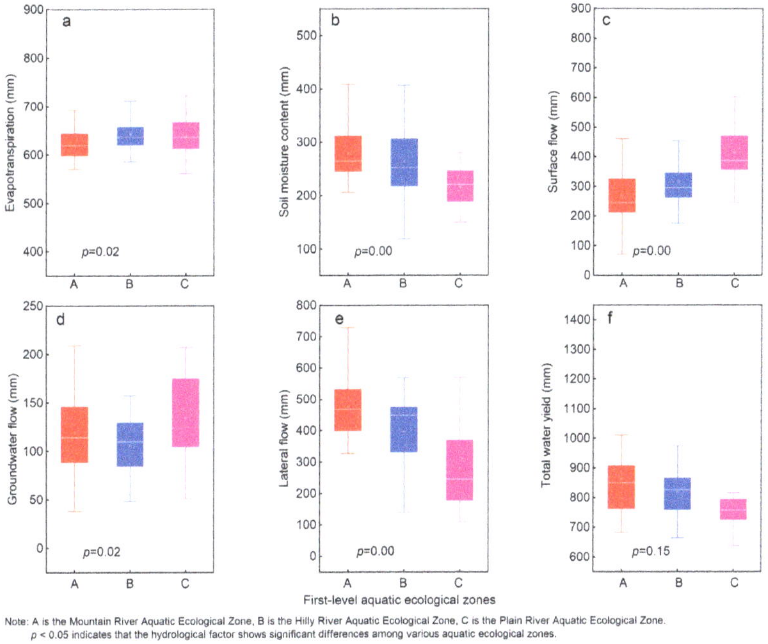

Figure 6. Comparison of evapotranspiration (**a**), soil moisture content (**b**), surface flow (**c**), groundwater flow (**d**), lateral flow (**e**) and total water yield (**f**) in three types of first-level aquatic ecological zones.

These findings indicate that the upstream mountainous river ecological zone plays a vital role in water conservation, significantly regulating runoff and protecting aquatic habitats in the middle and lower reaches of the basin. This observation also underscores the validity of the first-level aquatic ecological zoning to a considerable extent.

4.3. Secondary Aquatic Eco-Functional Zoning and Their Characteristics

Building upon the delineation of first-level water ecological zones, the Jinjiang Basin has been subdivided into 18 secondary water ecological functional zones. These zones are categorized into five distinct types based on indicators for secondary water ecological functional zoning, specifically the predominant land use types. The categories are as follows: (1) Forest Water Conservation Functional Zone, (2) Agricultural Development and Maintenance of Water Environment Functional Zone, (3) Orchard Soil and Water Retention

Functional Zone, (4) Reservoir Hydrological Regulation Functional Zone, and (5) Urban Support Development Functional Zone (Figure 7).

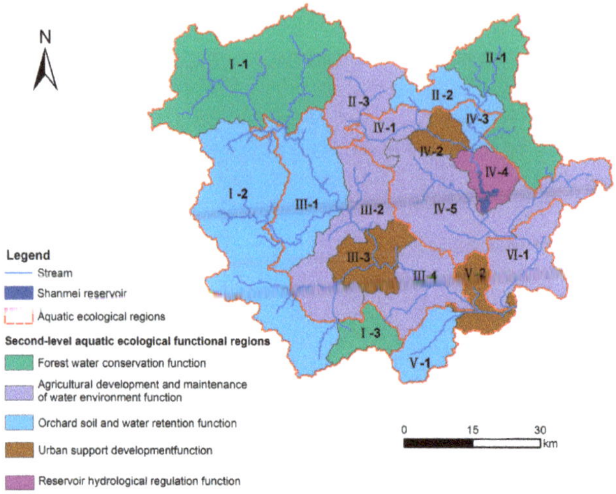

Figure 7. The secondary aquatic eco-functional regions in Jinjiang Basin.

The Forest Water Conservation Functional Zone encompasses three areas (I-1, I-3, II-1) (Figure 7), where forest land constitutes over 64.5% of the total area, significantly surpassing the basin-wide average of 54% (Figure 8a). These zones are situated within the mountainous river ecological regions of the first-level zones, characterized by high forest cover, minimal human disturbance, and abundant water resources. Their first-level ecological functions include runoff regulation and water source conservation.

The Agricultural Development and Maintenance of Water Environment Functional Zone consists of six areas: II-3, III-2, III-4, IV-1, IV-5, and VI-1 (Figure 7). A notable feature of these regions is the substantial area of arable land, with cultivated land surpassing the average proportion of arable land across the entire watershed, which is 16.87% (Figure 8b). Agricultural activities not only require significant water resources but also contribute to serious non-point source pollution challenges. The first-level ecological functions of these areas are centered on agricultural water supply and water quality enhancement.

The Orchard Soil and Water Retention Functional Zone consists of five areas (I-2, II-2, III-1, IV-3, V-1) (Figure 7), which have experienced extensive development of orchards and tea gardens. The proportion of orchard area in the upper, middle, and lower reaches is 27.9%, 21.8%, and 16.8% (Figure 8c), respectively. The large-scale development of these agricultural lands has raised significant concerns regarding soil erosion, leading to the first-level function of this zone being the control of soil erosion.

The Urban Support Development Functional Zone are primarily situated in the middle reaches of Yongchun County (IV-2) and Anxi County (III-3), as well as in the downstream region of Nanan City (V-2) (Figure 7), where the proportions of urban construction land are 16.37%, 20.35%, and 30.42% (Figure 8d), respectively. These ecological function areas are characterized by elevated levels of socio-economic activities. Their main functions include facilitating human habitation, providing recreational water landscapes, and treating urban wastewater generated from both industrial production and daily activities.

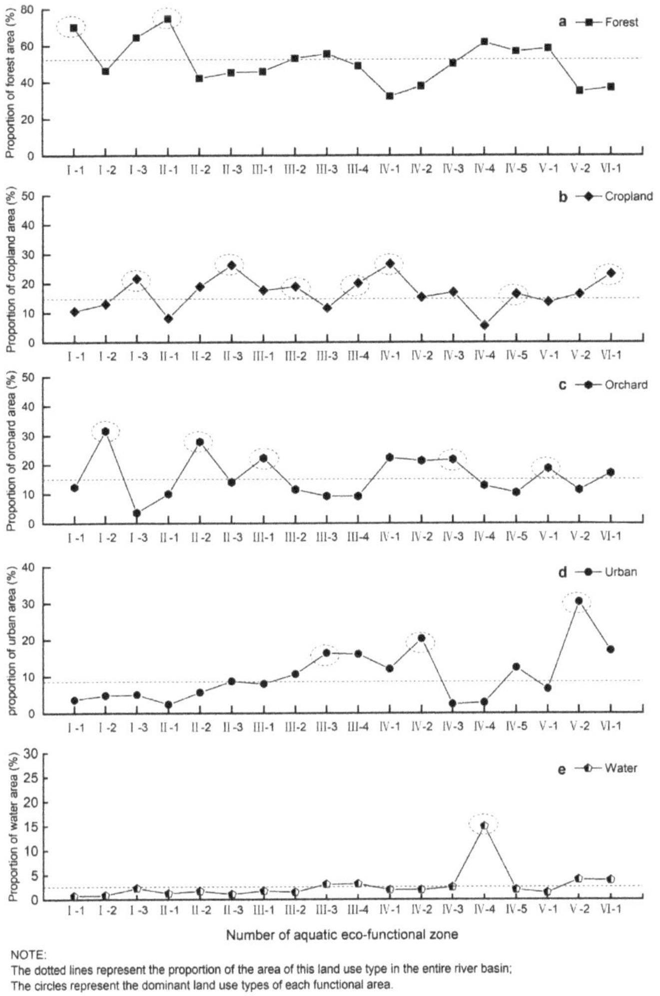

Figure 8. Proportions of forest (**a**), cropland (**b**), orchard (**c**), urban (**d**) and water (**e**) for the aquatic eco-functional regions in Jinjiang Basin.

The Reservoir Hydrological Regulation Functional Zone (IV-4) encompasses the Shanmei Reservoir, the only large reservoir within the Jinjiang watershed and the largest in the region (Figure 7), and the aquatic eco-function zone has the largest proportion of water area (Figure 8e). This reservoir, often referred to as the "Lifeblood of Quanzhou," plays a vital role in hydrological regulation. It ensures a continuous water supply for downstream agricultural irrigation, industrial activities, and residential drinking water through effective management. Additionally, it functions as a flood control mechanism by regulating water levels during flood seasons, thereby protecting the lives and properties of downstream residents. During dry seasons, it also provides ecological flow to downstream rivers.

5. Discussion

Water is a fundamental and dynamic component of ecosystems, serving as the essential medium for life. Hydrological processes are key drivers of material cycling and energy flow within these systems. Various environmental factors, including climate change within the watershed, surface characteristics, and anthropogenic activities, significantly influence the composition, structure, and function of aquatic ecosystems primarily by altering hydrological processes [16].

In this study, the Jinjiang watershed is delineated into five distinct water ecological functional zones, each defined by unique environmental factors and hydrological processes, resulting in diverse ecological functions. The extensive forest cover in the Forest Water Conservation Functional Zone facilitates the accumulation of precipitation during wet periods, thereby mitigating surface runoff. During dry periods, this forest cover contributes to river flow replenishment through soil moisture retention and groundwater recharge [24,33]. Additionally, this type of aquatic eco-function zone is mainly located in the upper reaches of the river basin, with the most abundant precipitation (Figure 5c) and water yield (Figure 6d). Therefore, it plays a crucial role in water conservation and the regulation of river runoff throughout the watershed.

In contrast, the extensive tea plantations established in the watershed, while providing significant economic benefits, have been shown to severely disrupt natural surface vegetation. This disruption reduces the surface's capacity to intercept rainfall, resulting in increased surface runoff, exacerbated topsoil erosion, and enhanced transport of soil particles, leading to serious soil erosion issues [34–36]. The Orchard Soil and Water Retention Functional Zone is mainly located in the upper and middle reaches of the river basin, with relatively high rainfall (Figure 5c) and water yield (Figure 6d). Soil erosion also has a significant impact on the aquatic ecosystem. Therefore, the first-level function of the aquatic ecological system in this region is to control soil erosion and safeguard soil resources. Conversely, in terraced farming areas, a considerable amount of rainfall is captured and infiltrated by the terraces during runoff events, ultimately reducing runoff efficiency [37,38].

The extensive use of chemical fertilizers and pesticides in agricultural practices can lead to non-point source pollution, particularly as a result of rainfall runoff. Research has demonstrated that agricultural runoff is a significant contributor to the elevated concentrations of nitrogen and phosphorus compounds in rivers [39,40]. In this region, the aquatic ecosystem is heavily dependent on agricultural water supply and the maintenance of water quality. Reservoirs, which are hydraulic engineering structures designed for flood control and runoff regulation, play a crucial role in the temporal and spatial redistribution of natural water resources. This is especially important for large reservoirs, as their hydrological regulation functions are vital for flood prevention, drought mitigation, and the efficient utilization of water resources within the watershed [41–43]. For instance, the construction of the Shanmei Reservoir in Jinjiang has led to a reduction in the average duration of droughts and a significant decrease in drought intensity at the Shilang hydrological station [26].

Urban development typically occurs in low-lying plain areas adjacent to major rivers, which are characterized by high flow and stable water volumes. The aquatic ecosystems in these regions can meet substantial water demands for urban industrial and agricultural production, residential needs, and urban landscaping, while also exhibiting strong capabilities for purifying wastewater generated by urban activities. Overall, it is imperative to recognize that hydrological processes not only connect and transmit elements of the natural environment and human activities but also serve as significant influencing factors themselves. Therefore, the spatial variability of these processes must be carefully considered in the functional zoning of aquatic ecosystems.

The SWAT model, a distributed hydrological model designed for watershed analysis, accounts for the spatial heterogeneity of meteorological variables and underlying surface conditions. This capability enhances the accuracy of simulating spatial variability in hydrological processes. Its application in delineating water ecological function zones effectively addresses the challenge of insufficient spatial distribution density of hydrological monitoring stations. This study has implications for future research, primarily in two areas. First, the SWAT model has demonstrated its ability to accurately simulate the spatial variability of water yield, sediment transport, and water quality at the watershed scale. This capability effectively addresses the challenge of obtaining hydrological process-related indicators for aquatic eco-functional zoning. Additionally, given the increasing focus on river health assessments across various aquatic ecoregions [44,45], the use of hydrological models in eco-functional zoning can provide comprehensive simulated data for evaluating river health in different zones. Furthermore, it facilitates the assessment and prediction of the impacts of human activities and climate change on river health through scenario modeling. Thus, this methodology serves as a valuable scientific reference for future research on river ecosystem health management.

However, this study also has certain limitations. First, due to the insufficient quantitative analysis of the relationships between hydrological processes and aquatic ecosystems, we relied solely on total water yield as the zoning indicator for hydrological processes. This approach may result in an inadequate representation of these processes and increase uncertainty. Future research should aim to quantify these relationships further to identify more representative hydrological indicators for zoning purposes. Second, the current lack of aquatic biological observation data limits the validation of zoning results. Subsequent studies should incorporate biological data (e.g., aquatic fauna, flora, and algae) to analyze biodiversity indices, species richness, endemic rarity, and biological integrity indices. Such data would not only enhance the zoning indicators but also provide a more robust foundation for verifying zoning results.

6. Conclusions

(1) The Jinjiang Basin can be categorized into three categories and six first-level aquatic ecological zones based on natural environmental factors such as topography, precipitation, vegetation, and hydrology. This classification reflects the spatial variability of these factors and their effects on river ecosystems.

(2) By incorporating predominant land use types, the Jinjiang Basin can be subdivided into five categories and eighteen second-level aquatic eco-functional zones, which build upon the first-level aquatic ecological zones. This subdivision highlights the extent to which human activities influence river ecosystems and the services they provide.

(3) The sub-basins delineated using the SWAT model serve as critical spatial units for the eco-functional zoning of the basin, thereby preserving the integrity of river ecosystems. By integrating hydrological simulation data from each sub-basin, this methodology effectively addresses the challenges associated with obtaining hydrological zoning indicators, thereby providing reliable technical support for the ecological functional zoning of the basin.

The zoning process primarily considers terrestrial environmental factors, encompassing both natural and anthropogenic influences. However, it does not analyze the spatial variability within river ecosystems nor does it investigate the relationship between terrestrial and riverine ecosystems. Future research should focus on enhancing the analysis of spatial differences in aquatic ecosystems and assessing the validity of the zoning outcomes.

Author Contributions: Data curation, M.L.; Formal analysis, B.L.; Funding acquisition, B.L. and L.Y.; Investigation, M.C.; Methodology, B.L. and Y.C.; Writing—original draft, B.L.; Writing—review and editing, B.L., Y.C., M.L., L.Y. and M.C. All authors have read and agreed to the published version of the manuscript.

Funding: This research was funded by the [Natural Science Foundation of Fujian Province] grant number [2024J01791], [Fujian Provincial Education and Scientific Research Project for Middle-aged and Young Teachers] grant number [JAT210301], [Key Project of the Education Department of Anhui Province] grant number [2022AH051108], and [Third Batch of "Integration of Specialties and Innovation" Demonstration Course Construction Project of Chuzhou University] grant number [2024zckc002].

Data Availability Statement: The raw data supporting the conclusions of this article will be made available by the authors on request.

Conflicts of Interest: The authors declare no conflicts of interest.

References

1. Omernik, J.M.; Bailey, G.B. Distinguishing between watersheds and ecoregions. *J. Am. Water Resour. Assoc.* **1997**, *33*, 935–949. [CrossRef]
2. Crowly, J.M. Biogeography. *Cannadian Geogr.* **1967**, *11*, 312–326. [CrossRef]
3. Omemik, J.M. Ecoregions of the conterminous United Stated. *Ann. Am. Assoc. Geogr.* **1987**, *77*, 118–125.
4. Biggs, B.J.; Duncan, M.J.; Jowett, I.G.; Quinn, J.M.; Hickey, C.W.; Davies-Colley, R.J.; Close, M.E. Ecological characterisation, classification, and modelling of New Zealand rivers: An introduction and synthesis. *N. Zealand J. Mar. Freshw. Res.* **1990**, *24*, 277–304. [CrossRef]
5. Harding, J.; Winterbourn, M. *New Zealand Ecoregions: A Classification for Use in Stream Conservation and Management*; Department of Conservation: Wellington, New Zealand, 1997; Volume 11.
6. Logan, P.; Furse, M. Preparing for the European Water Framework Directive making the links between habitat and aquatic biota. *Aquat. Conserv. Mar. Freshw. Ecosyst.* **2002**, *12*, 425–437. [CrossRef]
7. Moog, O.; Kloiber, A.S.; Thomas, O.; Gerritsen, J. Does the ecoregion approach support the typological demands of the EU "Water Frame Directive". *Hydrobiologia* **2004**, *516*, 21–33. [CrossRef]
8. Lorenz, A.; Feld, C.; Hering, D. Typology of streams in Germany based on benthic invertebrates: Ecoregions, zonation, geology and substrate. *Limnologica* **2004**, *34*, 379–389. [CrossRef]
9. Tison, J.; Park, Y.S.; Coste, M.; Wasson, J.G.; Ector, L.; Rimet, F.; Delmas, F. Typology of diatom communities and the influence of hydro-ecoregions: A study on the French hydrosystem scale. *Water Res.* **2005**, *39*, 3177–3188. [CrossRef]
10. Meyer, J.L. Stream health: Incorporating the human dimension to advance stream ecology. *J. N. Am. Benthol. Soc.* **1997**, *16*, 439–447. [CrossRef]
11. Meng, W.; Zhang, N.; Zhang, Y.; Zheng, B. Integrated assessment of river health based on water quality, aquatic life and physical habitat. *J. Environ. Sci.* **2009**, *8*, 1017–1027. [CrossRef]
12. Schroeder, W.; Pesch, R. Synthesizeing bioaccumulation data from the German metals in mosses surveys and relating them to ecoregions. *Sci. Total Environ.* **2007**, *374*, 311–327. [CrossRef]
13. Sheng, S.; Xu, C.; Wen, T.; Wan, Y.; An, S. Division Design of Water Eco-Functioning of the River Basin. *CLEAN–Soil Air Water* **2015**, *43*, 1640–1646. [CrossRef]
14. Yang, Y.; Chen, X.; Zhou, Y.; Li, W.; Chen, K. First-level and secondary aquatic ecological function division of the Yellow River Source Zone. *J. Hydroecology* **2023**, *44*, 1–9. (In Chinese)
15. Frissell, C.A.; Liss, W.J.; Warren, C.E.; Hurley, M.D. A hierarchical framework for stream habitat classification: Viewing streams in a watershed context. *Environ. Manag.* **1986**, *10*, 199–214. [CrossRef]
16. Liu, X.; Xu, Z.; Xu, C. A framework for aquatic ecoregion zoning. *Acta. Ecologica. Sinica.* **2010**, *30*, 4804–4814. (In Chinese)
17. Baker, D.B.; Richards, R.P.; Loftus, T.T.; Kramer, J.W. A new flashiness index: Characteristics and applications to midwestern rivers and streams. *J. Am. Water Resour. Assoc.* **2004**, *40*, 503–522. [CrossRef]
18. Gao, Y.; Gao, J.; Chen, J.; Xu, Y.; Zhao, J. Regionalizing aquatic ecosystems based on the river subbasin taxonomy concept and spatial clustering techniques. *Int. J. Environ. Res. Public Health.* **2011**, *8*, 4367–4385. [CrossRef] [PubMed]
19. Ma, B.R.; Ma, A.Q.; Chen, S.; Li, Z.; Wang, Z.K. Function regionalization of water ecosystem for the Daliaohe Drainage Area. *Period. Ocean Univ. China* **2014**, *44*, 093–099. (In Chinese)
20. Gao, Z.; Cao, X.; Huang, Y.; Li, F. Research of level I and II aquatic ecological function regionalization in Lake Dianchi basin. *J. Lake Sci.* **2015**, *27*, 175–182. (In Chinese)
21. Krysanova, V.; Arnold, J.G. Advances in ecohydrological modelling with SWAT-a review. *Hydrolog. Sci. J.* **2008**, *53*, 939–947. [CrossRef]
22. Zhang, D.; Chen, X.; Yao, H. Development of a prototype web-based decision support system for watershed management. *Water* **2015**, *7*, 780–793. [CrossRef]

23. Ricci, G.F.; Girolamo, A.M.D.; Abdelwahab, O.M.M.; Gentile, F. Identifying sediment source areas in a mediterranean watershed using the swat model. *Land Degrad. Dev.* **2018**, *29*, 1233–1248. [CrossRef]
24. Lin, F.; Chen, X.; Yao, H.; Lin, F. SWAT model-based quantification of the impact of land-use change on forest-regulated water flow. *Catena* **2022**, *211*, 105975. [CrossRef]
25. Lin, B.; Chen, X.; Yao, H.; Chen, Y.; Liu, M.; Gao, L.; James, A. Analyses of landuse change impacts on catchment runoff using different time indicators based on SWAT mode. *Ecol. Indic.* **2015**, *58*, 55–63. [CrossRef]
26. Wu, J.; Chen, X.; Yao, H.; Gao, L.; Chen, Y.; Liu, M. Non-linear relationship of hydrological drought responding to meteorological drought and impacts of a large reservoir. *J. Hydrol.* **2017**, *551*, 495–507. [CrossRef]
27. Arnold, J.G.; Srinivasan, R.; Muttiah, R.S.; Williams, J.R. Lager area hydrologic modeling and assessment part I: Modeling development. *JAWRA J. Am. Water Resour. Assoc.* **1998**, *34*, 91–101. [CrossRef]
28. Chaplot, V. Impact of spatial input data resolution on hydrological and erosion modeling: Recommendations from a global assessment. *Phys. Chem. Earth.* **2014**, *67–69*, 23–35. [CrossRef]
29. Tan, M.L.; Ramli, H.P.; Tam, T.H. Effect of DEM resolution, source, resampling technique and area threshold on SWAT outputs. *Water Resour. Manage* **2018**, *32*, 4591–4606. [CrossRef]
30. Lin, B.; Chen, X.; Yao, H. Threshold of sub-watersheds for SWAT to simulate hillslope sediment generation and its spatial variations. *Ecol. Indic.* **2020**, *111*, 106040. [CrossRef]
31. Zhang, D.; Chen, X.; Yao, H.; Lin, B. Improved calibration scheme of SWAT by separating wet and dry seasons. *Ecol. Model.* **2015**, *301*, 54–61. [CrossRef]
32. Moriasi, D.N.; Arnold, J.G.; Van Liew, M.W.; Bingner, R.L.; Harmel, R.D.; Veith, T.L. Model evaluation guidelines for systematic quantification of accuracy in watershed simulation. *Trans. ASABE* **2007**, *50*, 885–900. [CrossRef]
33. Rabbai, A.; Wendt, D.E.; Curioni, G.; Quick, S.E.; MacKenzie, A.R.; Hannah, D.M.; Kettridge, N.; Ullah, S.; Hart, K.M.; Krause, S. Soil moisture and temperature dynamics in juvenile and mature forest as a result of tree growth, hydrometeorological forcings, and drought. *Hydrol. Process.* **2023**, *37*, e14919. [CrossRef]
34. Gharibreza, M.; Bahrami Samani, A.; Arabkhedri, M.; Zaman, M.; Porto, P.; Kamali, K.; Sobh-Zahedi, S. Investigation of on-site implications of tea plantations on soil erosion in Iran using 137Cs method and RUSLE. *Environ. Earth Sci.* **2021**, *80*, 34. [CrossRef]
35. Chit, T.; Díaz-Pinés, E.; Butterbach-Bahl, K.; Marzaioli, F.; Valentini, F. Soil organic carbon changes following degradation and conversion to cypress and tea plantations in a tropical mountain forest in Kenya. *Plant Soil.* **2017**, *422*, 527–539. [CrossRef]
36. Hamzah, Z.; Amirudin, C.Y.; Saat, A.; Wood, A.K. Quantifying soil erosion and deposition rates in tea plantation area, Cameron Highlands, Malaysia using 137Cs. *Malays. J. Anal. Sci.* **2014**, *18*, 94–106.
37. Arnáez, J.; Lana-Renault, N.; Lasanta, T.; Ruiz-Flaño, P.; Castroviejo, J. Effects of farming terraces on hydrological and geomorphological processes. A review. *Catena* **2015**, *128*, 122–134. [CrossRef]
38. Meng, X.; Zhu, Y.; Shi, R.; Yin, M.; Liu, D. Rainfall–runoff process and sediment yield in response to different types of terraces and their characteristics: A case study of runoff plots in Zhangjiachong watershed, China. *Land Degrad. Dev.* **2024**, *35*, 1449–1465. [CrossRef]
39. Srinivas, R.; Singh, A.P.; Dhadse, K.; Garg, C. An evidence based integrated watershed modeling system to assess the impact of non-point source pollution in the riverine ecosystem. *J. Clean. Prod.* **2020**, *246*, 118963. [CrossRef]
40. Chang, R.; Wang, Y.; Liu, H.; Wang, Z.; Ma, L.; Zhang, J.; Li, J.; Yan, Z.; Zhang, Y.; Li, D. Optimizing Non-Point Source Pollution Management: Evaluating Cost-Effective Strategies in a Small Watershed within the Three Gorges Reservoir Area, China. *Land* **2024**, *13*, 742. [CrossRef]
41. Van Loon, A.F.V.; Gleeson, T.; Clark, J.; Van Dijk, A.I.J.M.; Stahl, K.; Hannaford, J.; Baldassarre, G.D.; Teuling, A.J.; Tallaksen, L.M.; Uijlenhoet, R.; et al. Drought in the anthropocene. *Nat. Geosci.* **2016**, *9*, 89–91. [CrossRef]
42. Gudmundsson, L.; Boulange, J.; Do, H.X.; Gosling, S.N.; Grillakis, M.G.; Koutroulis, A.G.; Leonard, M.; Liu, J.; Schmied, H.M.; Papadimitriou, L.; et al. Globally observed trends in mean and extreme river flow attributed to climate change. *Science* **2021**, *371*, 1159–1162. [CrossRef] [PubMed]
43. Wei, J.; Dong, N.; Fersch, B.; Arnault, J.; Wagner, S.; Laux, P.; Zhang, Z.; Yang, Q.; Yang, C.; Shang, S.; et al. Role of reservoir regulation and groundwater feedback in a simulated ground-soil-vegetation continuum: A long-term regional scale analysis. *Hydrol. Process.* **2021**, *35*, e14341. [CrossRef]
44. Growns, I.; Rourke, M.; Gilligan, D. Toward river health assessment using species distributional modeling. *Ecol. Indic.* **2013**, *29*, 138–144. [CrossRef]
45. Zhang, X.; Meng, Y.; Xia, J.; Wu, B.; She, D. A combined model for river health evaluation based upon the physical, chemical, and biological elements. *Ecol. Indic.* **2018**, *84*, 416–424. [CrossRef]

water

MDPI

Article

Shifts in Precipitation Variability near the Danube Delta Biosphere Reserve (1965–2019)

Alina Bărbulescu [1] and Cristian Ștefan Dumitriu [2,*]

[1] Department of Civil Engineering, Transilvania University of Brașov, 5 Turnului Str., 500152 Brașov, Romania; alina.barbulescu@unitbv.ro

[2] Faculty of Mechanical Equipment for Constructions and Robotics, Technical University of Civil Engineering, 59 Calea Plevnei, 021242 Bucharest, Romania

* Correspondence: cristian.dumitriu@utcb.ro

Abstract

Nowadays, climate change is one of the significant threats humanity faces. Many researchers have documented its effects on water availability and vulnerable systems. This study examines the long-term precipitation record (1965–2019) from the Tulcea station, located just 4 km from the Danube Delta Biosphere Reserve (DDBR), to evaluate the impact of climate change on precipitation variability, which can significantly affect biodiversity in this protected area. We integrated change point detection (CPD), stationarity tests, trend analysis, and series decomposition to characterize shifts and patterns in the time series. The Lee & Heghinian test detected a change point (CP) in all data series, whereas the Hubert segmentation methods and Cumulative Sum Method (CUSUM) found fewer series that present at least a CP. The Mann–Kendall (MK) trend test and Innovative Trend Analysis (ITA) indicated an increasing trend in the annual, monthly, and October precipitation series. The Seasonal-Trend decomposition using Loess STL decomposition found the highest seasonality indices in June and July. The Ensemble Empirical Mode Decomposition (EEMD) emphasizes a substantial difference in the seasonal cycle. The results indicate a high variability in the precipitation pattern, with periods of high precipitation followed by dry periods.

Keywords: change point (CP); trend; EEMD; STL decomposition

1. Introduction

Climate change has emerged as one of the most pressing environmental challenges of our time, profoundly influencing water resources and hydrological systems across the globe [1]. Understanding the induced changes' mechanisms, timing, and spatial patterns is essential for ensuring sustainable water management, safeguarding public access to fresh water, and minimizing adverse impacts on water quality and ecosystems [2,3]. The modification in precipitation regime directly affects water availability and quality, ecosystem health, agriculture, and human activities [4–7].

Over recent decades, observational data and climate model simulations have documented significant alterations in precipitation intensity, frequency, and spatial distribution, primarily driven by evolving atmospheric circulation and humidity patterns [8–10]. These changes have increased hydrological variability, with some regions experiencing intensified flooding and others enduring prolonged droughts. Such shifts present growing challenges for freshwater management and ecosystem resilience [11–13].

To better understand and respond to these threats, robust statistical tools are needed to analyze hydro-meteorological time series data. Time series analysis plays a central role in hydrology, climatology, and meteorology, particularly for studying long-term temperature and precipitation series, key indicators of climate variability and change. Identifying trends, seasonal cycles, and structural changes in these variables is critical for interpreting underlying climate dynamics [14–16].

An important concern when studying time series evolution is the determination of breakpoints or change points, i.e., detecting the moments where there are modifications of the series' statistical properties or distribution. They are widely utilized across disciplines, including hydrometeorology [17–20], healthcare [21], signal processing [22,23], etc.

Trend detection and seasonal decomposition methods enable the separation of intra-annual cycles from longer-term patterns and random noise, thereby enhancing insights into seasonal behavior and its evolution over time [24–27].

Precipitation variability has substantial environmental and socio-economic implications, especially in regions where water resources are already constrained or climate conditions are highly variable. One such area is DDBR, a UNESCO World Heritage Site and Ramsar Wetland of international importance, one of Europe's most dynamic and ecologically valuable wetland systems. Belonging to the Dobrogea region in southeastern Romania, DDBR is particularly sensitive to climatic fluctuations due to its low elevation, fragile ecosystems, and strong dependency on hydrological regimes [28–31].

Most of the previous research primarily focused on analyzing extreme weather events in Dobrogea in general. For instance, a Generalized Pareto Distribution was applied to the extreme monthly precipitation series [32], and 11 rainfall extreme indices were calculated using daily precipitation data from January 1961 to December 2005 [33]. The studies [34, 35] have indicated a significant increase in drought events in Romania, particularly in Dobrogea, from 1980 to 2000, with more than 50% of those years experiencing below-average rainfall, whereas SPI12 and SPEI12 readings at Sulina indicate an increase in the frequency and duration of droughts, especially since the early 1980s [36].

Despite existing analyses of the broader Dobrogea region, few have specifically addressed precipitation dynamics in the Danube downstream area, and none have utilized extended time series data reaching up to the recent years. Therefore, this research aims to address the long-term precipitation variability at the mouth of the Danube, with a focus on the DDBR, an area that has received limited attention in previous studies [37–39]. This paper seeks to fill that gap by examining long-term precipitation records from Tulcea, at the boundary of the Danube Delta, to detect structural breaks, trends, and seasonal shifts. These changes may indicate broader hydrological transformations, with significant implications for water resource management, ecosystem resilience, and flood or drought risk in the DDBR. In doing so, the research provides critical insights into the climate sensitivity of one of Europe's most ecologically vulnerable wetland systems.

2. Study Area and Data Series

Dobrogea is located in the southeastern part of Romania. It is bounded by the Danube River to the west and north, the Black Sea to the east, and Bulgaria to the south. It covers approximately 23,000 km^2 and includes Constanța and Tulcea counties [34] and the Danube Delta. Its topography is a combination of low-lying plains, hills, and the Măcin Mountains. Climatically, Dobrogea (Figure 1, left) lies in a transition zone between temperate-continental and maritime influences, with steppe characteristics particularly evident in its central and southern areas. Dobrogea is one of the driest regions in Romania.

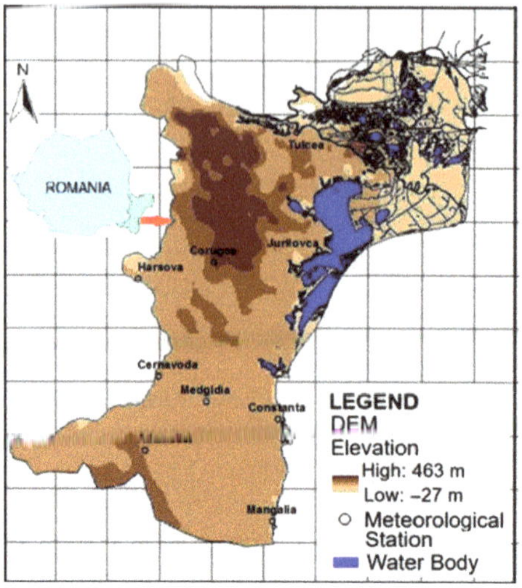

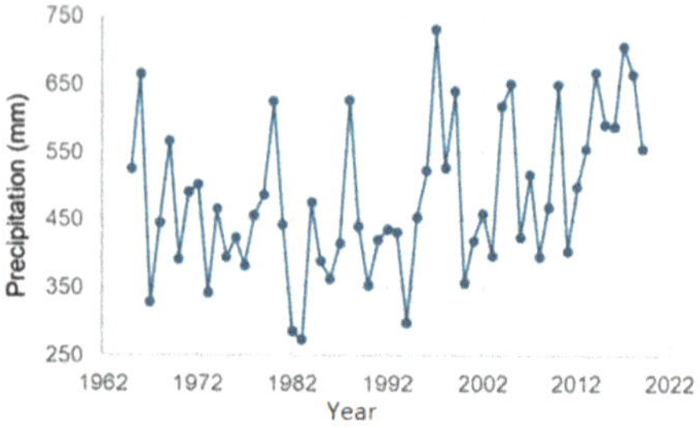

Basic statistics	Value
minimum [mm]	273.70
maximum [mm]	732.00
Mean [mm]	484.19
Standard deviation [mm]	113.63
skewness	0.35
kurtosis	-0.65

Figure 1. (**left**) Dobrogea region; (**top right**) Annual precipitation series spanning 1965–2019 recorded at Tulcea; (**bottom right**) Basic statistics of the annual series recorded at Tulcea.

The mean annual precipitation recorded at ten main hydrometeorological stations in the region varied between 350 and 450 mm during the period 1965–2005 [12,32]. Rainfall is irregular, seasonally concentrated. The climate is often marked by prolonged dry spells and sudden, intense convective events, posing significant challenges to agriculture, water resource management, and ecosystem stability [40–43].

The Danube Delta is the Europe's second largest and one of the most intact deltaic systems. It is a biodiversity hotspot that hosts over 5000 species, including numerous rare and endangered taxa. Its complex mosaic of freshwater, brackish, and terrestrial habitats, offers a unique natural laboratory for studying ecosystem function, species interactions, and environmental change. The Delta is a critical migratory corridor for over 300 avian species (including threatened taxa such as the white-tailed eagle, red-breasted goose, and Dalmatian pelican). It also contains the world's largest continuous reed bed system, which plays a vital role in water filtration, habitat provision, and sediment stabilization. Hydrology is a key driver of ecological processes in the Delta. Seasonal flooding, particularly in spring, reshapes habitats, enriches nutrient availability, and sustains the reproductive cycles of many species. These temporal dynamics underpin the Delta's exceptional biological productivity. Recognized as a UNESCO World Heritage Site and Ramsar Wetland, the Danube Delta deserves detailed investigation to sustain its conservation and management efforts in the face of ongoing climate change and anthropogenic pressures [44].

Tulcea is in inland Dobrogea, at the western edge of the DDBR. The distance to ecologically sensitive wetland zones is 3 to 5 km, and that to the strictly protected areas around 10 km from. Tulcea serves as the administrative center and logistical gateway to the DDBR. It is designed as the primary observation site for monitoring climate-induced changes in water systems [45]. It offers long-term climate records to the European Climate Assessment & Dataset. This makes it strategically important for understanding regional climate variability and hydrological dynamics in the Danube Delta. Due to its geographical position, Tulcea is exposed to maritime and deltaic influences. In 2019, at Tulcea, about 271 days were without rain. The number of rainy days was between 3 and 8 per month, with the highest average precipitation in June (about 80 mm) and the lowest in August (23 mm).

Although there are other main meteorological stations in Dobrogea, all are situated inland, at least 65 km from the Danube Delta. Therefore, the climate at those stations is not the same as in the DDBR. Extrapolating the results of the analysis performed for the precipitation series from those stations to the Danube Delta would not be correct. So, the study of the data collected at Tulcea is essential for understanding the variability in the DDBR. The analyzed series consists of precipitations recorded at Tulcea from 1965 to 2019 as follows:

- the total annual precipitation series, denoted by ANTOT (Figure 1 (top, right));
- the total monthly precipitation series that spans January 1965 to December 2019, denoted by MTOT (Figure 2a);
- the monthly series recorded each month (e.g., January series, formed by the precipitation recorded in January from 1965 to 2019, February series, etc.) (Figure 2b–e).

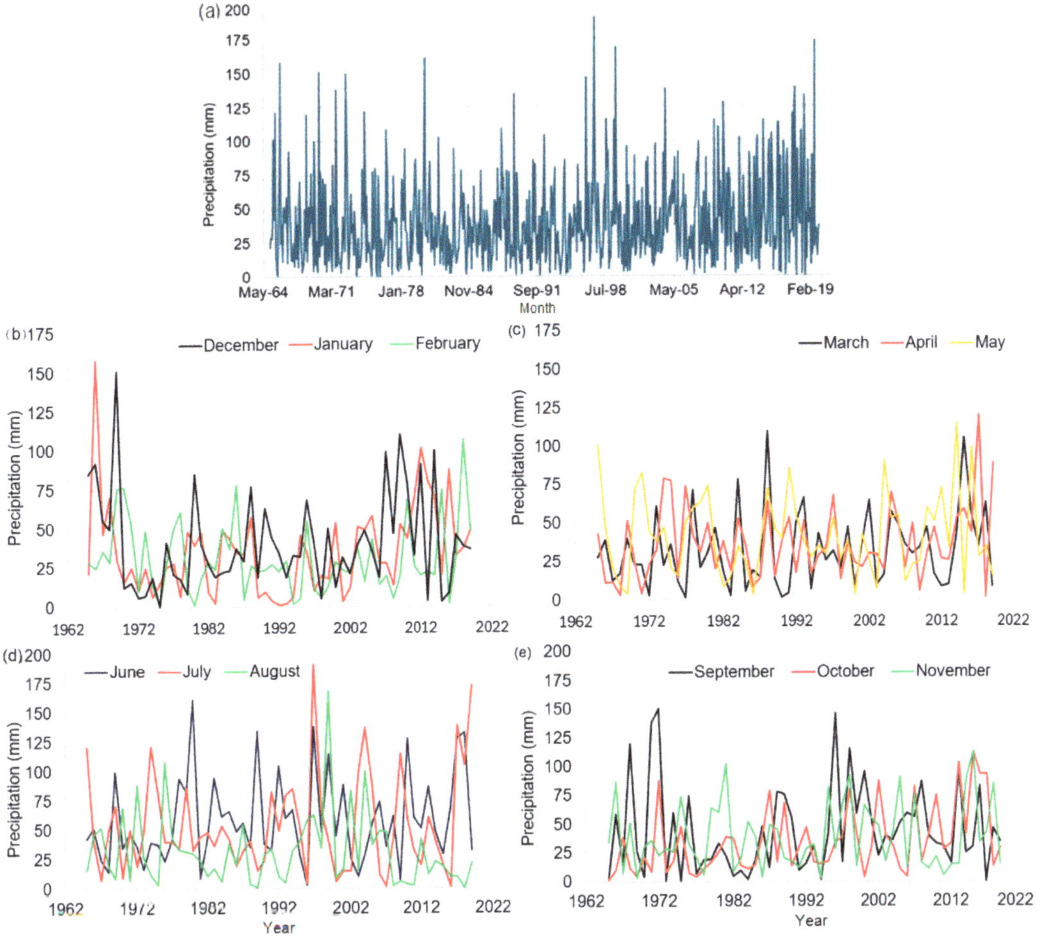

Figure 2. (**a**) MTOT, (**b**) Precipitation series recorded in December, January, and February. (**c**) Precipitation series recorded in March, April, and May. (**d**) Precipitation series recorded in June, July, and August. (**e**) Precipitation series recorded in September, October, and November.

The series are sourced from the National Institute of Meteorology, where they were checked for accuracy. No series contains missing values.

EMD enables the extraction of oscillatory components from nonstationary signals and reveals how these variations evolve across different time scales.

4. Results

The basic statistics of the monthly series are presented in Table 1. The highest amplitude of the monthly series corresponds to July, June, and January. The highest standard deviation was found in July, followed by September and June.

Table 1. Basic statistics of the monthly series.

	Jan.	Febr.	March	April	May	June	July	Aug.	Sept.	Oct.	Nov.	Dec.
min	1.20	1.00	1.90	1.60	3.50	2.80	1.20	0.00	0.40	1.10	0.00	0.70
max	158.10	107.00	109.50	120.00	115.00	160.90	191.10	168.60	149.80	113.00	113.00	151.20
mean	35.64	30.77	32.59	36.86	42.04	58.73	55.24	31.92	44.23	35.46	39.63	41.09
stdev	29.11	23.07	24.79	24.08	26.97	37.66	43.12	31.76	38.86	29.55	29.28	31.84
skew	1.67	1.17	1.11	1.03	0.71	0.91	1.22	2.05	1.13	1.06	0.76	1.25
kurt	4.69	1.27	1.29	1.42	0.05	0.16	1.33	5.70	0.73	0.10	−0.43	1.55

By comparison, ANTOT and MTOT have amplitudes (i.e., maximum–minimum) of 458.30 mm and 190.40 mm, means of 484.19 mm and 40.35, and standard deviations of 113.63 mm and 32.28 mm, respectively.

The highest number of outliers was detected in October, indicating years with high precipitation compared to similar months in other years. All series are right-skewed, and all but November are leptokurtic. November is the only series without aberrant values. MTOT also has outliers, indicating a high variability of the entire monthly series. The boxplots are presented in Figure 4.

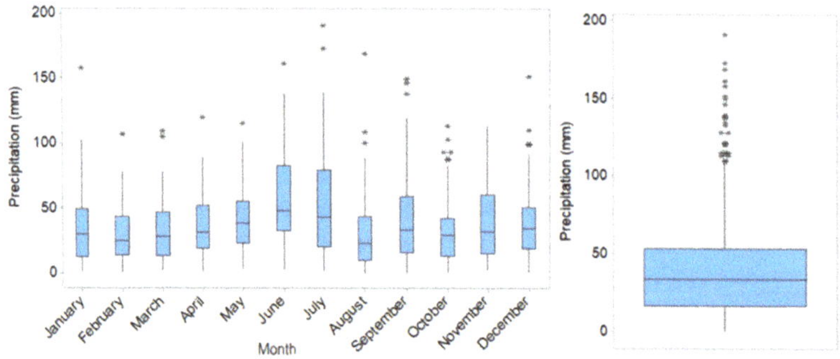

Figure 4. The boxplot of the series by month (**left**) and the box of MTOT (**right**).

The presence of numerous outliers in the October precipitation series in Tulcea, in contrast to the absence of outliers in November, can be attributed to the distinct atmospheric dynamics characterizing the two months. October represents a transitional period between the warm season and the onset of winter, during which residual high sea surface temperatures in the Black Sea and incursions of cold air from northern latitudes frequently interact. This interaction fosters conditions favorable for convective activity and cyclogenesis, leading to occasional episodes of intense rainfall and high variability in daily or monthly precipitation totals. These extreme events contribute to a broader distribution of values and the emergence of statistical outliers. In contrast, November is typically characterized by more stable atmospheric conditions, lower temperatures, and reduced convective potential, resulting in more uniform and moderate precipitation. Consequently, the November precipitation series tends to exhibit lower variability and lacks the extreme values necessary to generate statistical outliers. This seasonal contrast underscores the impact of regional climatic dynamics and transitional weather patterns on precipitation variability in the lower Danube and Black Sea regions.

The distribution of ANTOT is almost symmetric, is slightly platykurtic (Figure 1), and has no outliers. MTOT is right-skewed and leptokurtic. The hypothesis of series normality was rejected for all series. The series distribution was almost Gaussian after various transformations. The third column in Table 2 contains the type of transformation and the corresponding parameter, if applicable.

Table 2. Autocorrelation, transformations to reach normality, and results of stationarity tests.

	Autocorrelation	Normality-Transformation	ADF	KPSS-Level	KPSS-Trend
January	no	Box–Cox 0.77	Yes (0.0208)	No (0.1000)	Yes (0.0254)
February	no	Box–Cox 0.62	No (0.1391)	No (0.1000)	Yes (0.0442)
March	no	Box–Cox 0.81	Yes (0.0100)	No (0.1000)	No (0.1000)
April	no	Box–Cox 0.92	No (0.1154)	No (0.1000)	No (0.1000)
May	no	Box–Cox 0.97	Yes (0.0100)	No (0.1000)	No (0.1000)
June	no	Box–Cox 0.65	Yes (0.0107)	No (0.1000)	No (0.1000)
July	no	Box–Cox 0.67	Yes (0.0181)	No (0.1000)	No (0.1000)
August	no	Logarithm	No (0.3191)	No (0.1000)	No (0.1000)
September	no	Box–Cox 0.71	No (0.1558)	No (0.1000)	No (0.1000)
October	yes	Logarithm	Yes (0.0100)	Yes (0.0100)	No (0.1000)
November	no	Box–Cox 0.64	Yes (0.0265)	No (0.1000)	No (0.1000)
December	no	Box–Cox 0.65	No (0.1399)	No (0.1000)	No (0.1000)
MTOT	yes	Box–Cox 0.41	Yes (0.0100)	Yes (0.0100)	No (0.1000)
ANTOT	yes	Square root	Yes (0.03691)	Yes (0.0236)	No (0.1000)

According to the article of Wang et al. [62], the transformations improve the power of detection methods and increase the accuracy in identifying shifts in data patterns. The raw data series were used for other tests, so that the physical significance of the data was not impacted.

The last three columns of the same table contain the ADF and KPSS tests' results and the corresponding p-values (inside the brackets) for trend- and level-stationarity. In the second column of Table 2, 'yes' indicates the existence of the series autocorrelation. In the third column, 'Box–Cox' followed by a number, indicates the parameter in the Box–Cox transformation utilized to reach the series' normality. 'Yes' in the last three columns means the rejection of the hypothesis.

The ADF test did not reject the unit root hypothesis for February, April, August, September, and December, for which the non-stationarity was expected. Taking the first-order differences and performing the same tests for these five series, the unit-root hypothesis could be rejected. The KPSS test rejected the level-stationarity only for October, MTOT, and ANTOT. So, these series are expected to be non-stationary around a certain level. The January series is not trend stationary.

The output of stationarity tests shows that the series that require attention when studying the CP in the mean are mainly those that are level nonstationary. When studying the change in trend, the series for which the ADF did not reject the null hypothesis, and the KPSS-trend test rejected the null hypothesis simultaneously should also be considered.

Pettitt and Buishand's tests rejected the hypothesis of CPs' existence for all but the January, October, ANTOT, and MTOT series. In Table 3, 'yes' in the column of the Buishand test means that the test could not reject the hypothesis of a CP existence.

Table 3. The results of the CP tests.

	Pettitt	Buishand	Lee & Heghinian	Hubert
January	2002	yes	2011	1965, 1966, 2009
February	-	-	2017	2017
March	-	-	2013	-
April	-	-	2013	2013
May	-	-	1965	-
June	-	-	1977	-
July	-	-	2016	2016
August	-	-	2007	-
September	-	-	1965	-
October	1985	yes	1965	2012, 2017
November	-	-	2013	-
December		-	1969	1970
Monthly	July 1996	yes	July 1996	June 1991, July 1991
Annual	1995	yes	2012	2012

The numbers in the table's columns indicate the year (month) when the CPs were found. It is no surprise that different tests detected different CPs, given their different mathematical background on which the computation relies. For example, 1965 is the first year in the data series, so detecting it as a CP means that the corresponding value is an outlier. Similarly, June 1991 and July 1991 (the row before the last one, the last column) show that the values recorded in those months are aberrant, masking the real CPs. The Lee & Heghinian algorithm detected the most probable CPs of the series March, April, and November in 2011–2013 (specifically, January 2011, March, April, and November 2013) and 2016–2017 (July and February). They are reflected in the CP of the annual series (in 2012). July 1996 was found as a CP of MTOT by the same test. This finding concurs with those of the Hubert procedure and Lee & Heghinian for February, April, July, December, and ANTOT.

Since the autocorrelation hypothesis was rejected for October, ANTOT, and MTOT, the non-parametric CUSUM was applied when searching for CPs. The CUSUM advantage is that it can also detect CPs in the standard deviation and the coefficient of variation. The algorithm identified CPs in the mean for January, October, December, MTOT, and ANTOT Series, and in variance for March, August, October, and ANTOT (Table 4), with confidence levels exceeding 94%.

Table 4. CPs of the Tulcea series.

	CP in Mean				CP in Variance			
Month	Year or Month	Confidence Level	From	To	Year or Month	Confidence Level	From	To
January	1969 2009	96% 100%	74.325 25.325	25.325 59.091				
March					2003 2013	93% 90%	25.790 8.596	8.596 34.596
August					1999 2005	96% 91%	28.411 88.691	88.691 11.532
October	2007	94%	28.198	58.923	2007	95%	28.198	58.923
December	1970	98%	86.740	36.524				
MTOT	August 1996	100%	36.394	45.684	March 2017	97%	25.161	72.337
ANTOT	1996 2014	94% 94%	438.93 513.67	513.67 629.58				

We remark that some CPs found with the previous methods are among those found by CUSUM. We note that in this last case, using the non-parametric version (which confirmed the finding of the parametric version), which is less sensitive to outliers than the other methods, improved the finding accuracy. The CUSUM charts for the monthly series are presented in Figure 5.

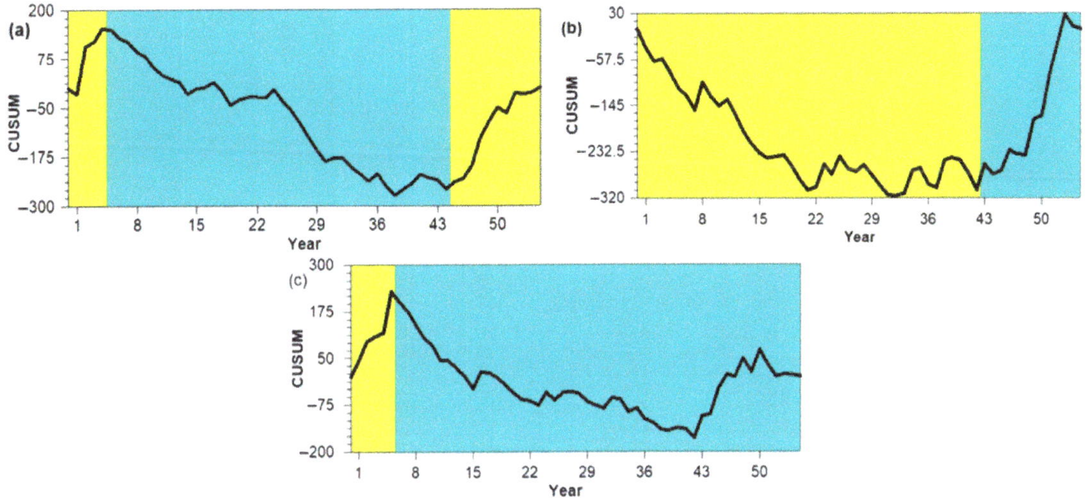

Figure 5. CUSUM charts for (**a**) January, (**b**) October, and (**c**) December series.

As a general remark, in for Gaussian data with a single CP, the most powerful choices are Lee & Heghinian or CUSUM. For a single change in non-Gaussian data, Pettitt's test provides robustness at some cost in power. For multiple change points, either CUSUM-based segmentation algorithm (especially rank-based) or Hubert's segmentation is most appropriate.

The MK test followed by the Sen procedure detected an increasing trend in the October series (slope = 0.667), ANTOT (slope = 2.555), and MTOT (slope = 0.015). The Hirsch–Slack test provided a p-value of 0.00514 and a positive z-value, confirming the existence of a positive trend in the MTOT series. The results of the ITA are presented in Figure 6. The position of the dots around the first bisector of the coordinate axes (represented by the red lines in Figure 6) for all but the October series indicates no trend in the data series.

The position of the dots above the first bisector indicates a positive trend in the October series. This means that the single month for which an increase in precipitation over time is October, whereas for the rest of the series, no increasing or decreasing tendency was detected. These results confirm the output of the MK test.

Figure 7 presents MTOT, its trend, seasonal factors, and the remainder obtained by the STL decomposition.

The trend obtained by Lowess (i.e., by a more general method than that of Sen) has an undulating pattern showing the long-term behavior of the series. The rising trend around 2007–2015 suggests increased rainfall during that period. The period 1980–1995 is characterized by some regularity in precipitation increase and decrease, with the lowest minima and maxima compared to other periods. The trend attenuation is noticed after 2015. These modifications indicate a non-stationarity in the series trend, which can be due to climatic change and modifications in land use.

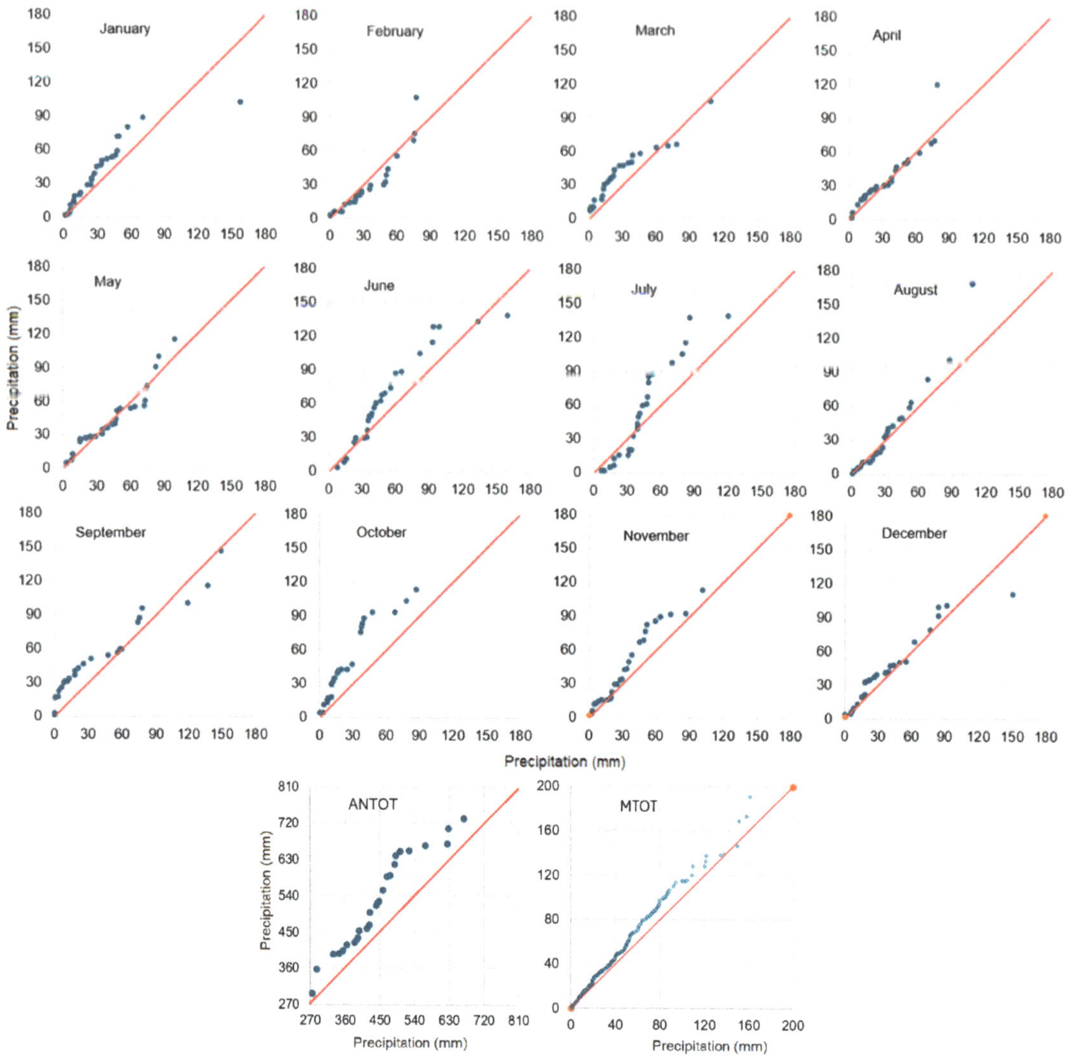

Figure 6. Results of the ITA trend test.

The Seasonal Component exhibits consistent peaks and troughs, characteristic of climates that experience distinct wet and dry seasons. The seasonal amplitude, measuring approximately 10–15 mm, is notably smaller than the overall rainfall variability, indicating that while seasonality is present, it is not the predominant factor.

The highest indices of seasonality correspond to June and July, showing the highest seasonal variability in precipitation during these months. In contrast, the indices' absolute values are lowest in November, December, and May, with values ranging from 0.68 to 1.64, suggesting much lower seasonal variability during these months. Consequently, the seasonal pattern highlights a rainy season in the summer and drier weather in November and December.

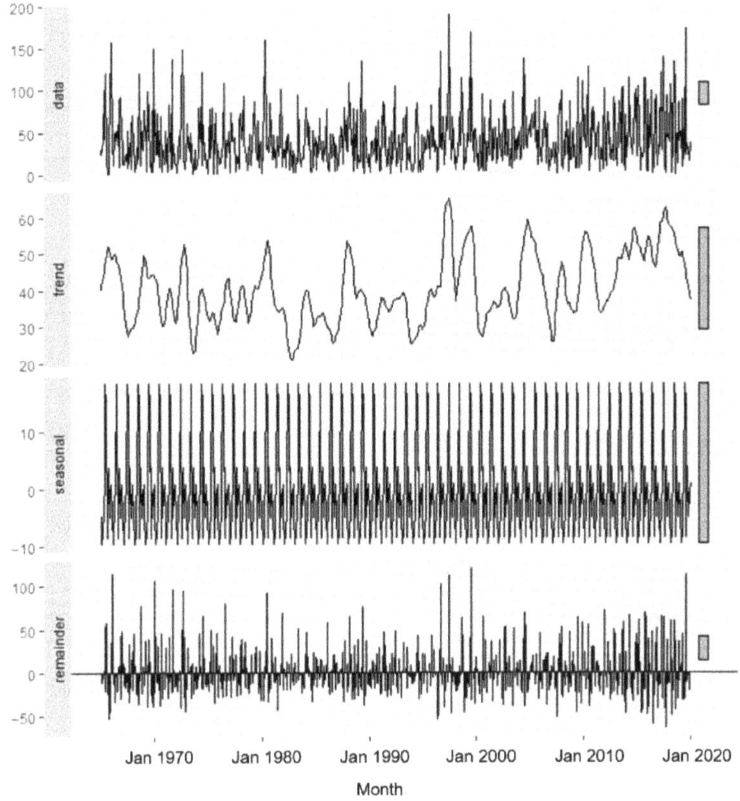

Figure 7. STL decomposition of MTOT series. From top to bottom, the rectangles contain data series, trend, seasonality, and remainder.

The Residual reflects the unexplained variability within the data series, which may include unusual droughts, isolated dry months, or extreme rainfall events. Notably, spikes in this variability are more frequent and intense following the year 2000, indicating an increase in climate volatility. In the context of the MTOT analysis, EMD decomposes the series into IMFs that represent the precipitation behavior at different scales (Figure 8).

IMF 1 and IMF 2 present high frequency. IMF 1 captures very short-term rainfall variability (1–2 months), characterized by unusually wet or dry months. Such fluctuations often arise from transient weather systems and are not linked to specific seasonal patterns, appearing more as random noise in the climate data. IMF 2 displays short-term swings (2–4 months) in rainfall amounts, which may indicate intra-seasonal oscillations. These oscillations can lead to noticeable changes in precipitation within a single season, affecting local weather patterns.

IMFs 3 and 4 present a medium frequency. IMF3, with a time scale of about 4–8 months, starts to reveal seasonal variability, shedding light on the patterns of wet and dry periods throughout the year. It captures the rhythmic changes in rainfall associated with different times of the year, reflecting the climate seasonality. There is a clear sine-decrease in this IMF's values, followed by an opposite behavior from month 80 (September 1971) to about 356 (August 1994), followed by a diminishing variation amplitude until 2015. IMF 4 reflects the annual cycle of rainfall, capturing consistent patterns tied to wet and dry seasons. It provides a clear indication of seasonal behavior, illustrating the predictable shifts in precipitation that occur as the year progresses. The highest seasonal variability was noticed in August 1976–March 1988 (a subperiod September 1971–August 1994, reflected in the variation in IMF 3).

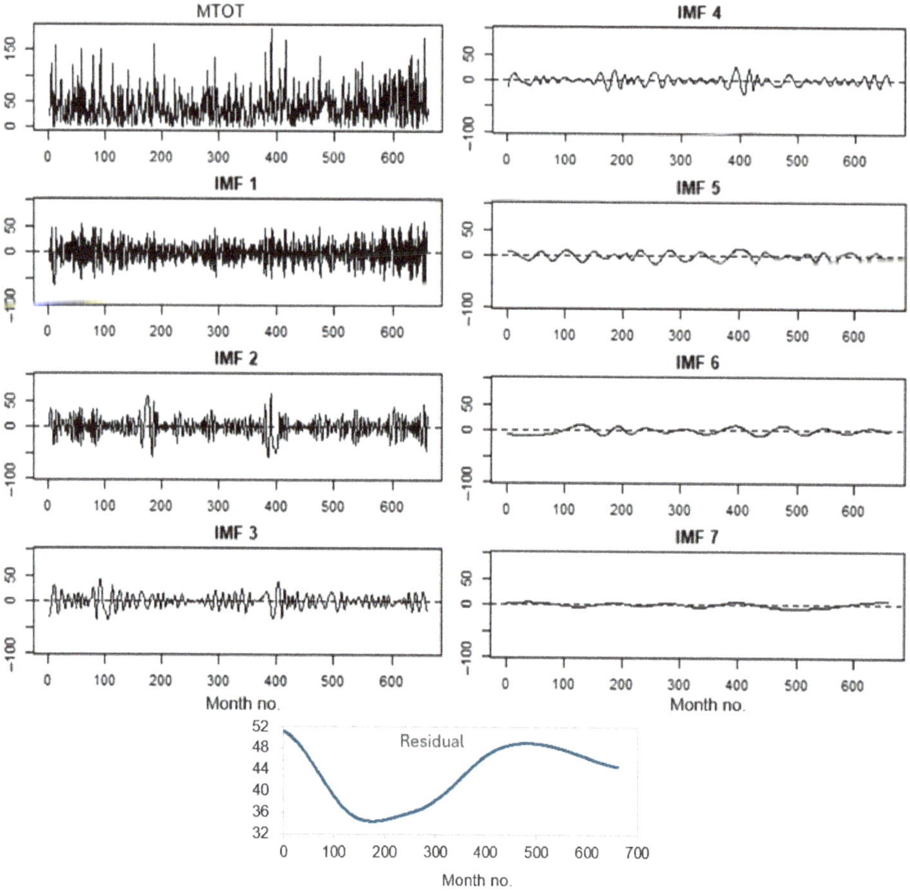

Figure 8. EMD decomposition of MTOT series: IMFs 1–7 and the residual.

The IMF 5 shows the interannual variability in rainfall reflecting longer-term climate influences. We talk about a period of 1.5 to 3 years when the frequency appears low. An almost regular sine wave is still present, highlighting trends that extend beyond single seasons.

The last two IMFs, with lower and much lower frequencies than those of IMF 5, capture stronger interannual signals for 3–7 years and 8–12 years, respectively. In the case of IMF 6, the timeframe allows for analysis of regional climate variability, revealing how rainfall patterns may shift over several years. In the case study, two periods of significant increase were followed by a sine-damped decrease. IMF 7 reflects decadal-scale variability, illustrating long-term climate cycles or persistent drought or wet conditions patterns that can last for many years. For the Tulcea series, the amplitude of these periods is higher after 2000, with the lengths moving toward 12 years.

Residual represents long-term climate trends, such as gradual drying or wetting phenomena, and exhibits the same sine pattern. These shifts may stem from climate change (at least after 1990) or land use transformations.

5. Discussion

5.1. Discussions of the Results

This study provides evidence of statistically significant and non-stationary behavior in precipitation time series, particularly for October and for the aggregated series, MTOT and ANTOT. These findings align with regional studies reporting enhanced autumn precipitation across southeastern Europe and the Romanian territory in recent decades. For example, Spinoni et al. [63] documented increases in heavy precipitation events and precipitation totals during autumn, especially after the 1990s, consistent with broader climate shifts in the Balkans and Mediterranean.

The upward trends in ANTOT and MTOT, in addition to October-specific changes, indicate that precipitation increases are not limited to isolated months but may reflect a broader shift in the annual precipitation regime. These can be interpreted as a manifestation of enhanced climate variability, particularly due to increased contributions from extreme precipitation events rather than uniform increases in daily or monthly rainfall. This distinction is crucial: rising totals may result from a lower number but more intense rainfall events, a dynamic frequently observed in Mediterranean and Balkan climates [64].

Unlike Western Europe, where atmospheric teleconnections like the North Atlantic Oscillation (NAO) have heavy climatological influence, in the Dobrogea and the Danube Delta, regional and local climatic drivers are more important in shaping extreme weather and hydrological patterns. One such driver is the frequent formation of Eastern Mediterranean cyclonic systems, especially during spring and autumn. These low-pressure systems transport moist air masses from the Aegean and Levantine basins into southeastern Europe, where they can produce heavy precipitation and flash flooding, particularly in the low-lying regions [65].

The Black Sea plays a central role in modulating regional climate. During periods of anomalously warm sea surface temperatures, evaporation rates increase, contributing to greater atmospheric instability over the coastal region. This can enhance local convective activity, resulting in heavy rainfall events or localized storms. In late summer and early autumn, this mechanism is especially influential when warm marine air interacts with polar air intrusions descending over Eastern Europe [66–68].

The STL decomposition of MTOT revealed a high variability in the seasonal components. At the same time, the Remainder showed augmented irregularity and volatility post-2000, suggesting increasing influence from high-frequency disturbances and short-duration extremes. These findings align with those of Irașoc et al. [69], who reported rising maximum rainfall events in Dobrogea, especially during the autumn months. The presence of more outliers and shifts in distribution skewness in October's monthly precipitation supports the hypothesis that climate variability in the region is intensifying, even when mean values show only moderate increases. The concentration of significant changes in October may point to a seasonal reorganization of precipitation regimes, with autumn becoming increasingly important in annual water budgets. Given the location of Tulcea near the Danube Delta, a sensitive socio-ecological system, such shifts may have implications for flood risk, agriculture, and water resource management. Similar concerns have been raised by Zaharia et al. [70], who projected that stronger seasonal contrasts and unpredictable rainfall patterns will mark the future hydrological regime of southeastern Romania.

The change-point detected by multiple statistical tests further confirmed an abrupt shift around the early 2000s in precipitation behavior. This period has also been identified in other Romanian hydrological studies as a pivot for changes in streamflow regimes and the frequency of hydrological extremes. These abrupt changes are often linked to broader atmospheric teleconnections, including shifts in the Mediterranean Oscillation, and sea surface temperature anomalies in the Black Sea and eastern Mediterranean [71–73].

Although only October and the aggregated series (ANTOT and MTOT) exhibited statistically significant trends, other months (March and August) showed alterations in the variance that trend analysis alone cannot fully capture. This reinforces the idea that climate change in southeastern Europe is increasingly expressed through greater intra-seasonal and interannual variability [72,74].

The decomposition via Empirical Mode Decomposition (EEMD) indicated the influence of both interannual variability and longer-term low-frequency patterns. These oscillations are consistent with previous findings in the Balkan and Mediterranean domains, where climate signals are known to be modulated by multi-decadal oscillations and synoptic-scale variability [75].

5.2. Comparative Discussions with the Results for the Monthly Sulina Series

To complement the actual study, we compared the CPs detected for the MTOT with the monthly precipitation series spanning the period January 1965–December 2019 collected at Sulina, a station which is situated 13 km offshore. This new data series was retrieved from the European Climate Assessment & Dataset (https://www.ecad.eu/ (accessed on 12 May 2025)). This comparison will give an idea about the monthly precipitation variability at different points in the DDBR.

The interquartile range method (IQRM) detected 50 outliers of the Sulina monthly series (denoted STOT in the following). The Hubert method detected as CPs the months August 1992, September 1992, June 2013, and May 2014. The parametric CUSUM found three CPs for the same series (Table 5), compared to only one (August 1996) for Tulcea. The same method identified two CPs (November 1972 and July 2009) in standard deviation of STOT, compared to only one CP (March 2017) for MTOT.

Table 5. CPs in mean and standard deviation for Sulina STOT series determined by the nonparametric CUSUM.

CP in Mean				CP in Standard Deviation			
Month	**Confidence Level**	**From**	**To**	**Month**	**Confidence Level**	**From**	**To**
August 1982	96%	26.13	19.57	November 1972	98%	23.80	11.64
January 2000	96%	19.57	15.64	July 2009	99%	11.64	17.82
October 2006	100%	15.64	28.59				

The existence of such a high number of outliers could bias the CPs' detection. For example, in the Hubert method, two consecutive months were determined as CPs (August 1992, September 1992) given that they are outlies. However, the nonparametric CUSUM, which is less sensitive to the aberrant values, detected only two CPs in mean and one in standard deviation for STOT.

The MK test did not find a significant trend for any of the segments delimited by the CPs of STOT. Therefore, the CPs are due to a high precipitation variability and the intensification of the extreme events, and not to a slow increase/decrease in the precipitation quantity.

A summary of CPs in mean determined for the monthly and annual series are presented in Table 6. No matter which version of CUSUM was used, the number of CPs and the moment of their apparition are not the same for the series registered at Sulina and Tulcea. The difference between the CPs detected at the two stations situated at only 15 km distance can be explained only by the most pregnant climatic influences—more accentuated continental for Tulcea, and more humid maritime at Sulina—as well as to the atmospheric circulation. They will be deeply investigated in a future work.

Table 6. CPs in mean and for Sulina and Tulcea monthly series determined by the nonparametric CUSUM.

Month	January	February	March	April	May	June	July	August	September	October	November	December	Annual Series
Sulina	1969 2003	1972			1981				1973	2009		1970	1982 2013
Tulcea	1969 2009									2007		1970	1996 2014

5.3. *Implications of the Study for Regional Planning*

5.3.1. Implications for Planning in the Danube Delta Resulted from the Statistical Analysis

The Danube Delta, with its complex network of water channels, wetlands, and biologically rich habitats, is uniquely vulnerable to changes in hydrological input and precipitation variability.

The statistical analysis indicated that the DDBR is likely to experience increasing hydrological irregularity, as evidenced by the break points, seasonal shifts, and rising variability detected in precipitation records. These changes can lead to fluctuating water levels, altered sedimentation dynamics, and disrupted aquatic habitats, which threaten both ecological integrity and human activities such as fishing, agriculture, and navigation.

The DDBR requires proactive and integrated environmental management tailored to increasing precipitation variability and changing water dynamics. Planning mechanisms must be flexible and adaptive, incorporating feedback from ecological monitoring and climate trend analyses to adjust priorities over time. Climate variability must be treated not as an episodic disturbance but as a central consideration in all levels of territorial development forecast.

The spatial planning should incorporate climate scenarios and hydrological projections derived from advanced data decomposition methods, which will enable authorities to prepare for long-term shifts in water regimes rather than respond only to immediate pressures. Policies must prioritize long-term monitoring and research, particularly into the decadal and interannual cycles revealed by EMD analysis. The STL and EMD results can inform predictive models that forecast rainfall anomalies and guide infrastructure lifespan planning.

The results of this study, particularly the identification of variance shifts and CPs in key months, highlight the growing complexity of water regime changes that cannot be addressed through static management approaches. Therefore, the Delta's water management strategy should be guided by:

- Dynamic hydrological models that integrate local precipitation trends and Danube River discharge;
- Ecological flow maintenance policies to support wetland biodiversity during dry periods;
- Flexible flood control mechanisms that prioritize nature-based solutions over hard infrastructure alone;
- Real-time hydrometeorological monitoring systems that provide actionable intelligence for basin-scale coordination

5.3.2. Restoring and Rewilding Solutions for European River Delta

Europe's deltas and estuaries are remarkable landscapes, home to rich biodiversity, productive ports, and vital wetlands. Yet, these areas face mounting pressures from urban growth, infrastructure, and climate-related risks. Balancing human activity with ecological preservation is a challenge. Different projects, among which DeltaNet [76], offered practical examples of how to achieve it. Active from 2010 to 2013, DeltaNet brought together nine delta regions across seven EU countries to share knowledge and develop sustainable policies for sensitive areas. The project focused on restoring and maintaining healthy delta ecosystems, improving flood and sediment management, and fostering public engagement. The key approaches were:

- Integrated Delta Management: Combining economic development in port areas with ecological conservation.
- Flood and Sediment Control: Restoring natural sediment flows and reducing flood risks.
- Wetland and Habitat Restoration: Reconnecting rivers to floodplains and reviving wetlands.
- Public Awareness and Policy Coordination: Engaging stakeholders and harmonizing regional strategies.

Notable good practices included:

- Elbe Estuary, Germany, and Severn Estuary, UK: Developed integrated management plans that merge ecological and economic objectives.
- Ebro River, Spain: Restored sediment flux to the delta, revitalizing wetlands and river habitats.
- Danube Delta, Romania: Reconnected abandoned polders to the river, enhancing floodplain wetlands and ecological function.

Nature-based solutions in European river deltas demonstrate the potential of ecosystem restoration to simultaneously mitigate climate risks, enhance biodiversity, and sustain local livelihoods [77]. Hydrological reconnection projects, such as the restoration of the Babina polder in the Danube Delta, have shown rapid recovery of ecological functions and socio-economic activities once natural processes are re-established [78]. Rewilding initiatives in the Danube Delta, including the reintroduction of water buffalo, horses, and deer, have revived natural grazing dynamics, enhanced habitat diversity, and contributed to fire risk reduction [79]. In the Ebro Delta, sediment augmentation and lagoon reconnection have increased resilience against sea-level rise while improving habitats for endangered species [80,81].

Across all cases, integrated and participatory governance emerges as essential, with stakeholder engagement and adaptive legal frameworks providing flexibility and societal support for long-term management [77]. Complementary initiatives in the Po Delta (Italy) [82], the Nida River Inland Delta (Poland) [83], and under the DaWetRest [84] program in the Danube Delta highlight the added value of coupling ecological restoration with cultural heritage preservation and innovative business models. Collectively, these examples confirm that leveraging natural processes is a cost-effective and sustainable pathway for the resilient management of European deltas under accelerating environmental change.

From a planning perspective, the DDBR requires a tailored, hydro-ecologically sensitive approach. Critical infrastructure (e.g., sluices, levees, and embankments) must be evaluated and upgraded to respond to both flash flooding risks and extended low-water periods. Strategic planning should preserve and enhance the Delta's natural capacity to regulate water through restoration and protection of flood retention areas, reedbeds, and natural levees. It must also reconcile its ecological significance with the socio-economic roles. Settlement expansion must be strictly limited in flood-prone zones, while tourism infrastructure should be subject to strict environmental assessments. Conservation zoning, aligned with the seasonal and decadal hydrological dynamics revealed in this study, can support the long-term resilience of wetlands and aquatic ecosystems.

Considering the Danube Delta's exposure to upstream influences, transboundary water governance and coordinated cross-sectoral policy are indispensable pillars of an effective climate adaptation framework. Participation in the International Commission for the Protection of the Danube River should be strengthened to ensure regional coherence in managing flow variability and ecological pressures.

6. Conclusions

In this study, we examined the long-term precipitation record (1965–2019) from the Tulcea station, located at the mouth of the Danube Delta in Romania, to provide a comprehensive evaluation of temporal precipitation variability. This comprehensive analysis reveals a complex interplay of seasonal, interannual, and long-term climatic dynamics. High variability, especially in the summer months, alongside significant skewness and kurtosis in most monthly series, highlights the irregular nature of rainfall events in the region. The presence of CPs, detected by various statistical methods, underscores modifications in precipitation regimes, notably around the years 1965, 1996, and 2012. The application of STL and EMD confirms the presence of multi-scale oscillations and long-term trends, with increasing climate volatility observed particularly after 2000. These findings indicate a departure from stationarity in both mean and variance, likely driven by climate change. The detection of positive trends in series such as October, MTOT, and ANTOT highlights a gradual intensification of rainfall, underscoring the need for adaptive water resource management and further investigation into regional climate drivers. EMD, while powerful, lacks a formal statistical framework for significance testing, and the subjective interpretation of IMFs may introduce bias.

Despite the robustness of the multi-method approach, some limitations should be acknowledged. First, change point detection methods have various degrees of sensitivity to outliers and sample length, leading to inconsistencies in CP identification across techniques. They also rely on various mathematical or statistical methodologies. The biased CP detection can significantly affect the quality of the models that could be built for the data series, and, especially, the forecast based on them. Therefore, the analysis should also be complemented by the knowledge of climate evolution on the region and metadata, which should support the acceptance of the CP existence. Second, the narrow database limits the results extrapolation to the entire Danube Delta. Therefore, for a full understanding of climate variability in the DDBR, the precipitation evolution (e.g., trend, CPs in mean and variance) should be analyzed considering the data series from other meteorological points in the region.

High-resolution spatial data and remote sensing could help assess the changes. Moreover, future work should focus on integrating physically based climate models to better interpret the statistical findings in the context of atmospheric dynamics. A deeper analysis of the connections with physical mechanisms of atmospheric circulation (e.g., NAO) could strengthen the interpretation of the statistical results. Furthermore, the use of more advanced machine learning techniques, such as deep learning-based time series models, could enhance trend and anomaly detection. Finally, extending the analysis to include other data series (e.g., temperature and evapotranspiration), updated to 2025, would provide a better view of regional water cycle shifts in response to climate change. The results would support the water resources management and planning.

Author Contributions: Conceptualization, A.B. and C.Ș.D.; methodology, A.B. and C.Ș.D.; software, A.B.; validation, C.Ș.D.; formal analysis, C.Ș.D.; investigation, A.B. and C.Ș.D.; resources, A.B.; data curation, A.B.; writing—original draft preparation, A.B.; writing—review and editing, C.Ș.D.; visualization, A.B.; supervision, A.B.; project administration, A.B.; funding acquisition, A.B. and C.Ș.D. All authors have read and agreed to the published version of the manuscript.

Funding: This research received no external funding.

Data Availability Statement: Data can be freely downloaded from: European Climate Assessment & Data: https://www.ecad.eu/dailydata/index.php (accessed on 12 May 2025).

Conflicts of Interest: The authors declare no conflicts of interest.

Abbreviations

The following abbreviations are used in this manuscript:

ANTOT	Annual total precipitation series RECORDED AT Tulcea
CP	Change Point
CPD	Change Point Detection
CUSUM	Cumulative Sum
DDBR	Danube Delta Biosphere Reserve
ITA	Innovative trend analysis
MK	Mann–Kendall
MTOT	Tulcea monthly precipitation series
SMK	Seasonal Mann–Kendall
STOT	Sulina monthly precipitation series

References

1. IPCC Sixth Assessment Report. Impacts, Adaptation and Vulnerability. 2022. Available online: https://www.ipcc.ch/report/sixth-assessment-report-working-group-ii/ (accessed on 22 June 2025).
2. The Global Water Monitor. 2024. Available online: https://www.globalwater.online/globalwater/report/index.html (accessed on 22 June 2025).
3. Do, H.X.; Mei, Y.; Gronewold, A.D. To what extent are changes in flood magnitude related to changes in precipitation extremes? *Geophys. Res. Lett.* **2020**, *47*, e2020GL088684. [CrossRef]
4. Impact of Climate Change on Water Availability and Quality. Available online: https://www.unicef.org/serbia/en/impact-of-climate-change-on-water-availability-and-quality (accessed on 22 June 2025).
5. van Vliet, M.T.H.; Thorslund, J.; Strokal, M.; Hofstra, N.; Flörke, M.; Macedo, H.E.; Nkwasa, A.; Tang, T.; Kaushal, S.S.; Kumar, R.; et al. Global river water quality under climate change and hydroclimatic extremes. *Nat. Rev. Earth Environ.* **2023**, *4*, 687–702. [CrossRef]
6. Barbulescu, A.; Barbes, L.; Dumitriu, C.S. Assessing the Water Pollution of the Brahmaputra River Using Water Quality Indexes. *Toxics* **2021**, *9*, 297. [CrossRef] [PubMed]
7. Sjerps, R.M.A.; ter Laak, T.L.; Zwolsman, G.J.J.G. Projected impact of climate change and chemical emissions on the water quality of the European rivers Rhine and Meuse: A drinking water perspective. *Sci. Total Environ.* **2017**, *601–602*, 1682–1694. [CrossRef]
8. Bărbulescu, A.; Maftei, C. Modeling the climate in the area of Techirghiol Lake (Romania). *Rom. J. Phys.* **2015**, *60*, 1163–1170.
9. Berg, A.; Findell, K.; Lintner, B.; Giannini, A.; Seneviratne, S.I.; Van Den Hurk, B.; Lorenz, R.; Pitman, A.; Hagemann, S.; Meier, A.; et al. Land–atmosphere feedbacks amplify aridity increase over land under global warming. *Nat. Clim. Change* **2016**, *6*, 869–874. [CrossRef]
10. Brunner, M.I.; Hingray, B.; Zappa, M.; Favre, A.-C. Future Trends in the Interdependence Between Flood Peaks and Volumes: Hydro-Climatological Drivers and Uncertainty. *Water Resour. Res.* **2019**, *55*, 4745–4759. [CrossRef]
11. Slater, L.J.; Anderson, B.; Buechel, M.; Dadson, S.; Han, S.; Harrigan, S.; Kelder, T.; Kowal, K.; Lees, T.; Matthews, T.; et al. Nonstationary weather and water extremes: A review of methods for their detection, attribution, and management. *Hydrol. Earth Syst. Sci.* **2021**, *25*, 3897–3935. [CrossRef]
12. Bărbulescu, A.; Maftei, C. Evaluating the Probable Maximum Precipitation. *Rom. Rep. Phys.* **2023**, *75*, 704. [CrossRef]
13. Sun, W.; Li, J.; Yu, R.; Li, N.; Zhang, Y. Exploring changes of precipitation extremes under climate change through global variable-resolution modeling. *Sci. Bull.* **2024**, *69*, 237–247. [CrossRef]
14. Knoben, W.J.M.; Raman, A.; Gründemann, G.J.; Kumar, M.; Pietroniro, A.; Shen, C.; Song, Y.; Thébault, C.; van Werkhoven, K.; Wood, A.W.; et al. Technical note: How many models do we need to simulate hydrologic processes across large geographical domains? *Hydrol. Earth Syst. Sci.* **2025**, *29*, 2361–2375. [CrossRef]
15. Alifujiang, Y.; Abuduwaili, J.; Maihemuti, B.; Emin, B.; Groll, M. Innovative Trend Analysis of Precipitation in the Lake Issyk-Kul Basin, Kyrgyzstan. *Atmosphere* **2020**, *11*, 332. [CrossRef]

16. Cao, S.; Li, X.; You, R.; Diao, Y.; Wang, J.; Liu, Y. Investigation of the trend and cycles using the long-term records of annual precipitation in Shandong, China. *J. Water Clim. Change* **2022**, *13*, 3761–3775.
17. Bărbulescu, A.; Deguenon, J. Change point detection and models for precipitation evolution. Case study. *Rom. J. Phys.* **2014**, *59*, 590–600.
18. Conte, L.C.; Bayer, M.D.; Bayer, F.M. Bootstrap Pettitt test for detecting change points in hydroclimatological data: Case study of Itaipu Hydroelectric Plant, Brazil. *Hydrol. Sci. J.* **2019**, *64*, 1312–1326. [CrossRef]
19. Valderrama Balaguera, J.C. Precipitation forecast estimation applying the change point method and ARIMA. *Cogent Eng.* **2024**, *11*, 2340191. [CrossRef]
20. Getahun, Y.S.; Li, M.-H.; Pun, I.-F. Trend and change-point detection analyses of rainfall and temperature over the Awash River basin of Ethiopia. *Heliyon* **2021**, *7*, e08024. [CrossRef] [PubMed]
21. Fotoohinasab, A.; Hocking, T.; Afghah, F. A Graph-constrained Changepoint Detection Approach for ECG Segmentation. In Proceedings of the 2020 42nd Annual International Conference of the IEEE Engineering in Medicine & Biology Society (EMBC), Montreal, QC, Canada, 20–24 July 2020; pp. 332–336.
22. Shvetsov, N.; Buzun, N.; Dylov, D.V. Unsupervised Non-Parametric Change Point Detection in Quasi-Periodic Signals. 2020. Available online: https://arxiv.org/abs/2002.02717 (accessed on 22 June 2020).
23. Alawadhi, F.A.; Alhulail, D. Bayesian change points analysis for earthquakes body wave magnitude. *J. Appl. Stat.* **2016**, *43*, 1567–1582. [CrossRef]
24. Jiqin, H.; Gelata, F.T.; Gemeda, S.C. Application of MK trend and test of Sen's slope estimator to measure impact of climate change on the adoption of conservation agriculture in Ethiopia. *J. Water Clim. Change* **2023**, *14*, 977–988. [CrossRef]
25. Kliengchuay, W.; Mingkhwan, R.; Kiangkoo, N.; Suwanmanee, S.; Sahanavin, N.; Kongpran, J.; WaiAung, H.; Tantrakarnapa, K. Analyzing temperature, humidity, and precipitation trends in six regions of Thailand using innovative trend analysis. *Sci. Rep.* **2024**, *14*, 7800. [CrossRef]
26. Cleveland, R.B.; Cleveland, W.S.; McRae, J.E.; Terpenning, I.J. STL: A seasonal-trend decomposition procedure based on loess. *J. Off. Stat.* **1990**, *6*, 3–33.
27. Bush, E.R.; Abernethy, K.A.; Jeffery, K.; Tutin, C.; White, L.; Dimoto, E.; Dikangadissi, J.-T.; Jump, A.S.; Bunnefeld, N. Fourier analysis to detect phenological cycles using long-term tropical field data and simulations. *Methods Ecol. Evol.* **2017**, *8*, 530–540.
28. Bandoc, G.; Prăvălie, R. Climatic water balance dynamics over the last five decades in Romania's most arid region, Dobrogea. *J. Geogr. Sci.* **2015**, *25*, 1307–1327. [CrossRef]
29. Cheval, S.; Dumitrescu, A.; Birsan, M.-V. Variability of the aridity in the South-Eastern Europe over 1961–2050. *Catena* **2017**, *151*, 74–86. [CrossRef]
30. Bădăluță, C.A.; Haliuc, A.; Bădăluță, G.; Scriban, R.E. Spatiotemporal variability of drought in Romania during 1901–2021 using the Standardized Precipitation Evapotranspiration Index (SPEI). *Ann. Univ. Oradea Ser. Geogr.* **2024**, *34*, 33–44.
31. Petre, M.J. Assessment of the UTCI and HSI Biometeorological Indices for Constanța and Tulcea. Available online: https://riscurisicatastrofe.reviste.ubbcluj.ro/Volume/XXIV_Nr_34_2024/PetreMJ_RCJ2024_5.pdf (accessed on 20 June 2025).
32. Deguenon, J.; Bărbulescu, A.; Sarr, M. GPD Models for extreme rainfall with cyclic trend detection over ten stations in Dobrudja. *Comput. Eng. Syst. Appl.* **2011**, *2*, 131–136.
33. Deguenon, J.; Bărbulescu, A. Trends of extreme precipitation events in Dobrudja. *Ovidius Univ. Ann. Ser. Civ. Eng.* **2011**, *7*, 73–80.
34. Maftei, C.; Dobrica, G.; Cerneaga, C.; Buzgaru, N. Drought Land Degradation and Desertification—Case Study of Nuntasi-Tuzla Lake in Romania. In *Water Safety, Security and Sustainability. Advanced Sciences and Technologies for Security Applications*; Vaseashta, A., Maftei, C., Eds.; Springer: Cham, Switzerland, 2021; pp. 583–597. [CrossRef]
35. Preoteasa, L.; Vespremeanu-Stroe, A.; Hanganu, D.; Katona, O.; Timar-Gabor, A. Coastal changes from open coast to present lagoon system in Histria region (Danube delta). *J. Coast. Res.* **2013**, *65*, 564–569. [CrossRef]
36. Nagavciuc, V.; Ionita, M.; Roibu, C.-C. Extreme Aridity in the South-Eastern Part of Romania. In Proceedings of the EGU General Assembly 2021, Online, 19–30 April 2021. [CrossRef]
37. Bărbulescu, A.; Băutu, E. Mathematical models of climate evolution in Dobrudja. *Theor. Appl. Climatol.* **2010**, *100*, 29–44.
38. Bărbulescu, A.; Dumitriu, C.S.; Dragomir, F. Detecting Aberrant Values and Their Influence on the Time Series Forecast. In Proceedings of the 2021 International Conference on Electrical, Computer, Communications and Mechatronics Engineering (ICECCME), Mauritius, Mauritius, 7–8 October 2021. [CrossRef]
39. Saliba, Y.; Bărbulescu, A. A Multi-Method Approach to Analyzing Precipitation Series and Their Change Points in Semi-Arid Climates: The Case of Dobrogea. *Water* **2025**, *17*, 391. [CrossRef]
40. Bărbulescu, A.; Maftei, C.; Dumitriu, C.S. The modeling of the climatic process that participates at the sizing of an irrigation system. *Bull. Appl. Comput. Math.* **2002**, *CII-2048*, 11–20.
41. Trif, A.; Mihai, D. Effect of climate change on Dobrogea agricultural area. *Sci. Pap. Ser. E. Land Reclam. Earth Obs. Surv. Environ. Eng.* **2014**, *3*, 137–140.
42. Bărbulescu, A.; Dumitriu, C.S.; Maftei, C. On the Probable Maximum Precipitation Method. *Rom. J. Phys.* **2022**, *67*, 801.

43. Prăvălie, R.; Bandoc, G. Aridity Variability in the Last Five Decades in the Dobrogea Region, Romania. *Arid Land Res. Manag.* **2015**, *29*, 265–287. [CrossRef]
44. RESTORE4CS. Available online: https://www.restore4cs.eu/case-pilots/danube-delta/ (accessed on 27 July 2025).
45. Tătui, F. Potential usage of met mast datasets for climatic parameters monitoring in Tulcea County, Romania. *Forum Geogr.* **2015**, *14*, 22–29. [CrossRef]
46. Shapiro, S.S.; Wilk, M.B. An analysis of variance test for normality (complete samples). *Biometrika* **1965**, *52*, 591–611. [CrossRef]
47. Dickey, D.A.; Fuller, W.A. Distribution of the Estimators for Autoregressive Time Series with a Unit Root. *J. Am. Stat. Assoc.* **1979**, *74*, 427–431. [PubMed]
48. Kwiatkowski, D.; Phillips, P.C.B.; Schmidt, P.; Shin, Y. Testing the null hypothesis of stationarity against the alternative of a unit root. *J. Econ.* **1992**, *54*, 159–178. [CrossRef]
49. Buishand, T.A. Tests for detecting a shift in the mean of hydrological time series. *J. Hydrol.* **1984**, *73*, 51–69. [CrossRef]
50. Pettitt, A.N. A non-parametric approach to the change point problem. *J. R. Stat. Soc. Ser. C. Appl. Stat.* **1979**, *28*, 126–135. [CrossRef]
51. Lee, A.F.S.; Heghinian, S.M. A Shift of the Mean Level in a Sequence of Independent Normal Random Variables—A Bayesian Approach. *Technometrics* **1977**, *19*, 503–506. [CrossRef]
52. Hubert, P. Segmentation Procedure as a Tool for Discrete Modeling of Hydro-Meteorological Regimes. *Stoch. Environ. Res. Risk Assess.* **2000**, *14*, 297–304.
53. Stapenhurst, T. *Mastering Statistical Process Control*, 1st ed.; Routledge: London, UK, 2005. [CrossRef]
54. Scheffe, H. *The Analysis of Variance*; Wiley: New York, NY, USA, 1959.
55. Taylor, W.A. Change-Point Analysis: A Powerful New Tool for Detecting Changes. 2000. Available online: https://variation.com/wp-content/uploads/change-point-analyzer/change-point-analysis-a-powerful-new-tool-for-detecting-changes.pdf (accessed on 4 June 2025).
56. Hirsch, R.M.; Slack, J.R. A Nonparametric Trend Test for Seasonal Data With Serial Dependence. *Water Resour. Res.* **1984**, *20*, 727–732. [CrossRef]
57. Gilbert, R.O. *Statistical Methods for Environmental Pollution Monitoring*; Van Nostrand Reinhold Company: New York, NY, USA, 1987.
58. Şen, Z. Innovative trend analysis methodology. *J. Hydrol. Eng.* **2012**, *17*, 1042–1046. [CrossRef]
59. Hyndman, R.J.; Athanasopoulos, G. *Forecasting: Principles and Practice*, 3rd ed.; OTexts: Melbourne, Australia, 2021; Available online: https://otexts.com/fpp3/ (accessed on 25 June 2025).
60. Huang, N.E.; Wu, Z. A review on Hilbert-Huang Transform: Method and its applications to geophysical studies. *Rev. Geophys.* **2008**, *46*, RG2006.
61. Kim, D.; Kim, K.-O.; Oh, H.-S. Extending the Scope of Empirical Mode Decomposition using Smoothing. *EURASIP J. Adv. Signal Process.* **2012**, *168*, 168. [CrossRef]
62. Wang, X.L.; Chen, H.; Wu, Y.; Feng, Y.; Pu, Q. New Techniques for the Detection and Adjustment of Shifts in Daily Precipitation Data Series. *J. Appl. Meteor. Climatol.* **2010**, *49*, 2416–2436. [CrossRef]
63. Spinoni, J.; Naumann, G.; Vogt, J.V. Pan-European Seasonal Trends and Variability of Precipitation and Temperature in Europe (1951–2010). *Glob. Planet. Change* **2017**, *148*, 113–130. [CrossRef]
64. Zittis, G. Observed rainfall trends and precipitation uncertainty in the vicinity of the Mediterranean, Middle East and North Africa. *Theor. Appl. Climatol.* **2018**, *134*, 1207–1230. [CrossRef]
65. Birsan, M.V. Trends in Monthly Natural Streamflow in Romania and Linkages to Atmospheric Circulation in the North Atlantic. *Water Resour. Manag.* **2015**, *29*, 3305–3313. [CrossRef]
66. Meredith, E.; Semenov, V.; Maraun, D.; Park, W.; Chernokulsky, A. Crucial role of Black Sea warming in amplifying the 2012 Krymsk precipitation extreme. *Nat. Geosci.* **2015**, *8*, 615–619. [CrossRef]
67. Mohamed, B.; Ibrahim, O.; Nagy, H. Sea Surface Temperature Variability and Marine Heatwaves in the Black Sea. *Remote Sens.* **2022**, *14*, 2383.
68. Gavrikov, A.V.; Zolina, O.G.; Razorenova, O.A.; Aleksandrova, M.P.; Gulev, S.K.; Stenchikov, G.L. Extreme Precipitation in June 2021 over the Black Sea in the Context of Long-Term Climate Change. *Oceanology* **2022**, *62*, 303–309. [CrossRef]
69. Irașoc, A.; Ionac, N.Ș.; Dumitrescu, A.; Beteringhe, A. Extreme Rainfall Intensities at Sub-Hourly Temporal Scale in Dobrudja, Romania. *Geogr. Tech.* **2024**, *19*, 103–122. [CrossRef]
70. Zaharia, L.; Ioana-Toroimac, G.; Perju, E.R. Hydrological Impacts of Climate Changes in Romania. In *Water Resources Management in Romania*; Negm, A., Romanescu, G., Zeleňáková, M., Eds.; Springer: Cham, Switzerland, 2020; pp. 309–351. [CrossRef]
71. Ionita, M.; Chelcea, S.; Rimbu, N.; Adler, M.-J. Spatial and temporal variability of winter streamflow over Romania and its relationship to large-scale atmospheric circulation. *J. Hydrol.* **2014**, *519*, 1339–1349. [CrossRef]
72. Ioniță, M.; Lohmann, G.; Rimbu, N.; Chelcea, S.; Dima, M. Interannual to decadal summer drought variability over Europe and its relationship to global sea surface temperature. *Clim. Dyn.* **2012**, *38*, 363–377. [CrossRef]

73. Cramer, W.; Guiot, J.; Fader, M.; Garrabou, J.; Gattuso, J.–P.; Iglesias, A.; Lange, M.A.; Lionello, P.; Llasat, M.C.; Paz, S.; et al. Climate change and interconnected risks to sustainable development in the Mediterranean. *Nat. Clim. Change* **2018**, *8*, 972–980. [CrossRef]
74. Danilovich, I.S.; Loginov, V.F.; Groisman, P.Y. Changes of Hydrological Extremes in the Center of Eastern Europe and Their Plausible Causes. *Water* **2023**, *15*, 2992. [CrossRef]
75. Zampieri, M.; Toreti, A.; Schindler, A.; Scoccimarro, E.; Gualdi, S. Atlantic multi-decadal oscillation influence on weather regimes over Europe and the Mediterranean in spring and summer. *Glob. Planet. Change* **2017**, *151*, 92–100. [CrossRef]
76. Network of European Delta Regions-Sustainable Delta Governance. Available online: https://keep.eu/projects/423/Network-of-European-Delta-Reg-EN/ (accessed on 8 September 2025).
77. European Commission: Directorate-General for Research and Innovation. *Evaluating the Impact of Nature-Based Solutions—A Handbook for Practitioners*; Publications Office of the European Union: Luxembourg, 2021; Available online: https://data.europa.eu/doi/10.2777/244577 (accessed on 15 May 2025).
78. Natural Water Retention Measures. The Babina Restoration Project (Danube Delta, Romania). 2014. Available online: https://www.nwrm.eu (accessed on 8 September 2025).
79. Rewilding Danube Delta. Available online: https://rewilding-danube-delta.com/ (accessed on 8 September 2025).
80. Spanish Climate Change Adaption Platform. Habitat Restoration and Management in the Ebro Delta Coastal Lagoons. Available online: https://adaptecca.es/en/casos-practicos/habitat-restoration-and-management-ebro-delta-coastal-lagoons (accessed on 8 September 2025).
81. Rodríguez-Santalla, I.; Navarro, N. Main Threats in Mediterranean Coastal Wetlands. The Ebro Delta Case. *J. Mar. Sci. Eng.* **2021**, *9*, 1190. [CrossRef]
82. Conservation Programme for the Po Delta Park Geographical Areas (Second Phase). Available online: https://econservation.jrc.ec.europa.eu/project/2000338 (accessed on 8 September 2025).
83. Project LIFE4DELTA_PL. Available online: https://life4delta.pl/?lang=en (accessed on 8 September 2025).
84. DaWetRest. Available online: https://dawetrest.eu/ (accessed on 8 September 2025).

water

MDPI

Article

Stable Isotope Monitoring in a Semi-Arid Olive Orchard Suggest Changes in Ecohydrological Dynamics from Contrasting Drip Irrigation Regimes

Taha Attou [1,2,*], M. H. Kharrou [1], S. Kuppel [2], Y. Ait Brahim [1], L. Bouchaou [1,3], V. Demarez [4], M. M. Lehmann [5], F. Raibi [1,6], T. Elghali [6], A. Elazhari [1,7], N. Rhoujjati [8], H. Bouimouass [9] and A. Chehbouni [9]

[1] International Water Research Institute, Mohammed VI Polytechnic University, Benguerir 43150, Morocco; mohamed.kharrou@um6p.ma (M.H.K.); yassine.aitbrahim@um6p.ma (Y.A.B.); lhoussaine.bouchaou@um6p.ma (L.B.); fatima.raibi@um6p.ma (F.R.); ahmed.elazhari@um6p.ma (A.E.)

[2] Géosciences Environnement Toulouse, Université de Toulouse/CNRS/IRD/CNES, 31400 Toulouse, France; sylvain.kuppel@ird.fr

[3] Laboratory of Applied Geology and Geo-Environment, Faculty of Science, Ibn Zohr University, Agadir 80000, Morocco

[4] Université de Toulouse, CNRS, CNES, INRAe, IRD, CESBIO, Centre d'Etudes Spatiales de la Biosphère, 31400 Toulouse, France; valerie.demarez@utoulouse.fr

[5] Forest and Soil Ecology, Swiss Federal Institute for Forest, Snow and Landscape Research WSL, Zürcherstrasse 111, 8903 Birmensdorf, Switzerland; marco.lehmann@wsl.ch

[6] Centre National de l'Énergie, des Sciences et des Techniques Nucléaires (CNESTEN), Rabat 14000, Morocco; elghali.tibari@gmail.com

[7] Geotop-UQAM, Hydro Sciences, Department of Earth and Atmospheric Sciences, University of Quebec at Montreal (UQAM), Montreal, QC H3C 3P8, Canada

[8] Geoenvironment and Civil Engineering (L3G), Laboratory of Georessources, Faculty of Science and Technology, Cadi Ayyad University (UCA), Av. A. El Khattabi, Marrakech 40000, Morocco; nadiarhoujjati@gmail.com

[9] Center for Remote Sensing Applications, Mohammed VI Polytechnic University, Benguerir 43150, Morocco; houssne.bouimouass@um6p.ma (H.B.); abdelghani.chehbouni@um6p.ma (A.C.)

* Correspondence: taha.attou@um6p.ma or taha.attou.th@gmail.com; Tel.: +212-661809887

Abstract

In semi-arid regions of Morocco, where the majority of water withdrawals are devoted to irrigation, optimizing irrigation practices in agriculture is a national priority in the face of recurring droughts and growing pressure on groundwater resources. However, the hydrological impacts of different drip-irrigation systems in the soil–plant–atmosphere continuum remain insufficiently understood. We monitored the stable isotope composition (δ^2H, $\delta^{18}O$) across the two agricultural plots in Marrakech (Morocco) with surface drip and subsurface drip irrigation treatments for a complete hydrologic year (June 2022 to June 2023). Weekly to daily samples of rainfall, irrigation water, groundwater, and soil at various depths (5–50 cm) were sampled, and water from branch xylem was extracted using the cryogenic vacuum distillation method. We found that the subsurface irrigation treatment, which delivered water directly to the root zone, maintained narrow isotopic ranges in water of soils beyond 30 cm, as well as in branch xylem and leaf water. By contrast, surface irrigation treatment plots showed pronounced evaporative isotopic enrichment: summer topsoil water $\delta^{18}O$ peaked at −1.1‰ (vs. −8.7‰ in subsurface irrigation treatment), and leaf water reached +13‰ (vs. +8‰ in subsurface). Despite this larger isotopic heterogeneity in surface irrigation site, branch xylem water $\delta^{18}O$ remained within −6 to 2.5‰ across all soil depth, similar to subsurface irrigation treatment, which ranged between −5 and 0‰. This suggests that olive roots accessed soil water uniformly from the upper 50 cm under both irrigation treatments. Seasonal xylem isotopic enrichment in spring and midsummer mirrored shifts towards shallow, evaporatively altered soil water under surface irrigation, but not under the subsurface. The results suggest that subsurface drip irrigation can significantly improve drought resilience and water-use efficiency in the expanding olive sector of the Maghreb, while continuous isotope monitoring serves as a practical approach to enhance sustainable and adaptive water management in water-limited regions.

Keywords: soil–plant–atmosphere continuum (SPAC); irrigation management; stable isotopes of water; isotope hydrology; cryogenic vacuum distillation

1. Introduction

Water is a critical resource for agriculture, industry, and the environment, particularly in semi-arid and arid regions where increasing water scarcity poses serious challenges [1]. Factors such as population growth, agricultural intensification, and climate change have intensified stress on freshwater supplies, making it difficult to balance water allocations among competing sectors [2]. In agriculture, specifically, a consistent water supply is crucial for maintaining robust crop yields and ensuring global food security [3]. However, due to the changing climate, weather patterns have become more erratic, characterized by a higher frequency of extreme events especially droughts, caused by higher temperature and reduced rainfall in semi-arid regions [4], where a significant share of the irrigation expansion is precisely taking place [5]. This situation underscores the necessity of understanding the dynamics of water and its movement through soil, absorption by crops, and subsequent evaporation into the atmosphere. Mastering this concept is crucial for ensuring efficient water usage in agriculture in areas that are currently experiencing significant water scarcity [6].

In semi-arid regions like Morocco, improving irrigation efficiency is crucial for sustainable agriculture [7]. Traditionally, flood irrigation has been the most common method, where this technique results in high water losses due to evaporation and deep percolation, making it unsuitable for water-scarce regions [8]. Recognizing the need for a more sustainable approach, the Moroccan government launched the Green Morocco Plan in 2008, a comprehensive agricultural strategy aimed at modernizing farming practices and optimizing water resources, followed by the Green Generation plan in 2020. A key initiative within this framework was the National Irrigation Water Saving Program (PNEEI), which set an ambitious goal of converting from flood irrigation to localized irrigation systems, particularly drip irrigation. By reducing water losses and increasing crop productivity, the plan played a transformative role in Morocco's agricultural landscape.

As a result, many studies have compared irrigation systems in Morocco in terms of water savings and irrigation efficiency. For instance, a recent study [9] demonstrated that seasonal deep percolation losses under flood irrigation reached up to 53% of total water input, with an associated irrigation recharge efficiency of only 44%, compared to 21% losses and up to 78% efficiency under drip irrigation in olive orchards of the Haouz Plain.

However, while conventional water balance approaches are useful for estimating gross irrigation performance, they often lack the resolution to track how water is distributed and used within the root zone and beyond [10]. This is particularly relevant when comparing irrigation strategies such as surface versus subsurface drip irrigation, where water delivery depth may alter plant uptake patterns and soil moisture retention in subtle ways. In this context, stable isotopes of hydrogen (δ^2H) and oxygen ($\delta^{18}O$) have become valuable tools for investigating water fluxes within the soil–plant atmosphere continuum (SPAC) [11,12]. By analyzing the isotopic composition of water at different points in the SPAC such as in the soil profile, plant xylem, and atmospheric vapor, researchers can trace the origins and pathways of water with a level of detail not easily achievable by other methods [13,14]. This approach facilitates the assessment of key processes, including evapotranspiration, soil moisture distribution, and deep percolation. Global studies leveraging these techniques have shown their effectiveness in diverse agro-ecological settings, including semi-arid regions of Morocco, where efficient irrigation strategies are urgently needed to conserve limited water resources [15]. At the same time, the successful application of isotopic methods depends on robust sampling protocols and reliable extraction methods that minimize artifacts such as evaporation fractionation and sample contamination, while also accounting for the inherent limitations and biases introduced by laboratory constraints [16].

For instance, the stable isotope method provides a promising tool for partitioning evapotranspiration (ET). This method has been successfully used in many studies related to water movement in the soil–vegetation–atmosphere continuum [17–19] since the measurements of stable isotopic compositions are now relatively easy and robust due to the rapid progress in development of mass and laser spectrometry devices during the last few years. Water uptake by plants was treated as essentially non-fractionating, so the isotope ratio that moves from soil through roots, stems, and into the xylem is assumed to match that of the source water [20]. More recent studies nonetheless show that minor isotopic offsets can occur along the soil–root–stem–leaf continuum. But these deviations are an order of magnitude smaller than the enrichment produced by direct evaporation and are therefore often neglected in practice [21]. By contrast, water evaporated from soil surfaces is strongly fractionated and depleted in heavy isotopes [22]. This "isotopic fingerprinting" may also help to understand groundwater recharge and dynamics in arid environments [23], including the impact or agricultural practices [24]. For example, Aouade et al. [15] combined δ^2H measurements of soil water, xylem sap, and canopy-layer vapor with concurrent eddy covariance fluxes over an irrigated winter wheat field near Marrakech. Their SPAC-based mass balance showed that transpiration dominated post-irrigation water losses (about 70% of total ET) and rose to nearly 80% under drier pre-irrigated soil, translating to roughly 4–5 mm d^{-1} of plant water use versus 1–2 mm d^{-1} of soil evaporation. This field scale demonstration confirms that integrating isotopic profiles with micrometeorological data can allow partitioning ET into both soil evaporation and plant transpiration, offering a pathway for improving irrigation scheduling in semi-arid agriculture.

Therefore, the present study examines the hydrological performance of two irrigation systems: surface drip irrigation (DI) and subsurface drip irrigation (SDI) in a semi-arid agricultural setting. Specifically, it compares their water fluxes within the SPAC through an isotopic analysis. Such an integrative strategy reveals not only the overall water balance but also the spatial and temporal nuances of water uptake by plants, including the depth at which roots predominantly acquire water under varying irrigation practices. In this study we aim to add information on the gap and interconnect between these previous studies conducted in the Moroccan context. The central hypotheses of this work are the following:

1. Isotopic range of the soil profile is narrower in SDI than in DI, depending on the seasonal cycle in the isotopic composition of rainwater?
2. Soil and plant water experience more evaporative, non-productive loss of water under DI than under SDI, as indicated by the isotopic enrichment in topsoil.
3. Plants under SDI take consistently water from the subsurface irrigation layer throughout the season, while under DI it takes mainly water from the topsoil.

By testing these hypotheses through a careful synthesis of isotopic data, this study seeks to elucidate the complex interactions among soil, plant, and atmospheric components under contrasting irrigation regimes. The finding of DI will help refine our understanding of water partitioning in agricultural systems, informing the development of water-saving strategies, improving irrigation scheduling, and ultimately contributing to more sustainable management of scarce water resources in regions facing escalating climatic and hydrological pressures.

2. Materials and Methods

2.1. Study Area

The study was conducted in a semi-arid olive orchard located near Marrakech, called Agdal Domain, where planted with 240-year-old olive trees, grown in an orchard of about 275 ha. The density of olive trees was about 225 trees per hectare, which corresponds to an area of about 45 m^2 occupied by each tree. The area, (Latitude 31.587491 | Longitude: −7.980781 | 489 m elevation) is characterized by flat terrain, warm temperatures, and notably scarce and erratic rainfall. The average annual temperature in this region ranges from 12 °C in winter to 30 °C in summer, with occasional peaks exceeding 40 °C during heatwaves. Rainfall in Marrakech is typically between 200 mm and 250 mm per year, occurring mainly in the winter months, with long dry periods during the summer. These conditions contribute to the region's semi-arid climate. This region's climate is further marked by moderate to strong winds, with average wind speeds ranging from 15 to 30 km/h, and high evapotranspiration rates reaching up to 1600 mm per year, depending on seasonal variations in hot and windy conditions [25]. The orchard's soils, composed of roughly 30% clay, 44% sand, and 25% silt, reflect a texture that can retain a reasonable amount of water (field capacity of about 32%) but is still vulnerable to drying (wilting point around 19%), particularly under the hot and windy conditions [26]. These soil properties were determined using the pedotransfer function described by Wösten et al. [27], providing valuable insights into how the orchard's soil stores and releases water across seasons. The water table in the region varies, but in general, groundwater is found at depths of 30 to 260 m, depending on the proximity to natural water sources. The depth of the groundwater table and its recharge rates vary seasonally, influenced by a complex interplay between natural precipitation patterns, the demands of agricultural water use, and the impact of various on-farm practices. For instance, natural rainfall contributes significantly to the replenishment of groundwater, though its spatially and temporally heterogeneous distribution makes it difficult to rely on as a sole source of recharge. Agricultural water use also plays a key role, particularly during peak irrigation periods when the demand for water increases. On-farm practices, such as fertilization, tillage, and crop rotation, further influence the efficiency of water usage and the sustainability of water resources [28].

Water for irrigation in the studied olive orchard is primarily sourced from wells, with additional reserves provided by nearby dams and seguias, creating a dual water supply system that is essential for coping with the region's fluctuating and heterogenous rainfall patterns. This combined use of different water sources is common and reflects regional efforts to adapt to the uncertainties of semi-arid climates and ensure a consistent and reliable water supply for agricultural purposes. Two irrigation systems: surface drip (DI) and subsurface drip (SDI) are used to optimize water-use efficiency for olive production under challenging climate conditions (Figure 1). In the surface drip irrigation system, pierced plastic lines deliver water to the soil surface near the base of each tree. In contrast, in the subsurface system the drip lines are buried at 30 cm depth, therefore supplying water closer to the root system, aiming at reducing direct evaporation and improving moisture retention (especially during dry spells). Both the surface (DI) and subsurface (SDI) drip irrigation systems were equipped with self-compensating drippers, each delivering an average flow rate of 2.3 L h^{-1} under a pressure of 1 bar. These systems represent both classical and improved drip irrigation methods, allowing local farmers to adapt to water scarcity. The choice between surface and subsurface drip systems is influenced by soil type and water availability. Subsurface drippers, effectively conserve water and deliver moisture to the root zone. The sampling stations were distributed over various emitter distances, soil profiles, and phenological stages to represent the spatial and temporal variability of the orchard (see Section 2.2). This approach embodies daily orchard management, when farmers continuously adjust practices in response to fluctuating weather, crop demand, and water availability. These elements create a dynamic environment for evaluating water fluxes in the soil–plant–atmosphere continuum and for identifying strategies that balance olive production with water conservation in semi-arid regions.

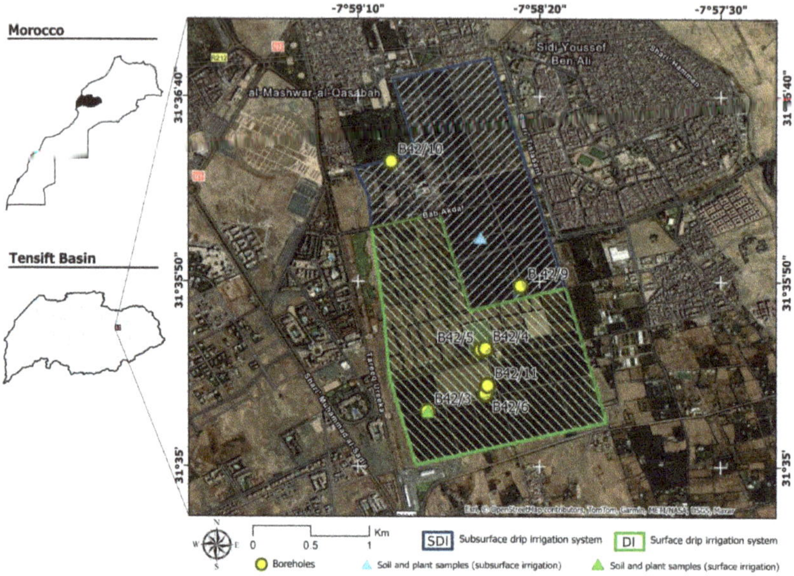

Figure 1. Location of the study area (Royal Domain of Agdal) east of Marrakech (Tensift Basin) highlighting the two experimental fields: surface drip irrigation DI (green hatching) and subsurface-drip irrigation SDI (blue hatching). Wells with groundwater access are shown as yellow circles; soil- and plant-sampling points are marked within each plot.

2.2. Sampling Protocol

The fieldwork was conducted in the Agdal olive orchards, where olive trees are irrigated using two distinct drip systems: surface (DI) and subsurface (SDI). These sites were selected to examine how irrigation methods influence water fluxes within the soil–plant–atmosphere continuum (SPAC). To capture the full range of water inputs and outputs, a comprehensive sampling strategy was implemented over a hydrological year (23 June 2022–23 June 2023).

The DI site used water from the 42/3 well, the primary source for surface irrigation, with 81 samples collected regularly to monitor variations in groundwater isotopic composition. This allowed for detailed analysis of temporal fluctuations potentially linked to irrigation schedules, rainfall, or broader hydrological processes. In contrast, the SDI site's 42/10 well was sampled only once, offering a limited but still valuable isotopic profile of the subsurface irrigation source.

Thirty precipitation samples were collected daily on rainy days using automatic water samplers, which were filled with paraffin oil to prevent evaporation before they reached the laboratory [29]. Irrigation water was sampled every other day to match its application schedule, allowing for the detection of short-term isotopic fluctuations. During this study, sixty-seven soil samples were taken monthly at four depths: 5 cm, 10 cm, 30 cm, and 50 cm using a hand auger. These depths were chosen to represent near-surface processes, as well as deeper infiltration and potential root uptake zones, with particular relevance to comparing surface and subsurface irrigation. By maintaining fixed depth increments, the sampling protocol ensured consistent and comparable snapshots of soil water content and isotope ratios throughout the year. The soil samples were collected in very thin small plastic bags, then stored in a field cooler at a cold temperature before being transferred to the refrigerator at 4 °C until the day of extraction. Monthly plant samples were also collected: 18 branches and 16 leaves of the olive trees were harvested destructively, sealed promptly, and transported under chilled conditions. In the laboratory, the plant material was likewise placed in glass tubes and stored frozen (−0 °C). For both soil and plant samples, this procedure ensured sealed storage without grinding or open-air handling, thereby minimizing evaporation and isotopic fractionation prior to extraction.

2.3. Laboratory Analysis

All soil and plant samples underwent water extraction at the National Center for Energy Sciences and Nuclear Techniques (CNESTEN) laboratory using a cryogenic vacuum distillation system, a technique widely regarded as a gold standard for preserving original isotopic signatures [30]. In this method, the samples were first ground to increase surface area, then covered with a small amount of glass wool to prevent contamination, sealed in airtight glass flasks, labeled, tightly sealed with screw caps, and frozen until processing. For extraction, the sample tubes were assembled with empty trapping tubes and then shielded for 3 min in a cold trap, cooled with liquid 2-propanol at approximately −60 °C to reduce water vapor in the gas phase during evacuation. The entire setup was connected to a high-capacity vacuum pump to lower the pressure, effectively reducing the boiling point of water and allowing it to evaporate at relatively low temperatures (90–120 °C). This careful temperature control ensures that water is liberated without breaking down organic molecules or causing significant isotopic fractionation. Afterward, the sample tube was placed on a block heater while the water collection tube remained immersed in the cold trap. Distillation was carried out at 100 °C for 2 h. At the end of the process, the distillation assembly was vented, opened, and closed to allow any ice to melt. Finally, the collected water was pipetted into 2 mL sample vials, with a 0.3 mL insert used, if necessary, for stable isotopic analysis. To ensure reliability, soil extractions were performed in duplicate, which allowed us to assess the consistency of the isotopic signatures obtained. While full triplicates were not conducted for all samples, duplicates provided sufficient reproducibility, and any variability was within the analytical precision of the isotope measurements. The reported extraction efficiency (>98%) is based on earlier validation studies of the CVD [31], and our duplicate runs confirmed that our results remained within this range of accuracy for both $\delta^{18}O$ and $\delta^{2}H$ [31].

Water samples from precipitation, irrigation, soil, and groundwater were analyzed at the UM6P (University Mohammed 6 Polytechnic) laboratory using a Picarro L2140i cavity ring-down spectrometer (CRDS) (Picarro, Santa Clara, CA, USA). The measured $\delta^{18}O$ and $\delta^{2}H$ values were expressed in per mil with respect to Vienna Standard Mean Ocean Water (VSMOW) as follows in Equation (1):

$$\delta = \left(\frac{R_{sample}}{R_{standart}} - 1\right) \times 1000‰ \quad (1)$$

where R_{sample} and $R_{standard}$ are the ratios of $^{18}O/^{16}O$ or $^{2}H/^{1}H$ in the sample and the standard (VSMOW), respectively. The precision of the measurements was ±1.0‰ for $\delta^{2}H$ and ±0.1‰ for $\delta^{18}O$. To avoid isotopic biases in plant water extracts (i.e., branch xylem and leaves), isotopic analyses were conducted at WSL Birmensdorf (Swiss Federal Institute for Forest, Snow, and Landscape Research) isotope laboratory using a thermal conversion elemental analyzer coupled to an DeltaXP isotope ratio mass spectrometer (TC/EA-IRMS, all Thermo Finnigan, Bremen, Germany). The precision of the WSL laboratory analysis is ±0.8‰ for δ^2H and ±0.1‰ for δ^{18}O. From these paired isotopic measurements, we also calculate the deuterium-excess parameter, d = δ^2H − 8·δ^{18}O. Because d-excess is sensitive to non-equilibrium processes such as evaporation, it provides a complementary line of evidence for distinguishing between source water signatures and evaporative enrichment along the soil–plant–atmosphere continuum. Together, δ^{18}O, δ^2H, and their derived d-excess constitute the core dataset that will be used to trace water movement and quantify evaporation in the two irrigation layouts examined here.

3. Results and Discussion

3.1. Description of δ^{18}O and δ^2H for Different Waters

3.1.1. Rainwater

The stable isotopic composition of the rainfall collected in the study area during the period of the sampling varies within a wide range from −13.1‰ to −2.66‰ and from −91.2‰ to −0.2‰ for δ^{18}O and δ^2H, respectively (Figure 2). The lowest isotope values are recorded during February 2022. Figure 2 shows that the rainwater samples fell above the global meteoric water line (GMWL: δ^2H = 8 × δ^{18}O + 10) [22], and around the local meteoric water line (LMWL: δ^2H = 8 × δ^{18}O + 13.7) performed by Raibi in 2006. This indicates a mixed Atlantic and Mediterranean origin of the air masses that generate the precipitation in the study site with a little evaporation. This interpretation is further supported by the d-excess values, which remain close to +13‰, consistent with moisture sources influenced by both marine and continental conditions, and more specifically pointing to a predominantly Mediterranean origin.

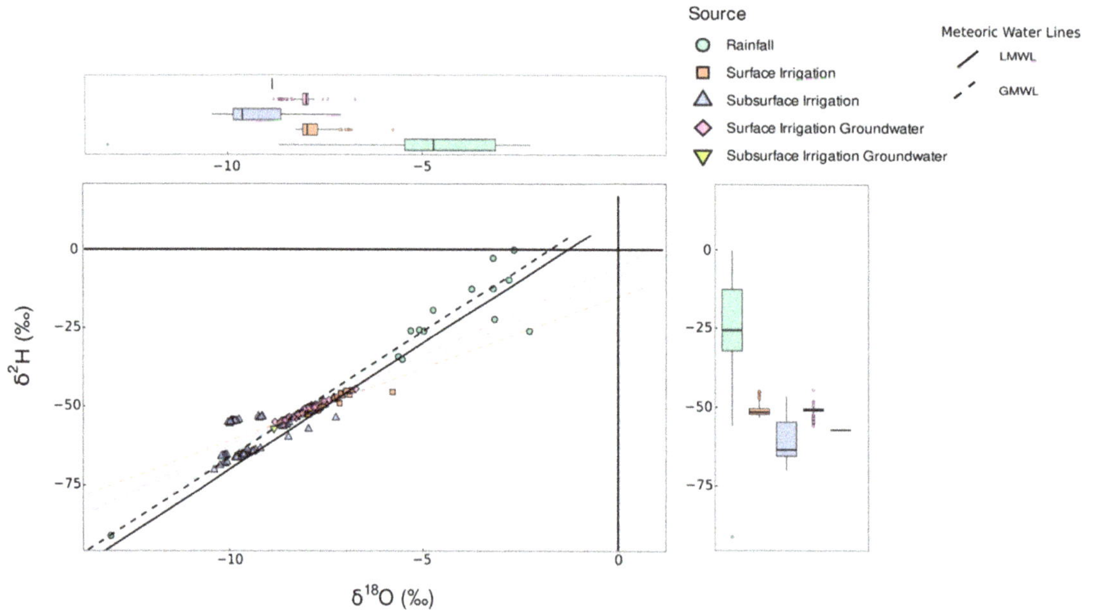

Figure 2. Dual-isotope plot (δ^{18}O vs. δ^2H) illustrating the isotopic composition of multiple hydrological compartments (rainfall, groundwater, irrigation water) relative to the global meteoric water line (GMWL) and local meteoric water line (LMWL).

3.1.2. Irrigation Water

Irrigation water samples were collected from two open storage basins, each one fed by the aforementioned wells and dedicated to one of the surface (DI) or subsurface drip (SDI) irrigated plots. The basins occasionally receive river water from a traditional canal called *Taseltant* (Figure 3), which diverts water from the Ourika river at the mountain-front.

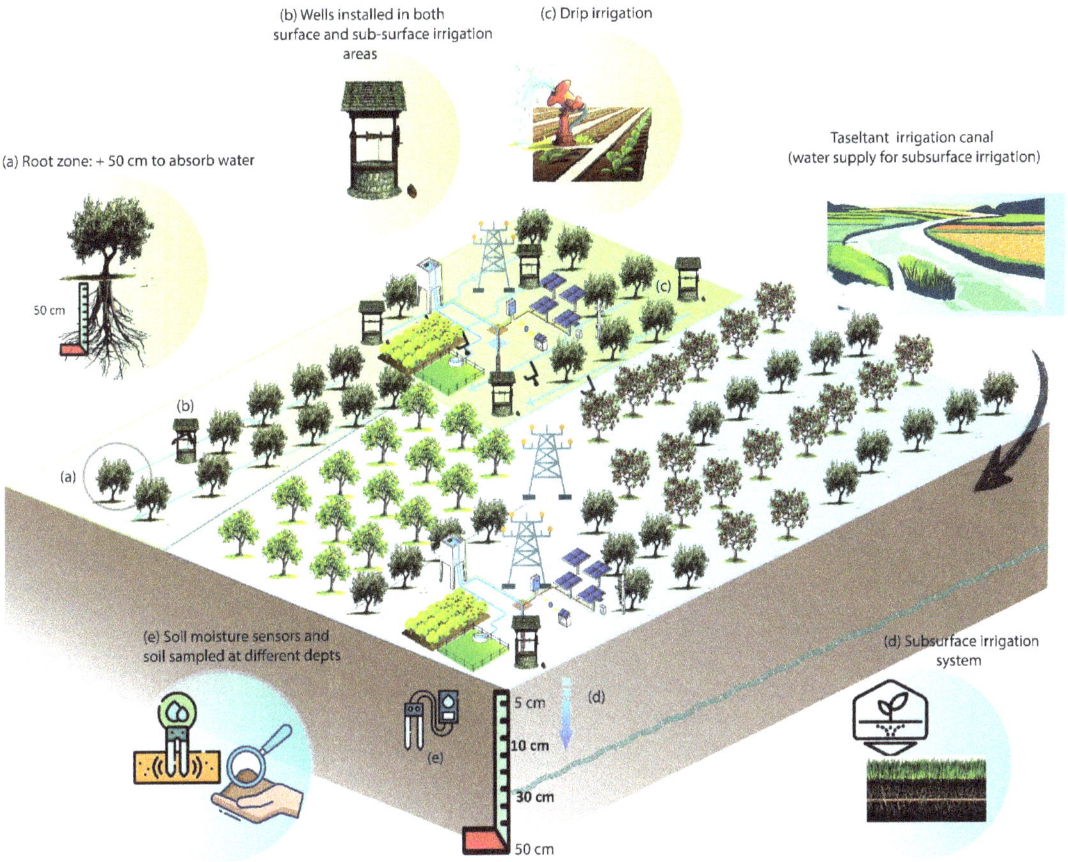

Figure 3. Schematic overview of the experimental olive orchard setup showing key components for monitoring water dynamics under surface and subsurface irrigation systems. The design includes olive trees, drip and subsurface irrigation systems, and soil isotope sampling at multiple depths.

The isotopic composition of surface irrigation water ranged from −8.23‰ to −5.76‰ and from −53.3‰ to −45.4‰ for δ^{18}O and δ^{2}H, respectively, while subsurface irrigation samples ranged from −10.41‰ to −7.07‰ and from −70.2‰ to −46.4‰, for δ^{18}O and δ^{2}H, respectively (Figure 2). The isotopic composition of irrigation water splits into two distinct groups with the subsurface irrigation group being more depleted in heavy isotopes than the surface irrigation group (Figure 4). Both groups fell between the global and local meteoric water lines with a slight evaporation effect. The surface irrigation group had a similar isotopic signature to its corresponding groundwater samples (Figure 2) with a slight evaporation effect probably caused by the residence time at the storage basin after pumping. Subsurface irrigation group is more depleted than its groundwater source (Figure 2) indicating that it may come from the other well where we only had one sample (42/10), or that may originate from the Taseltant canal (Figure 3).

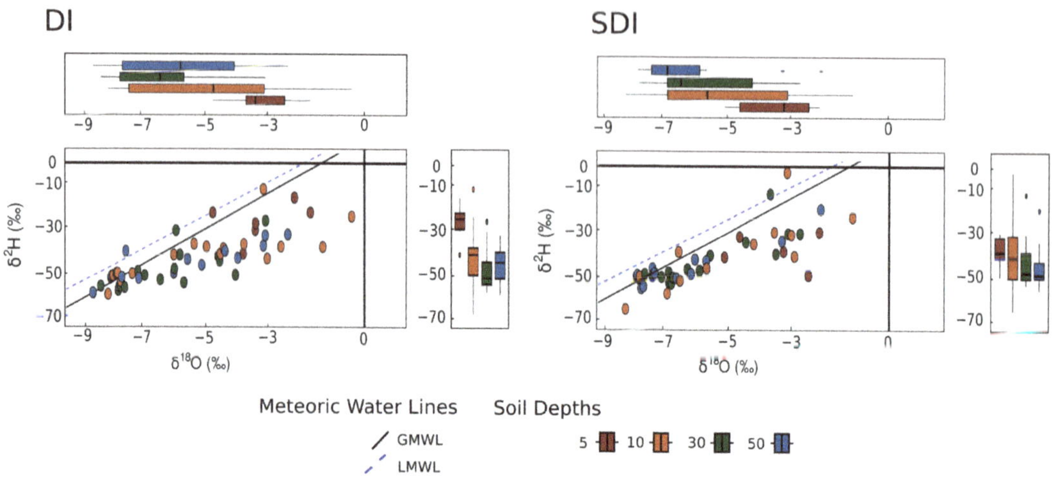

Figure 4. Dual-isotope plot ($\delta^{18}O$ vs. δ^2H) of soil water from different depths (5, 10, 30, and 50 cm) for (**DI**) surface irrigation site (on the (**left**)) and (**SDI**) subsurface irrigation site (on the (**right**)).

3.1.3. Groundwater

The isotopic composition of groundwater supplying in DI (well 42/3) ranged from −8.8‰ to −6.7‰ and from −55.1‰ to −44.5‰ for $\delta^{18}O$ and δ^2H, respectively (Figure 2), The groundwater isotope signatures are mainly scattered toward depleted values, with slight evaporation, that could be explained by recharge seasonality. For SDI, we only had one sample of the 42/10 well, with values of −57.1‰ and −8.85‰ for δ^2H and $\delta^{18}O$, respectively. This depletion could be attributed to the altitude of recharge, which controls groundwater in the Haouz aquifer because of water sourced from the High-Atlas Mountain. Indeed, the difference between the groundwater signature and the local rain signature confirms that groundwater recharge at the study site is not direct but rather occurs indirectly through high-elevation sourced water [32].

3.1.4. Soil Water

Soil samples were collected at four depths (5 cm, 10 cm, 30 cm, and 50 cm) to capture $\delta^{18}O$ and δ^2H across the soil profile, especially the shallow layers (<10 cm) where strong seasonal dynamics and evaporative effects are expected. Sampling at deeper layers (30 and 50 cm) allowed the assessment of more stable soil water pools, although collecting samples beyond 50 cm was constrained due to the use of a hand auger in the arid soil conditions. Nevertheless, significant isotopic variation was anticipated due to the strong vertical evaporation gradient typical of arid and semi-arid soils. The isotopic composition varied with both soil depth and irrigation type. Within the soil profile, $\delta^{18}O$ ranged from −8.41‰ to −1.15‰, and δ^2H from −59.8‰ to −19.6‰, in the DI (surface irrigation) site, while $\delta^{18}O$ ranged from −8.66‰ to −2.15‰ and δ^2H from −59.2‰ to −23.3‰, in SDI (subsurface irrigation). When considered separately (Figure 4), both sites display vertical isotopic depletion with increasing depth, consistent with decreasing evaporative enrichment. Average $\delta^{18}O$ values decreased from −3.39‰ at 5 cm to −6.48‰ at 50 cm across all samples, reflecting the effect of kinetic fractionation, whereby lighter isotopes preferentially evaporate, enriching the remaining soil water in heavier isotopes [33].

Despite the general trend, substantial isotopic variability between the various depths is observed in both the irrigation systems, particularly at 10 cm, where $\delta^{18}O$ ranges from less than −8‰ to more than −1‰, and δ^2H ranges from below −60‰ to above −20‰This wide spread suggests that evaporation strongly influenced the upper soil layers, likely extending beyond the surface into the top 10–20 cm, regardless of irrigation method. Previous studies have reported similar patterns, with isotopic fractionation influencing the top 50 cm of soil in Mediterranean climates [34] and up to 3 m in arid environments [35,36].

The large isotopic range is also attributable to the mixing of waters with contrasting isotopic signatures: enriched precipitation (e.g., $\delta^{18}O \approx -4.9‰$, $\delta^{2}H \approx -26.6‰$) and more depleted groundwater and surface irrigation water (e.g., $\delta^{18}O \approx -8.0‰$, $\delta^{2}H \approx -51.1‰$). Notably, the SDI treatment shows slightly more depleted values at depth compared to DI, likely due to reduced surface evaporation from buried water inputs and possibly greater reliance on deeper, less fractionated water sources.

3.1.5. Plant Water

The isotopic composition of plant water from the olive trees, including both branch xylem and leaf samples, differed between irrigation treatments and reflects the influence of evaporative processes and water source availability (Figure 5). DI Branch xylem water $\delta^{18}O$ ranged from −5.79‰ to 9.03‰ and $\delta^{2}H$ from −51.44‰ to 14.29‰ under the DI treatment, while leaf water values ranged from −4.79‰ to 13.02‰ for $\delta^{18}O$ and from −46.11‰ to 30.33‰ for $\delta^{2}H$. In contrast, the SDI treatment showed narrower, more depleted isotopic ranges with bch xylem $\delta^{18}O$ ranging from −2.76‰ to 6.12‰ and $\delta^{2}H$ from −27.33‰ to 11.68‰, and leaf water $\delta^{18}O$ from −2.16‰ to 7.44‰ and $\delta^{2}H$ from −22.92‰ to 16.59‰.

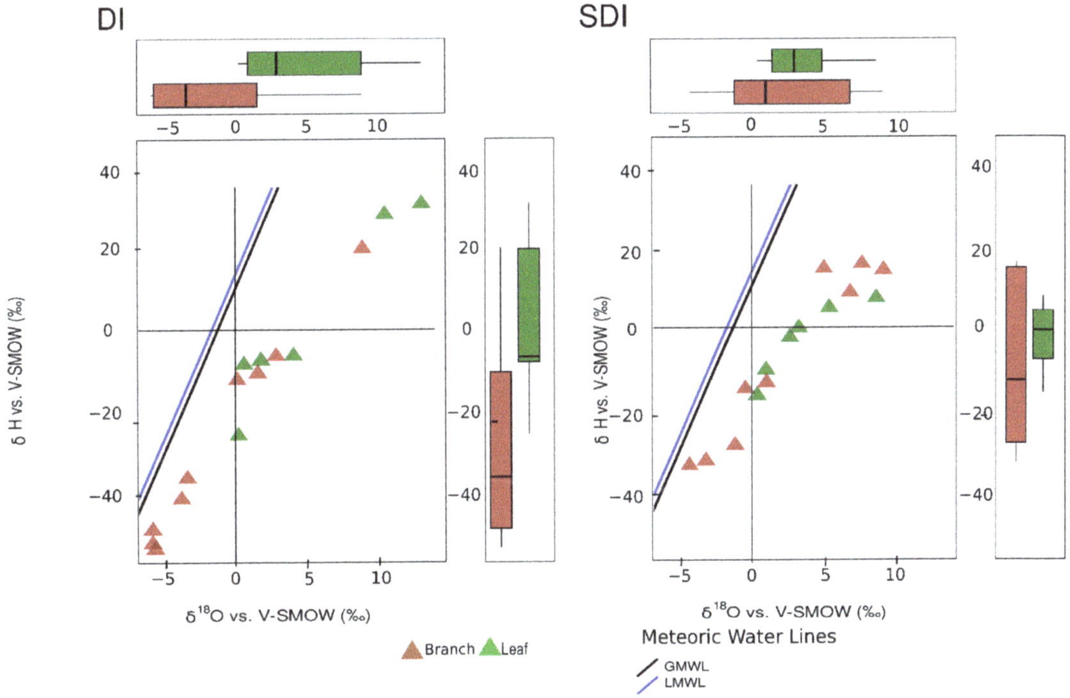

Figure 5. Dual-isotope plot ($\delta^{18}O$ vs. $\delta^{2}H$) of branch xylem and leaf water for DI (**left**) and SDI (**right**).

In both treatments, leaf water was on average more enriched in heavy isotopes than branch water, consistent with transpirative enrichment due to kinetic fractionation occurring at the leaf level. This is particularly evident in DI, where the greater isotopic range and higher enrichment suggest stronger evaporative demand and a higher degree of post-uptake fractionation. The evaporation trend of plant water intersects the LMWL both in

DI and SDI treatment, but the slope is more acute in SDI, indicating that under subsurface irrigation, plant uptake is dominated by isotopically more stable water likely sourced from deeper, less evaporated zones. This is consistent with findings by Orlowski et al. [37], who reported that irrigation method and water residence time in the soil can significantly influence isotopic source water signatures. In contrast, DI shows greater deviation from the LMWL and a wider spread in both $\delta^{18}O$ and $\delta^{2}H$, pointing to greater evaporative enrichment and likely root uptake of near-surface water affected by fractionation. The observed isotopic variability between DI and SDI reflects the physical configuration of the irrigation systems. In SDI, water is delivered at 30 cm, reducing surface exposure and evaporation, thus maintaining lower isotopic values and tighter clustering. In DI, the shallow application of irrigation water increases its exposure to atmospheric demand, enhancing isotopic enrichment before uptake. Similar effects have been documented in controlled experiments by Barbeta et al. [38], who demonstrated that soil surface evaporation can strongly influence the isotopic signature of xylem and leaf water, especially in trees accessing shallow soil water. The isotopic distinction between plant compartments (leaf vs. branch) and irrigation treatments illustrates the integrated effects of plant water uptake strategy, irrigation configuration, and soil evaporation.

3.2. Relationships Between Soil Water and Plant Water: Implications for Root Water Uptake of Olive Trees

The similarities between the isotopic composition of soil water and plant xylem can be used to infer the plant root water uptake depth, thus, assessing plant water use strategies under different environmental conditions [36]. This approach rests on two key assumptions: (i) minimal or no isotopic fractionation during water transport from roots to xylem, and (ii) a discernible isotopic gradient in the soil profile. In the DI treatment with surface emitters, the soil gradient is faint: $\delta^{18}O$ values at 10 cm and 50 cm overlap within analytical uncertainty (Figure 4), suggesting intense mixing of surface-applied water. Branch xylem nevertheless spans $-6‰$ to $0‰$ $\delta^{18}O$ and overlaps the soil field (Figure 5), so olives still rely on the first half meter; however, the blurred gradient masks a single dominant uptake depth. Leaves are again more enriched than branches, but the offset is smaller than under SDI, reflecting the lower evaporative demand during the DI sampling window.

The SDI layout, where emitters are buried at ~30 cm, tells a different story. The soil profile displays a depletion of heavy isotopes: median $\delta^{18}O$ falls from roughly $-3‰$ at 5 cm to almost $-7‰$ at 50 cm, with a parallel decline in $\delta^{2}H$ (Figure 4). Branch xylem samples plot inside the same envelope, confirming that roots tap the entire 0–50 cm layer but most actively the 5–30 cm zone around the emitters, as indicated by two autumn clusters centered on $-5.7‰$ and $-3.6‰$ $\delta^{18}O$. Leaves collected at the same times are markedly more enriched (0 to +10‰ $\delta^{18}O$, up to +30‰ $\delta^{2}H$), falling well above the local meteoric water line; this is the classic evaporative signature of water that has traveled from xylem to the leaves. A third, midsummer branch cluster averaging +0.7‰ $\delta^{18}O$ lies even further along the evaporation line than any sampled soil, implying that under peak atmospheric demand some roots shift to the highly evaporated 0–5 cm layer rather than any fractionation within the plant, a process shown to be minimal in olives [39].

But these three clusters are also shown in DI, in which we can deduct other assumptions: (i) these samples are either sourced from the soil layer less than 5 cm, subject to higher evaporation influence; (ii) have been subject to additional evaporation during transport from roots to branches; or (iii) the pattern could simply reflect an artifact introduced during sampling, transport, or extraction.

Figure 6 compresses the entire soil–plant–atmosphere continuum into a single dual-isotope view, revealing that every water pool, rainfall, irrigation, soil layers, branch xylem and leaves, lies along one shallow-slope evaporation trajectory that branches off the local meteoric water line whether on DI or SDI; the kinetic process is thus common across treatments, while the distance each pool travels along that line records how much evaporation it has endured. Rainfall and groundwater cluster tightly on the meteoric line, confirming negligible pre-infiltration enrichment; SDI water coincides with this meteoric cluster, whereas DI is already about 1–2‰ heavier in $\delta^{18}O$, nudging all subsequent DI points rightward before field evaporation even begins. Shallow DI soils (0–10 cm) extend furthest along the trajectory to nearly +5‰ $\delta^{18}O$, signaling repeated wet–dry cycles at the surface, while SDI soils, wetted at 30 cm depth, remain within +2‰ of rainfall and therefore occupy an up-gradient, less-evaporated position. Branch xylem samples from each treatment sit comfortably inside the isotopic envelope of their respective 0–50 cm soil profiles, indicating that olive roots draw a depth-integrated blend of available soil water without additional fractionation, and leaves then shift a further 3–5‰ along the same line through transpiration, with the DI canopy reaching the extreme enrichment seen in its topsoil whereas SDI leaves never attain such values. This stepped progression—rainfall → irrigation → soil → branch → leaf, forms a visible isotopic "ladder" whose rungs are widely spaced under surface drip but compressed under subsurface drip, illustrating that burying emitters shortens the evaporation path, limits non-productive water loss, and ultimately supplies the trees with isotopically cooler, more efficient water, beyond simply showing the expected enrichment of plant water, underscoring the actual sources of water used by trees under different irrigation regimes.

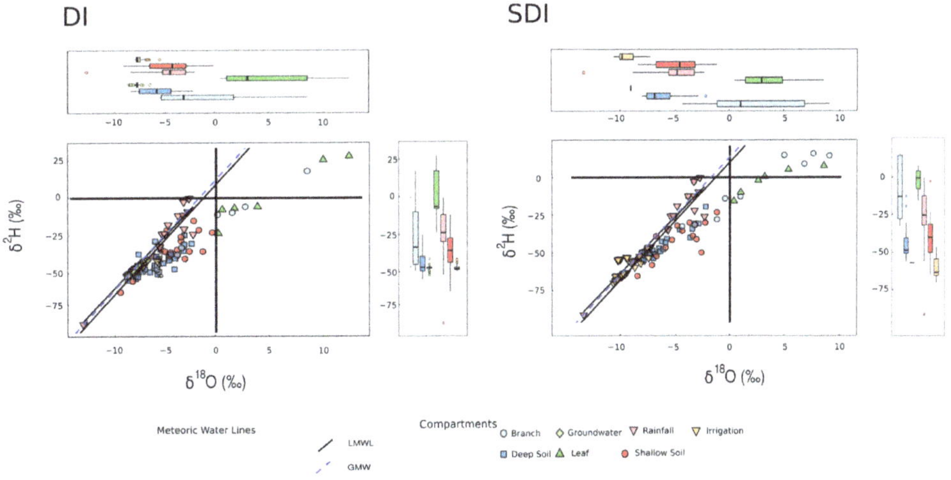

Figure 6. Dual-isotope plot ($\delta^{18}O$ vs. $\delta^{2}H$) displaying the isotopic composition of rainfall, groundwater, surface irrigation water, subsurface irrigation water, soil water (various depths), and olive plant water (leaves, branches). The black line (GMWL) and dark blue dashed line (LMWL) represent meteoric water lines, while the green line indicates the soil evaporation line.

3.3. Seasonal Variation in Plant Xylem and Soil Water

Seasonal changes in plant xylem and soil water $\delta^{18}O$ exhibit distinct patterns in DI and SDI systems (Figure 7). Winter values remain stable in both systems, reflecting infiltration of isotopically depleted precipitation [40]. Summer brings significant enrichment, particularly in DI surface soils due to evaporative fractionation, with xylem water mirroring this trend. SDI systems maintain more depleted values through subsurface water delivery [20]. The spring peak in DI plant (branches and leaves) $\delta^{18}O$ (exceeding summer values by 1–2‰) likely results from combined ecohydrological and methodological factors: (1) transient preferential uptake of evaporatively enriched shallow water during root system activation, and (2) potential analytical artifacts from cryogenic extraction that may amplify early-season enrichment [16]. Autumn shows intermediate values as trees transition to deeper sources. The more significant $\delta^{18}O$ variability in DI (15‰ range vs. 10‰ in SDI) highlights its more substantial evaporation exposure and shallower water sourcing compared to stable isotopic profile of SDI. These patterns demonstrate how irrigation methods fundamentally alter soil–plant water dynamics in semi-arid ecosystems.

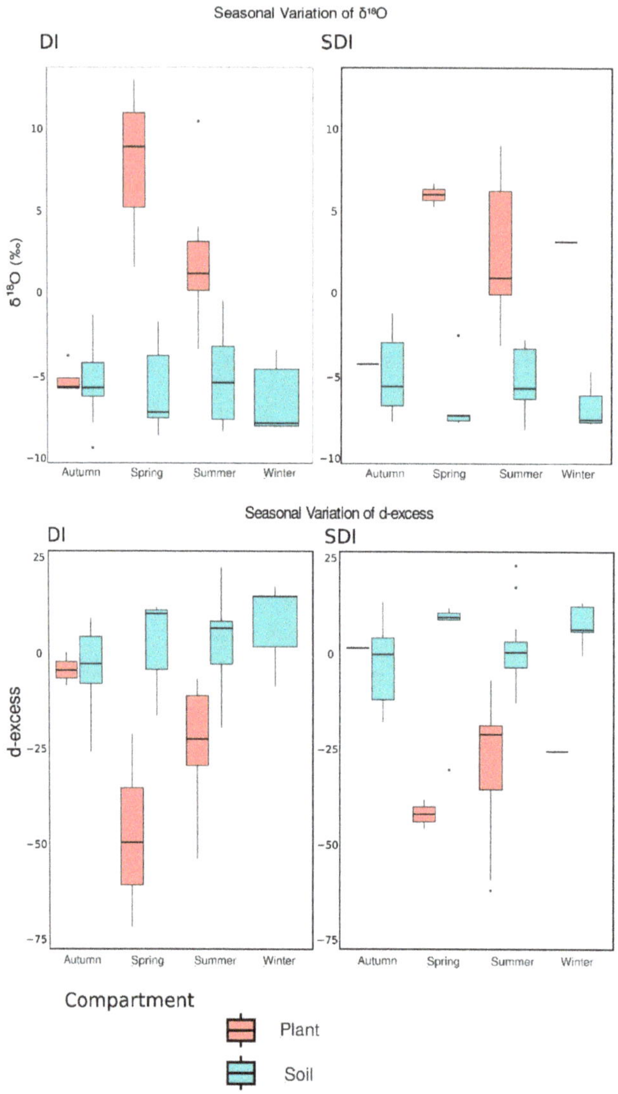

Figure 7. Seasonally grouped distributions of in $\delta^{18}O$ (‰) (**top**) and d-excess (**bottom**) values measured in plant water (red) and soil water (blue, merging all depths) in the surface (**left panels**) and subsurface (**right panels**) irrigation systems.

In Figure 7, the subsurface irrigation system (SDI) shows soil water d-excess maintaining consistently positive values throughout the year, especially during winter and spring. This stability indicates minimal evaporative influence and a predominant contribution from meteoric water inputs. During summer, a slight decline in d-excess suggests limited evaporative enrichment, underscoring the effectiveness of subsurface water delivery in mitigating evaporation. Conversely, in the surface irrigation system (DI), soil water d-excess exhibits greater seasonal variability. Moderately elevated positive values in winter and spring reflect recharge by precipitation that carries a typical Mediterranean d-excess of roughly +8 to +12‰, and infiltrates with minimal post-events evaporation, while a slight decrease during summer points to significant evaporative enrichment of surface-applied water.

The isotopic composition of branch water reflects the combined influence of source water uptake and post-uptake fractionation processes, though it is not possible to clearly distinguish between xylem and leaf water in the graph. In SDI, plant water d-excess remains relatively stable with a median of roughly −25‰ in summer, with a wide spread down to −60‰. This negative drop of d-excess means that the water has already undergone kinetic evaporation, where ^{18}O is retained preferentially over ^{2}H, either (i) before uptake in near-surface soil that partially dried, or (ii) after uptake through leaf-level fractionation that back-diffuses into the xylem. While in DI, plant water d-excess exhibits clear seasonal variation. During summer, markedly negative d-excess values point to increased reliance on shallow, evaporatively enriched soil water. The high level of enrichment, possibly influenced by leaf water signatures, reflects intensified transpiration under hot, dry conditions. These patterns are consistent with recent studies that report seasonal shifts in plant water isotopic signatures, highlighting the impact of soil enriched water in shaping water uptake and evaporative dynamics [41].

The inverse relationship between d-excess and $\delta^{18}O$ further elucidates the effects of evaporation and water source changes. During summer, increased $\delta^{18}O$ values coupled with increasingly negative d-excess in soil, branch, and leaf water under DI indicate substantial evaporative enrichment and a shift towards the use of shallower water sources. In SDI, the relatively stable $\delta^{18}O$ and d-excess values suggest that subsurface irrigation effectively buffers against seasonal isotopic shifts, maintaining access to deeper, isotopically consistent water. The stability of d-excess and $\delta^{18}O$ in SDI suggests that subsurface irrigation provides a more reliable water source, reducing evaporative stress and enhancing drought resilience. Similar stability has been documented in Alpine orchards irrigated with buried drip lines, where soil and xylem d-excess stayed within +2 to +9‰ despite summer heat waves [42]. Conversely, the pronounced seasonal isotopic shifts observed in DI mirror the large summer enrichment ($\delta^{18}O \approx +8‰$; d-excess $\approx -30‰$) found under surface-spray irrigation in European apple orchards [43], underscoring the evaporative losses inherent to surface delivery. These observations reinforce the broader European consensus that SPAC-scale isotope monitoring is a powerful tool for designing water-saving irrigation strategies [43].

3.4. Limitations and Perspective

Our research identified substantial alterations in isotopic behavior under surface (DI) and subsurface (SDI) drip irrigation in a semi-arid orchard system, notwithstanding numerous technical and conceptual challenges. The differentiation of plant and soil isotope data between the two irrigation treatments indicated that surface irrigation (DI) consistently yielded more enriched and variable $\delta^{18}O$ and $\delta^{2}H$ values in both soil and plant compartments, especially in upper soil layers and leaf water, thereby underscoring the significant impact of evaporative enrichment. Subsurface irrigation (SDI) had a more reduced and less variable isotopic signature, indicating a more stable water source from deeper, less evaporation-affected regions. Although the plot-scale isotope patterns clearly separate surface (DI) from subsurface (SDI) drip irrigation, several methodological constraints blur fine-scale interpretation. First, the point-source geometry of drip emitters promotes lateral redistribution and preferential flow; as a result, soil water collected 10 cm below an emitter can share the same $\delta^{18}O$–$\delta^{2}H$ signature as water sampled 50 cm deep even though the two parcels followed distinct flow paths [44]. Second, our fortnightly, bulk-core sampling relies on cryogenic vacuum extraction, a technique known to co-extract tightly bound pore water whose isotope composition differs from the mobile fraction accessed by roots, thereby dampening vertical gradients [37]. Third, recent syntheses show that xylem water may undergo within-plant fractionation or mix with leaf back-diffusion, meaning that "stem ≙ source" is not always valid and can further obscure depth assignment [45]. Moreover, there exists additional uncertainty due to potential isotopic alterations during sample preparation and storage. Despite the samples being chilled prior to cryogenic extraction, the interval between sampling and analysis likely resulted in significant isotopic fractionation, particularly in the drier topsoil samples characteristic of DI. Zhang et al. [46] and Amin et al. [39] emphasize that conventional cryogenic extraction methods are susceptible to bias in soils with low moisture or high organic content. Employing in situ vapor equilibration techniques or dual-method comparisons, such as cryogenic versus centrifugation procedures, may improve extraction accuracy [47,48]. To provide robust isotopic datasets, recent quality-control guidelines advocate for the assessment of extraction efficiency and reproducibility across varied samples [49]. Taken together, these factors explain the overlapping $\delta^{18}O$ and $\delta^{2}H$ ranges we observed between 10 cm and 50 cm and caution against using the present dataset to locate root water uptake zones. Future work combining in situ isotope probes, high-resolution soil-moisture sensing and tracer-pulse experiments would help disentangle these processes while retaining the robust treatment-level contrast documented here. Furthermore, isotope-based mixing models like MixSIAR or SIAR can offer valuable tools for estimating the proportions of shallow versus deep soil water contributing to plant uptake. However, the reliability of Bayesian mixing models hinges on how clearly separated the isotopic signatures of the potential sources are. In our study, there is significant overlap between the 10–50 cm soil horizons, which means it would be necessary to (i) group individual depths into broader categories representing 'shallow' and 'deep' sources, and (ii) incorporate informed priors derived from data on root distribution and soil moisture dynamics. When these conditions are met, MixSIAR has been shown to outperform simpler iso-source approaches in semi-arid cropping systems [12] and has proven effective in flooded rice fields when priors are applied [50].

In our context, these approaches may assist in ascertaining whether the observed enrichment in DI leaves signifies shallow water uptake or post-uptake fractionation, concerns particularly relevant in water-scarce agricultural settings. Prior research by Zhao et al. [51], and Rothfuss & Javaux [52] substantiates this approach by illustrating the application of plant and soil isotope signatures to model seasonal and depth-dependent root water uptake in Mediterranean and semi-arid environments where our findings confirm a broader consensus: isotope-based SPAC diagnostics must be more extensively integrated into routine agricultural monitoring to enhance irrigation efficiency and sustainability. Recent advancements in technology, particularly field-deployable laser spectroscopy devices like CRDS and OA-ICOS, enables high temporal resolution measurements of water isotopes in soil and plant compartments [45]. While these systems diminish the necessity for laboratory analysis and may detect quick variations in isotopic signals, they nevertheless demand substantial on-site infrastructure, frequent maintenance, and specialized technical knowledge. When combined with remote sensing and soil moisture data, these techniques offer potential for enhancing irrigation scheduling by delivering insights into actual plant water uptake dynamics, rather than depending exclusively on predicted soil moisture levels.

4. Conclusions and Implications

In this study, we examined the movement of irrigation water in the SPAC in olive tree plots in semi-arid conditions under two irrigation systems: surface drip (DI) and subsurface drip with drip laterals buried at 30 cm (SDI). We analyzed samples of precipitation, soil, plant, and irrigation water for stable water isotopes ($\delta^{18}O$, $\delta^{2}H$). By tracking the seasonal variation in distinct rainfall and irrigation water isotope signatures through the 0–50 cm profile and into xylem, we found that three consistent patterns emerged as follows: (i) the subsurface drip (SDI) maintained narrower and more depleted isotope ranges below approximately 20 cm, signaling deeper and more stable wetting fronts; (ii) branch water under SDI exhibited significantly less evaporative enrichment compared to that under DI, suggesting a closer connection between irrigation pulses and active root zones; (iii) under surface drip system (DI), isotope transients only propagated to about 15 cm, and branch water showed strong enrichment, indicating greater near surface evaporation and percolation losses.

Our findings underscore the diagnostic potential of stable isotopes to reveal changes in plant–soil interactions under various irrigation systems, despite the persistence of isotopic errors related to sampling depth and methodological consistency. The results showed that subsurface drip irrigation (SDI) mitigates seasonal isotopic fluctuations and limits evaporative enrichment in branch water compared to surface drip (DI), indicating a more stable and efficiently targeted root water supply in semi-arid orchards. Ultimately, this confirms SDI as a practical lever to reduce non-productive losses and to enhance irrigation to achieve deeper, better timed wetting fronts.

More broadly, our isotope approach turns "where plants actually drink" into a measurable management metric, paving the way for evidence-based irrigation design and scheduling. Future research should include routine isotope checks at additional strategic depths (1 m and beyond) during peak demand, paired with continuous soil moisture and sap flux monitoring to quantify real water savings under various irrigation regimes. Furthermore, to move beyond merely descriptive isotope patterns, isotope time series can be embedded within Bayesian mixing frameworks, such as MixSIAR, to statistically analyze depth-specific contributions to root water uptake. Concurrently, assimilating the same dataset into isotope-enabled process models such as EcH2o-iso for coupled plant–soil exchange would convert qualitative enrichment patterns into explicit estimates of seasonal evapotranspiration partitioning, deep percolation, and groundwater recharge. This study demonstrates how continuous, compartment-resolved isotope monitoring can be operationalized to verify the active rooting depths targeted by irrigation and to identify when irrigation water is lost by near surface evaporation. This experimental study advances isotope hydrology and opens the door to real time, isotope-informed irrigation control and policy tools tailored to semi-arid agriculture.

Author Contributions: T.A.: Writing—original draft, visualization, methodology, investigation, field experimentation, lab work, formal analysis, data curation, and conceptualization. M.H.K.: Writing—review and editing, visualization, supervision, methodology, investigation, formal analysis, data curation, conceptualization, and project administration. S.K.: Writing—review and editing, visualization, validation, supervision, methodology, investigation, and formal analysis. Y.A.B.: Writing—review and editing, visualization, supervision, methodology, investigation, formal analysis, data curation, and conceptualization. L.B.: Writing—review and editing, visualization, supervision, methodology, investigation, and formal analysis. V.D.: Writing—review and editing, validation, and supervision. M.M.L.: Writing—review and editing, methodology, investigation, lab work, formal analysis, and data curation. F.R.: Writing—review and editing, visualization, supervision, methodology, investigation, lab work, formal analysis, data curation, and conceptualization. T.E.: Writing—original draft, visualization, supervision, methodology, investigation, lab work, formal analysis, data curation, and conceptualization. A.E.: Writing—review and editing, visualization, investigation, field experimentation, and lab work. N.R.: Writing—review and editing, visualization, and conceptualization. H.B.: Writing—review and editing, visualization, methodology, investigation, and formal analysis. A.C.: Supervision, project administration, and funding acquisition. All authors have read and agreed to the published version of the manuscript.

Funding: This research was conducted in the frame of the ASSIWAT project (grant agreement no: AS_71) funded by the Office Chérifien des Phosphates (OCP Group S.A.). We would also like to thank the Moroccan Ministry of Higher Education, Scientific Research and Innovation and the OCP Foundation for their support through the APRD research program (GEANTech). M.M.L. is supported by the Swiss National Science Foundation (grants 213367 and 229377).

Data Availability Statement: The data presented in this study are not publicly available due to contractual and proprietary restrictions. The dataset is the property of OCP Group S.A., which funded this research, and its distribution is subject to confidentiality agreements. Access to the data may be granted upon reasonable request and with permission from OCP Group S.A.

Acknowledgments: We thank the managers of "Les domaines Agricoles" company for making their Agdal site available for this experiment and providing us with necessary data of the study site. Cryogenic vacuum distillation of soil water samples was conducted at CNESTEN (Centre National de l'Énergie, des Sciences et des Techniques Nucléaires), whose support is gratefully acknowledged. We warmly thank Tibari El ghali, Fatima Raibi, Manuela Oettli, and Matthias Saurer for their expert technical assistance in cryogenic vacuum distillation extraction of xylem and leaf samples and their subsequent isotopic analysis at the isotope laboratory of CNESTEN Rabat and WSL Birmensdorf.

Conflicts of Interest: The authors declare that this study received funding from OCP Group (AS_71). The funder was not involved in the study design, collection, analysis, interpretation of data, the writing of this article, or the decision to submit it for publication.

References

1. Matthews, J.B.R.; Möller, V.; van Diemen, R.; Fuglestvedt, J.S.; Masson-Delmotte, V.; Méndez, C.; Semenov, S.; Reisinger, A. Annex VII: Glossary. In *Climate Change 2021: The Physical Science Basis. Contribution of Working Group I to the Sixth Assessment Report of the Intergovernmental Panel on Climate Change*; Masson-Delmotte, V., Zhai, P., Pirani, A., Connors, S.L., Péan, C., Eds.; Cambridge University Press: Cambridge, UK; New York, NY, USA, 2021; pp. 2215–2256. [CrossRef]
2. Mekonnen, M.M.; Hoekstra, A.Y. Sustainability of the blue water footprint of crops. *Adv. Water Resour.* **2020**, *143*, 103679. [CrossRef]
3. Falkenmark, M.; Wang-Erlandsson, L. A water-function-based framework for understanding and governing water resilience in the Anthropocene. *One Earth* **2021**, *4*, 213–225. [CrossRef]
4. Ait Dhmane, L.; Saidi, M.E.; Moustadraf, J.; Rafik, A.; Hadri, A. Spatiotemporal characterization and hydrological impact of drought patterns in northwestern Morocco. *Front. Water* **2024**, *6*, 1463748. [CrossRef]
5. Mehta, P.; Siebert, S.; Kummu, M.; Deng, Q.; Ali, T.; Marston, L.; Xie, W.; Davis, K.F. Half of twenty-first century global irrigation expansion has been in water-stressed regions. *Nat. Water* **2024**, *2*, 254–261. [CrossRef]
6. Hamdy, A.; Ragab, R.; Scarascia-Mugnozza, E. Coping with water scarcity: Water saving and increasing water productivity. *Irrig. Drain.* **2003**, *52*, 3–20. [CrossRef]
7. Achli, S.; Ongoma, V.; Epule, T.E.; Dhiba, D.; Salih, W.; Ousayd, L.; Chehbouni, A. Exploring the key drivers of crop yields in Morocco—A systematic review. *Front. Agron.* **2025**, *7*, 1515938. [CrossRef]
8. Khabba, S.; Er-Raki, S.; Ezzahar, J.; Merlin, O.; Kharrou, M.H.; Chehbouni, A. Experimental assessment of irrigation efficiency over the dominant (flood and drip) irrigated crops in the Haouz plain, center of Morocco. *Acta Hortic.* **2025**, *1422*, 275–282. [CrossRef]
9. Attou, T.; Kuppel, S.; Kharrou, M.H.; Ezzahar, J.; Bouchaou, L.; Ait Brahim, Y.; Demarez, V.; Chehbouni, A. A process-based modelling of groundwater recharge under contrasting irrigation methods in semi-arid crops. *Agric. Water Manag.* **2025**, *316*, 109584. [CrossRef]
10. Modanesi, S.; Massari, C.; Bechtold, M.; Lievens, H.; Tarpanelli, A.; Brocca, L.; Zappa, L.; De Lannoy, G.J.M. Challenges and benefits of quantifying irrigation through the assimilation of Sentinel-1 backscatter observations into Noah-MP. *Hydrol. Earth Syst. Sci.* **2022**, *26*, 4685–4706. [CrossRef]
11. Ait Brahim, Y.; Bouchaou, L.; Sifeddine, A.; Khodri, M.; Reichert, B.; Cruz, F.W. Elucidating the climate and topographic controls on stable isotope composition of meteoric waters in Morocco, using station-based and spatially-interpolated data. *J. Hydrol.* **2016**, *543*, 305–315. [CrossRef]

12. Liu, Y.; Zhu, G.; Zhang, Z.; Sun, Z.; Yong, L.; Sang, L.; Wang, L.; Zhao, K. Isotopic differences in soil–plant–atmosphere continuum composition and control factors of different vegetation zones on the northern slope of the Qilian Mountains. *Biogeosciences* **2022**, *19*, 877–889. [CrossRef]
13. Tetzlaff, D.; Buttle, J.; Carey, S.K.; Kohn, M.J.; Laudon, H.; McNamara, J.P.; Smith, A.; Sprenger, M.; Soulsby, C. Stable isotopes of water reveal differences in plant—Soil water relationships across northern environments. *Hydrol. Process.* **2021**, *35*, e14023. [CrossRef]
14. Ceperley, N.; Gimeno, T.E.; Jacobs, S.R.; Beyer, M.; Dubbert, M.; Fischer, B.; Geris, J.; Holko, L.; Kübert, A.; le Gall, S.; et al. Toward a common methodological framework for the sampling, extraction, and isotopic analysis of water in the Critical Zone to study vegetation water use. *WIREs Water* **2024**, *11*, e1727. [CrossRef]
15. Aouade, G.; Ezzahar, J.; Amenzou, N.; Er-Raki, S.; Benkaddour, A.; Khabba, S.; Jarlan, L. Combining stable isotopes, Eddy Covariance system and meteorological measurements for partitioning evapotranspiration, of winter wheat, into soil evaporation and plant transpiration in a semi-arid region. *Agric. Water Manag.* **2016**, *177*, 181–192. [CrossRef]
16. Orlowski, N.; Breuer, L.; Angeli, N.; Boeckx, P.; Brumbt, C.; Cook, C.S.; Dubbert, M.; Dyckmans, J.; Gallagher, B.; Gralher, B.; et al. Inter-laboratory comparison of cryogenic water extraction systems for stable isotope analysis of soil water. *Hydrol. Earth Syst. Sci.* **2018**, *22*, 3619–3637. [CrossRef]
17. Wang, X.-F.; Yakir, D. Using stable isotopes of water in evapotranspiration studies. *Hydrol. Process.* **2000**, *14*, 1407–1421. [CrossRef]
18. Helliker, B.R.; Ehleringer, J.R. Differential 18O enrichment of leaf cellulose in C3 versus C4 grasses. *Funct. Plant Biol.* **2002**, *29*, 435–442. [CrossRef] [PubMed]
19. Williams, D.G.; Cable, W.; Hultine, K.; Hoedjes, J.C.B.; Yepez, E.A.; Simonneaux, V.; Er-Raki, S.; Boulet, G.; de Bruin, H.A.R.; Chehbouni, A.; et al. Evapotranspiration components determined by stable isotope, sap flow and eddy covariance techniques. *Agric. For. Meteorol.* **2004**, *125*, 241–258. [CrossRef]
20. Ehleringer, J.R.; Dawson, T.E. Water uptake by plants: Perspectives from stable isotope composition. *Plant Cell Environ.* **1992**, *15*, 1073–1082. [CrossRef]
21. Kuppel, S.; Tetzlaff, D.; Maneta, M.P.; Soulsby, C. EcH2O-iso 1.0: Water isotopes and age tracking in a process-based, distributed ecohydrological model. *Geosci. Model Dev.* **2018**, *11*, 3045–3069. [CrossRef]
22. Murad, A.; Krishnamurthy, R.V. Factors controlling the stable hydrogen and carbon isotoperatios in regional ground water of the Eastern United Arab Emirates (UAE). *Hydrol. Process.* **2007**, *22*, 1922–1931. [CrossRef]
23. Mohammed, A.M.; Krishnamurthy, R.V.; Kehew, A.E.; Crossey, L.J.; Karlstrom, K.K. Factors affecting the stable isotopes ratios in groundwater impacted by intense agricultural practices: A case study from the Nile Valley of Egypt. *Sci. Total Environ.* **2016**, *573*, 707–715. [CrossRef] [PubMed]
24. Craig, H.; Gordon, L.I.; Horibe, Y. Isotopic exchange effects in the evaporation of water: 1. Low-temperature experimental results. *J. Geophys. Res.* **1963**, *68*, 5079–5087. [CrossRef]
25. le Page, M.; Toumi, J.; Khabba, S.; Hagolle, O.; Tavernier, A.; Kharrou, M.; Er-Raki, S.; Huc, M.; Kasbani, M.; Moutamanni, A.; et al. A Life-Size and Near Real-Time Test of Irrigation Scheduling with a Sentinel-2 Like Time Series (SPOT4-Take5) in Morocco. *Remote Sens.* **2014**, *6*, 11182–11203. [CrossRef]
26. Nassah, H.; Younes, F.; Salah, E.; Said, K.; Mougenot, B. Mapping Soil Clay Content and Hydraulic Properties over an Agricultural Semiarid Plain Using Remote Sensing and Interpolation Techniques. *Ecol. Eng. Environ. Technol.* **2024**, *25*, 164–179. [CrossRef]
27. Wösten, J.H.M. Chapter 10 Pedotransfer functions to evaluate soil quality. *Dev. Soil Sci.* **1997**, *25*, 221–245. [CrossRef]
28. Zhou, L.; Zhang, Z.; Zhang, W.; An, K.; Lei, X.; He, M. Real-Time Water Level Prediction in Open Channel Water Transfer Projects Based on Time Series Similarity. *Water* **2022**, *14*, 2070. [CrossRef]
29. Williams, M.R.; Lartey, J.L.; Sanders, L.L. Isotopic (δ^{18}O and δ^{2}H) Integrity of Water Samples Collected and Stored by Automatic Samplers. *Agric. Environ. Lett.* **2018**, *3*, 180009. [CrossRef]
30. Tsuruta, K.; Yamamoto, H.; Katsuyama, M.; Kosugi, Y.; Okumura, M.; Matsuo, N. Effects of cryogenic vacuum distillation on the stable isotope ratios of soil water. *Hydrol. Res. Lett.* **2019**, *13*, 1–6. [CrossRef]
31. Wang, D.; Risi, C.; Tian, L.; Yang, D.; Bowen, G.; Fan, S.; Su, Y.; Pang, H.; Li, L. Correcting for water vapor diffusion in air bag samples for isotope composition analysis: Cases studies with drone-collected samples. *Preprints* **2024**. [CrossRef]
32. Boukhari, K.; Fakir, Y.; Stigter, T.Y.; Hajhouji, Y.; Boulet, G. Origin of recharge and salinity and their role on management issues of a large alluvial aquifer system in the semi-arid Haouz plain, Morocco. *Environ. Earth Sci.* **2015**, *73*, 6195–6212. [CrossRef]
33. Yang, Y.; Zhang, M.; Wang, S.; Qu, D.; Zhang, Y.; Wang, J. Soil moisture variability affected by sand mulch: An isotope-based assessment of irrigated farmland in Northwest China. *Ecohydrology* **2023**, *16*, 2477. [CrossRef]
34. Oshun, J.; Dietrich, W.E.; Dawson, T.E.; Fung, I. Dynamic, structured heterogeneity of water isotopes inside hillslopes. *Water Resour. Res.* **2016**, *52*, 164–189. [CrossRef]
35. Singleton, M.J.; Sonnenthal, E.L.; Conrad, M.E.; DePaolo, D.J.; Gee, G.W. Multiphase Reactive Transport Modeling of Seasonal Infiltration Events and Stable Isotope Fractionation in Unsaturated Zone Pore Water and Vapor at the Hanford Site. *Vadose Zone J.* **2004**, *3*, 775–785. [CrossRef]

36. Sprenger, M.; Leistert, H.; Gimbel, K.; Weiler, M. Illuminating hydrological processes at the soil-vegetation-atmosphere interface with water stable isotopes. *Rev. Geophys.* **2016**, *54*, 674–704. [CrossRef]
37. Orlowski, N.; Seeger, S.; Mennekes, D.; de Boer, H.; Weiler, M.; Rinderer, M. Monitoring tree species-specific water uptake strategies via continuous in-situ stable water isotope measurements. In Proceedings of the EGU General Assembly, Online, 4–8 May 2020; EGU2020-16786. [CrossRef]
38. Barbeta, A.; Gimeno, T.E.; Clavé, L.; Fréjaville, B.; Jones, S.P.; Delvigne, C.; Wingate, L.; Ogée, J. An explanation for the isotopic offset between soil and stem water in a temperate tree species. *New Phytol.* **2020**, *227*, 766–779. [CrossRef] [PubMed]
39. Amin, A.; Zuecco, G.; Marchina, C.; Engel, M.; Penna, D.; McDonnell, J.J.; Borga, M. No evidence of isotopic fractionation in olive trees (*Olea europaea*): A stable isotope tracing experiment. *Hydrol. Sci. J.* **2021**, *66*, 2415–2430. [CrossRef]
40. Dansgaard, W. Stable isotopes in precipitation. *Tellus A Dyn. Meteorol. Oceanogr.* **2012**, *16*, 436. [CrossRef]
41. Nasta, P.; Todini-Zicavo, D.; Zuecco, G.; Marchina, C.; Penna, D.; McDonnell, J.J.; Amin, A.; Allocca, C.; Marzaioli, F.; Stellato, L.; et al. Quantifying irrigation uptake in olive trees: A proof-of-concept approach combining isotope tracing and Hydrus-1D. *Hydrol. Sci. J.* **2023**, *68*, 1479–1486. [CrossRef]
42. Brighenti, S.; Tagliavini, M.; Comiti, F.; Aguzzoni, A.; Giuliani, N.; Abdelkader, A.B.; Penna, D.; Zanotelli, D. Drip irrigation frequency leads to plasticity in root water uptake by apple trees. *Agric. Water Manag.* **2024**, *298*, 108870. [CrossRef]
43. Penna, D.; Geris, J.; Hopp, L.; Scandellari, F. Water sources for root water uptake: Using stable isotopes of hydro--gen and oxygen as a research tool in agricultural and agroforestry systems. *Agric. Ecosyst. Environ.* **2020**, *291*, 106790. [CrossRef]
44. Sprenger, M.; Allen, S.T. Commentary: What ecohydrologic separation is and where we can go with it. *Water Resour. Res.* **2020**, *56*, e2020WR027238. [CrossRef]
45. Beyer, M.; Kühnhammer, K.; Dubbert, M. In situ measurements of soil and plant water isotopes: A review of approaches, practical considerations and a vision for the future. *Hydrol. Earth Syst. Sci.* **2020**, *24*, 4413–4440. [CrossRef]
46. Zhang, X.; Wang, Y.; Li, P.; Zhou, X. Improving sampling protocols for stable isotope analysis in agro-ecosystems. *J. Hydrol.* **2023**, *617*, 129031. [CrossRef]
47. Koeniger, P.; Gaj, M.; Wassenaar, L.I. Evaporation-induced oxygen-18 and deuterium enrichment in water sample storage: Field guidance for precipitation collectors. *Hydrol. Process.* **2020**, *34*, 921–925.
48. Oerter, E.J.; Siebert, G.; Bowling, D.R.; Bowen, G. Soil water vapour isotopes identify missing water source for streamside trees. *Ecohydrology* **2019**, *12*. [CrossRef]
49. Johansen, T.; Crispin, M.; White, S. Advances in cryogenic vacuum distillation for stable isotope analysis. *Anal. Methods* **2021**, *14*, 1090–1102.
50. Mahindawansha, A.; Orlowski, N.; Kraft, P.; Rothfuss, Y.; Racela, H.; Breuer, L. Quantification of plant water uptake by water stable isotopes in rice paddy systems. *Plant Soil* **2018**, *429*, 281–302. [CrossRef]
51. Zhao, L.; Zhang, J.; Liu, H.; Yang, M. Using stable isotopes to trace soil water movement in a semi-arid watershed. *Hydrol. Process.* **2020**, *34*, 103–116. [CrossRef]
52. Rothfuss, Y.; Javaux, M. Reviews and syntheses: Isotopic approaches to quantify root water uptake: A review and comparison of methods. *Biogeosciences* **2017**, *14*, 2199–2224. [CrossRef]

water

MDPI

Article

Performance Evaluation of the SRM and GRxJ—CemaNeige Models for Daily Streamflow Simulation in Two Catchments with Snow and Rain Dominated Hydrological Regimes

Bastián Rivas [1], Víctor Osores [2], David González [1], Carlo Gualtieri [3] and Santiago Yépez [1,*]

[1] Departamento Manejo de Bosques y Medio Ambiente, Facultad de Ciencias Forestales, Universidad de Concepción, Concepción 4070386, Chile; brivas2019@udec.cl (B.R.); dgonzale@udec.cl (D.G.)
[2] Departamento de Matemáticas, Físicas y Estadística, Facultad de Ciencias Básicas, Universidad Católica del Maule, Talca 3460000, Chile; vosores@ucm.cl
[3] Department of Structures for Engineering and Architecture, University of Napoli Federico II, Via Claudio 21, 80125 Napoli, Italy; carlo.gualtieri@unina.it
* Correspondence: syepez@udec.cl

Abstract

This study evaluated the Snowmelt-Runoff Model (SRM) and the Génie Rural à X Paramètres Journalier (GRxJ) model family, analyzing the latter both independently and in combination with the CemaNeige snow module. SRM and GRxJ represent snowmelt-runoff and rainfall-runoff hydrological models, respectively. Accurate streamflow estimation in snow- and rain-dominated basins is crucial for water resource management, especially in the Andes where climate variability and glacier retreat threaten long-term water availability. The analysis was conducted in two Chilean watershed basins with contrasting regimes: the snow-dominated Aconcagua and the mixed rain–snow Duqueco basins. Daily data (2012–2020) of precipitation, temperature, evapotranspiration, snow cover (MODIS), and streamflow were used. Models were calibrated and validated with optimization algorithms and evaluated using NSE, $RMSE$, R^2, $PBIAS$, KGE, MAE, $\log NSE$ and $APFB$. The results show that SRM effectively reproduces variability and, in the case of the rain–snow regime basin, extreme events, with NSE ranging from 0.70 to 0.78 (Aconcagua) and 0.93 to 0.94 (Duqueco). Model selection should take into account the dominant hydrological processes. In this study, SRM showed the best performance in both analyzed catchments, although with limitations in reproducing extreme streamflow events. In contrast, the GRxJ models did not adequately capture the hydrological dynamics of the snow-dominated Aconcagua catchment. However, their performance improved considerably when applied to the mixed regime of the Duqueco River. These findings highlight the importance of adapting modeling strategies to local hydrological conditions and limited data availability, offering practical guidance for water management and climate change adaptation in Andean catchments.

Keywords: snowmelt-runoff model (SRM); GRxJ; rainfall-runoff model; CemaNeige; snow hydrology; streamflow simulation; Andes

1. Introduction

Accurate estimation of streamflow in river basins is essential for effective water resource management, supporting applications such as irrigation planning, hydropower generation, and reliable water supply for domestic and industrial users [1–5]. Achieving high precision, however, remains challenging due to multiple factors, including the increasing variability of precipitation driven by climate change and the complex interactions between surface water and groundwater systems [6,7].

Moreover, the dynamics of stormflow generation are significantly influenced by antecedent soil moisture, precipitation magnitude, and watershed hydrological connectivity, factors that complicate hydrological modeling [8]. Likewise, abrupt landscape disturbances, such as seismic events, can alter vegetation and geomorphology, modifying runoff patterns and increasing the challenge of accurately predicting risks associated with extreme events [9].

Globally, the growing pressure on freshwater resources is increasingly recognized as a critical threat to human well-being and long-term sustainability [10,11]. Over the past two decades, population growth, intensified human activities, and climate variability have exacerbated tensions between water supply and demand [12]. Mountainous catchments are particularly strategic in this context, as meltwater from glaciers provides essential water for downstream populations. Nevertheless, ongoing glacier retreat compromises this natural storage, resulting in net water losses and transient increases in runoff during the ablation season [13,14].

Hydrological models have become essential for simulating future streamflow and assessing the impacts of climate and land-use changes on river catchments [15,16]. Their importance is especially evident in data-scarce environments where understanding of hydrological processes is limited [17]. Such models simulate catchment responses under diverse climatic scenarios, supporting predictions of runoff generation [18]. Modeling approaches differ in their spatial representation, ranging from lumped to semi-distributed and fully distributed, and in conceptual complexity, from conceptual to physically based formulations. While complex models allow detailed process representation, they require extensive datasets and are subject to uncertainties related to equifinality, model structure, and future climate projections [19–21].

Among the most widely used approaches, the Snowmelt-Runoff Model (SRM) is a degree-day model developed to simulate daily streamflow in glacierized mountain catchments [22]. SRM has been extensively used to assess climate change impacts on streamflow in mountainous watersheds, including the Upper Rio Grande Basin [23,24].

In parallel, Pushpalatha et al. [25] introduced the GR6J model, a six-parameter lumped conceptual model for daily rainfall-runoff simulation, designed to improve low-flow performance relative to GR4J and GR5J. The later integration of the CemaNeige module enabled an explicit representation of snow accumulation and melt, enhancing the applicability of the model in snow-dominated basins [26]. These models have been successfully validated across more than 380 catchments with varying climatic and altitudinal characteristics [27].

In South America, recent applications have demonstrated the effectiveness of these models in mountainous environments. For instance, Escanilla-Minchel et al. [28] applied SRM+G to Chilean Andean catchments, achieving performance metrics exceeding 0.8 and a 7% annual difference between observed and simulated discharge. Similarly, Flores et al. [29] evaluated GR4J, GR5J, and GR6J in forested catchments in south–central Chile, emphasizing the influence of evapotranspiration formulations and the consistent performance of GR6J under diverse land-cover conditions.

Despite numerous studies applying hydrological models such as SRM and the GRxJ family across different regions, no comparison to date has evaluated the performance of SRM and GRxJ (with and without the CemaNeige snow module) in reproducing streamflow in basins with hydrological regimes as contrasting as those commonly observed in Chile. Addressing this gap is essential for understanding the suitability of conceptual models in Chilean catchments that exhibit distinct rainfall- and snow-dominated dynamics. Therefore, the objective of this study is to assess and compare the performance of the SRM model and the GRxJ family (GR4J, GR5J, GR6J), the latter with and without the CemaNeige snow module, in two representative basins of central Chile: (1) the snow-dominated Aconcagua basin and (2) the mixed rain–snow Duqueco basin. Using traditional statistical performance metrics during both calibration and validation periods, this study aims to identify which model performs best under each of the hydrological regimes considered. The results provide valuable insights for water resource management and climate change adaptation in Andean catchments with limited data availability.

2. Materials and Methods

2.1. Study Area

The Aconcagua River Basin, located between 32.3° and 33° S and approximately 50 km north of Santiago, Chile, extends 215 km from the Andes to the Pacific Ocean, with an elevation difference exceeding 6100 m. Covering an area of 7333 km^2, this arid basin is crucial to the national economy, contributing about 12% of Chile's agricultural, livestock, and forestry production. Large-scale irrigated agriculture in the basin relies heavily on seasonal meltwater from Andean snowpacks and glaciers [30].

Within the Valparaíso Region, the Aconcagua sub-basin exhibits a nival hydrological regime, characterized by snow accumulation in winter and runoff generated by snowmelt during spring and summer [31]. It covers 2109 km^2, with a perimeter of 334 km, a mean elevation of 3270 m a.s.l., and a maximum altitude of 5923 m a.s.l. Its flow is sustained by a network of high-Andean tributaries, including the Juncal, Blanco, Colorado, Riecillos, and Juncalillo rivers, as well as smaller streams such as Vilcuya, Los Chacates, El Maitén, Del Bolsillo, De Los Leones, and Navarro.

The hydrometric station "Aconcagua en Chacabuquito", located at 950 m a.s.l. in the upper Andean sector of the sub-basin, monitors runoff from headwater tributaries before water is diverted for irrigation in the valley. Historical data (1950–2000) indicate an average annual discharge of 387 m^3/s and a mean monthly discharge of 32 m^3/s [32]. The station's location is presented in Figure 1.

In contrast, the Duqueco River, located in the upper Biobío Basin, is classified as a mountain river [33], with a pluvio-nival regime governed by both winter precipitation and spring snowmelt. The river basin covers an area of 1433 km^2, with a perimeter of 274 km, a mean elevation of 541 m a.s.l., and a maximum altitude of 3531 m a.s.l. Within this system, the Cerrillos sub-basin is particularly important, and streamflow and water quality are monitored at the "Cerrillos" station, situated in the mid-to-upper sub-basin (Figure 1). The main tributaries of this river are the Canicura River, the Dimilhue River, the Arilahuen River, and the Coreo River. This station provides essential information on seasonal hydrodynamics. According to Chile's national water balance [34], the area is characterized by isothermal conditions ranging from 4 to 15°C and annual precipitation between 1200 and 4000 mm, largely driven by Andean orographic effects.

2.2. Hydrometeorological Data

This study used daily records of precipitation, mean air temperature, and mean daily discharge for the period from 1 June 2012, to 29 April 2020. The data were obtained from the Aconcagua River at Chacabuquito station, operated by the Dirección General de Aguas (DGA), identified by the BNA code 05410002-7. The station is located in the middle Aconcagua River sub-basin, between the Colorado River and the Seco Stream. It is situated in the Valparaíso Region, Los Andes Province and Municipality, at an elevation of 950 m a.s.l.

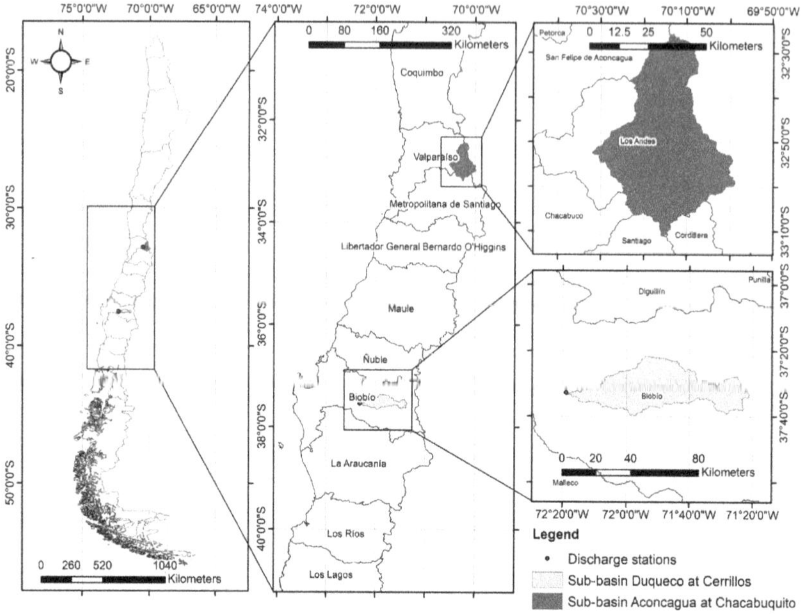

Figure 1. Geographical location of the Aconcagua and Duqueco catchments under study in central Chile.

Similarly, data for the Duqueco River were obtained from the Cerrillos station (BNA code 08323001-0), operated by the DGA. This station is located in the upper Biobío River basin, within the Biobío Region (Biobío Province, Los Ángeles Municipality), at an elevation of 129 m a.s.l.

Additionally, average daily potential evapotranspiration (PET) was used as a climatic variable for the GRxJ models, with and without the CemaNeige module analysis. These data were obtained from the CAMELS-CL (C) dataset (Catchment Attributes and Meteorology for Large-sample Studies—Chile), developed by Alvarez-Garreton et al. [35]. This dataset provides physiographic and climatic catchment attributes, along with daily time series of meteorological and hydrological variables for watersheds across the country.

In addition, four alternative models for estimating potential evapotranspiration were evaluated: Oudin et al. [36], Hargreaves and Samani [37], Priestley and Taylor [38] and Priestley-Taylor modified in Lhomme [39].

The Oudin (E_o) (Equations (1a) and (1b)):

$$E_o = \frac{R_e}{\lambda \rho} + \frac{t+5}{100}, \qquad t+5 > 0, \tag{1a}$$

$$E_o = 0, \qquad t+5 \leq 0. \tag{1b}$$

where E_0 Oudin's model estimates for potential evapotranspiration $(\mathrm{mm \cdot day^{-1}})$; R_e extraterrestrial radiation $\left(\mathrm{Mj \cdot m^2 \cdot day^{-1}}\right)$; t temperature (°C); λ latent heat flux $\left(\mathrm{Mj \cdot kg^{-1}}\right)$; ρ water density $\left(\mathrm{kg \cdot m^{-3}}\right)$.

The Hargraves-Samani (E_h) (Equation (2)):

$$E_h = 0.0135(t + 17.78)R_s, \tag{2}$$

where E_h Hargreaves' model estimates for potential evapotranspiration $\left(mm \cdot day^{-1}\right)$; R_s incident radiation $\left(Mj \cdot m^2 \cdot day^{-1}\right)$; t temperature (°C).

The Priestley-Taylor (E_p) (3a) and modified (3b):

$$E_p = \frac{\Delta}{\Delta + Y}(R_n - G), \quad (3a)$$

$$E_{pa} = \alpha \frac{\Delta}{\Delta + Y}(R_n - G). \quad (3b)$$

where E_p represents the equilibrium evaporation rate ($mm \cdot day^{-1}$), assuming no aerodynamic transfer; Δ is slope of the saturated steam heat curve $\left(Pa \cdot {}^{\circ}C^{-1}\right)$; Y psychometric constant ($Pa \cdot {}^{\circ}C^{-1}$); the equilibrium rate of actual evaporation is modified to give Priestley-Taylor PET as $E_{pa} = \alpha \cdot E_p$ [39]. The parameter "α" is associated with vegetation cover and reflects the ratio between actual evapotranspiration and the limiting evapotranspiration observed in the study area [38], α coefficient related to vegetation land cover.

2.3. Satellite Data

MODIS

In this study, snow cover area (SCA) was estimated to be using the MOD09A1.061 Terra Surface Reflectance product from the MODIS sensor, which provides 8-day composite surface reflectance data at a spatial resolution of 500 m. This product offers atmospherically corrected reflectance for MODIS bands 1 to 7 and is widely recognized for its reliability in snow monitoring applications [40]. Pixels included in this dataset are selected based on rigorous quality criteria, such as high observation frequency, low sensor viewing angles, minimal cloud or shadow contamination, and low aerosol presence, ensuring the consistency and accuracy of the input data, thus generating a satellite image composed of the best-quality, cloud-free pixels selected over the 8-day period.

The dataset was accessed and processed using Google Earth Engine (GEE), where the Normalized Difference Snow Index (NDSI) was applied to delineate snow and glacier-covered areas, following the methodology outlined by Riggs et al. [41]. The NDSI is calculated from the normalized difference between the green band green band (band 4) and the shortwave infrared (SWIR, band 6), using a threshold of 0.4, which is commonly applied to delineate snow and ice. This threshold has been shown to be adequate for satellite products with a spatial resolution of 500 m or more, reducing uncertainties in snow cover mapping [42].

The processing in GEE consisted of calculating the NDSI for each 8-day composite, applying the standard threshold of 0.4 to identify snow. Using this threshold, binary rasters were generated to represent only the snow-covered areas within the study basins. Subsequently, snow-covered area was quantified for each defined altitudinal band. Finally, the 8-day composite series was interpolated to produce a daily SCA time series.

2.4. Hydrological Models

2.4.1. SRM (Snowmelt-Runoff Model)

In [43] the authors developed the SRM model to calculate daily runoff mainly based on snow cover area, precipitation, and temperature. Equation (4) was used to estimate snowmelt-induced runoff in the study area. The model relies on daily data of snow cover, precipitation, and air temperature. It also incorporates parameters such as the runoff coefficient, degree-day factor, temperature lapse rate, and recession constant. Since runoff processes differ between snowmelt and rainfall, the runoff coefficient is separated into two values: C_S for snow and C_R for rain.

$$Q_{n+1} = [c_{Sn} \cdot a_n(T_n + \Delta T_n)S_n + c_{Rn}P_n](A \cdot 10000)/86400(1 - k_{n+1}) + Q_n k_{n+1}, \quad (4)$$

where Q mean daily discharge (in m^3/s), C_S snowmelt-runoff coefficient, C_R rainfallrunoff coefficient, a degree-day factor (in $cm \cdot {}^{\circ}C^{-1} \cdot d^{-1}$), which indicates the snowmelt depth due to 1 degree day, T the number of degree-days (in °Cd), ΔT temperature adjustment based on the temperature gradient from the measurement station to the basin average hypsometric elevation (in °Cd) and is given by $\Delta T = y \cdot (h_{st} - \bar{h}) \cdot \frac{1}{100}$ where y is the temperature lapse rate (°C per 100 m), h_{st} is the elevation at which the measurement station is located (m.s.n.m.) and $\bar{h}$ is the mean hypsometric elevation of the area (m.s.n.m.). S the ratio of the snow-covered area to the total area. P precipitation contributing to runoff (in cm), if the catchment is divided into zones, precipitation P increases by 3% for every 100 m of elevation. The parameter T_{crit} determines whether this precipitation is rainfall and turns into runoff immediately or it is snow and should be turned into runoff with a delay (once the melting condition is right), A the area of the basin (in km^2), k recession coefficient, which indicates how much the discharge declines in the absence of rainfall or snowmelt and is given by $k = Q_{n+1}/Q_n$ where n and $n+1$ are two consecutive days in a recession period. n the number of days in discharge calculations.

Since the temperature decreases with the increase in altitude, the probability of snowfall increases. If the basin elevation range is greater than 500 m, it is best to divide it into multiple zones, each with an elevation range of 500 m [44], for example, see Figure 2.

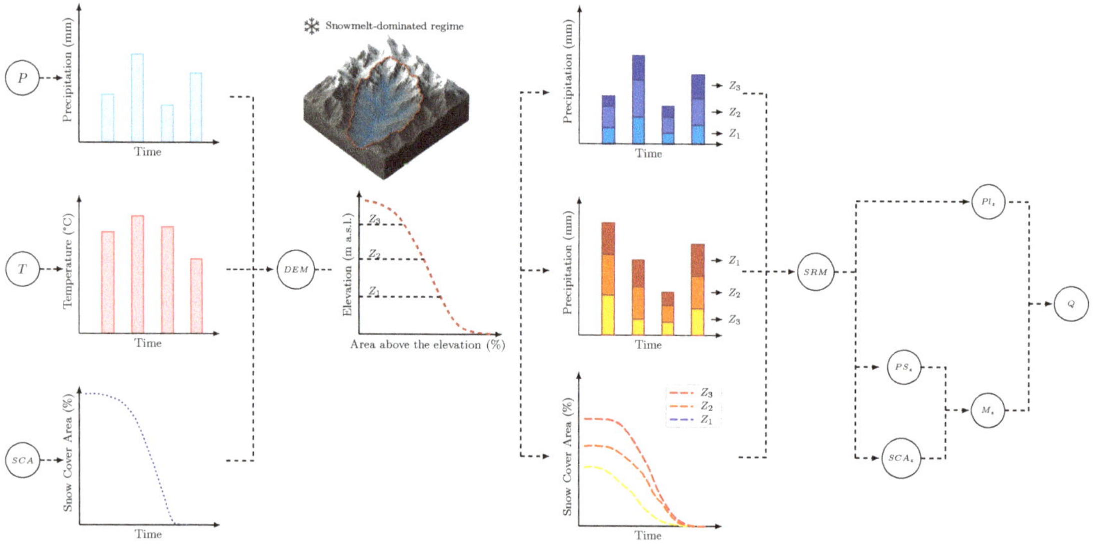

Figure 2. Conceptual diagram of the SRM model, where P is precipitation, T is temperature, SCA is snow cover area, DEM is the digital elevation model, and Z_1, Z_2, Z_3 represent elevation-based zones. PS_z is solid precipitation by zone, Pl_z is liquid precipitation by zone, SCA_z is snow cover area by zone, M_z is snowmelt by zone, and Q is the simulated streamflow.

2.4.2. GRxJ Models (GR4J, GR5J, GR6J) and the CemaNeige Module

The GR4J model is a lumped, conceptual rainfall-runoff model designed to simulate daily streamflow using four parameters. It requires daily inputs of precipitation and potential evapotranspiration (both in mm), along with observed streamflow data for model calibration. Over time, this model has been extended into more complex versions: GR5J [45], which introduces a fifth parameter, and GR6J [25], which adds a sixth parameter. The parameters of these models are described in Figure 3. These models belong to the family of soil moisture accounting models [46].

To simulate snow accumulation and melt processes, the CemaNeige snow accounting module (SAR) was implemented. This routine is based on a degree-day approach and incorporates the thermal inertia of the snowpack. It uses two parameters: weighting coefficient for snow pack thermal state (θ_{G1}) and degree-day melt coefficient (θ_{G2}). Although it uses lumped daily input data, CemaNeige allows for enhanced altitudinal representation by dividing the catchment into elevation bands of equal area based on the hypsometric curve. Meteorological inputs for each elevation zone are interpolated from basin-averaged values using temperature and precipitation lapse rates, following the methodology of Valéry et al. [27]. The implementation of CemaNeige and GR6J is shown in Figure 3.

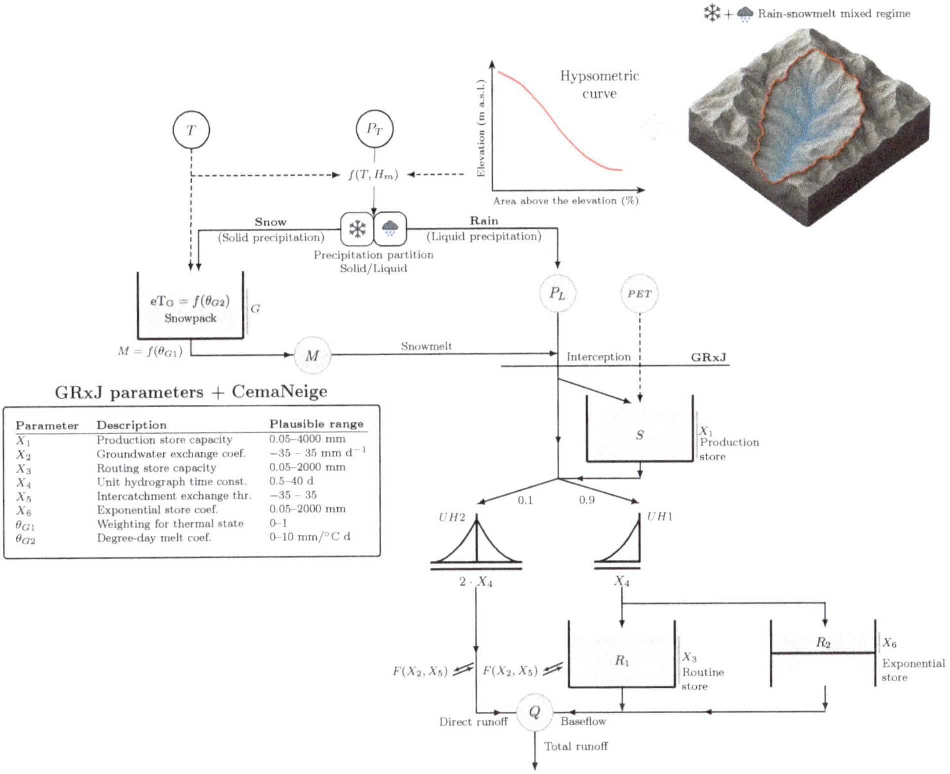

Parameter	Description	Plausible range
X_1	Production store capacity	0.05–4000 mm
X_2	Groundwater exchange coef.	−35 – 35 mm d^{-1}
X_3	Routing store capacity	0.05–2000 mm
X_4	Unit hydrograph time const.	0.5–40 d
X_5	Intercatchment exchange thr.	−35 – 35
X_6	Exponential store coef.	0.05–2000 mm
θ_{G1}	Weighting for thermal state	0–1
θ_{G2}	Degree-day melt coef.	0–10 mm/°C d

Figure 3. Diagram of the GR6J hydrological model and the snow module, based on [47]. T is temperature, P_T is total precipitation, H_m is a characteristic elevation from the hypsometric curve, eT_G is the cold-content of the snowpack, M is snowmelt, P_L is liquid precipitation, PET is potential evapotranspiration, $UH1$ and $UH2$ are unit hydrographs, and Q is total runoff.

2.5. Calibration Strategy

The calibration and validation of the hydrological models were carried out by comparing observed streamflow with the streamflow simulated by each model. The calibration period spanned from 1 January 2012, to 31 May 2016, while the validation period extended from 1 June 2016, to 29 April 2020. These periods were selected to include four complete hydrological cycles for both calibration and validation. The Nash-Sutcliffe Efficiency (NSE) was used as the objective function [48].

For the calibration of the Snowmelt-Runoff Model (SRM) in the Aconcagua River at Chacabuquito, a genetic algorithm-based optimization approach was implemented using the GA package in R [49]. This method relies on a stochastic optimization process inspired by the principles of natural selection, in which a population of candidate solutions evolves over successive generations with the aim of maximizing an objective function. The calibration involved the optimization of seven parameters: five of them (a, C_r, C_s, T_{crit}, and Lapse Rate [°C/100 m]) were defined as spatiotemporal variables (monthly and by elevation zone), resulting in a high-dimensional parameter space that includes 12 months and 9 elevation bands. Because this complexity could potentially increase the risk of overfitting, all parameters were constrained within physically plausible ranges reported in the literature and specifically aligned with the bounds proposed by Bhagwat et al. [50]. Additionally, model performance was evaluated using an independent validation period of equal length to the calibration period, which allowed verification of the temporal stability of the optimized parameters and ensured that the model did not fit only the specific characteristics of the calibration period The remaining two parameters (X and Y) were treated as fixed.

In the case of the Duqueco River at Cerrillos, a spatial-only optimization approach was applied to the same five parameters, with X and Y again held constant.

In the other case, the GRxJ models, with and without the CemaNeige module, were calibrated using the Mitchell optimization algorithm implemented in the airGR package (version 1.7.6) [51]. These models have a smaller number of parameters, all of which are aggregated in space and time, naturally reducing the risk of overfitting. In addition, the airGR calibration procedure combines an initial global search with a local refinement, allowing identification of parameter sets that maximize model performance without requiring additional measures like those applied in SRM.

2.6. Model Efficiency Indicators

To evaluate the performance of the simulations, commonly used metrics in hydrological modeling were applied, the Nash-Sutcliffe Efficiency (NSE), the Root Mean Square Error ($RMSE$), the Coefficient of Determination (R^2), the Percent Bias ($PBIAS$), the Kling-Gupta Efficiency (KGE), Mean Absolute Error (MAE), the logarithmic version of the NSE (log NSE) and The Annual Peak Flow Bias ($APFB$). The Nash-Sutcliffe efficiency (NSE), proposed by Nash and Sutcliffe [48] is a normalized statistic that estimates the relative magnitude of the residual variance compared to the measured data variance (Equation (5)) [52].

$$NSE = 1 - \frac{\sum_{i=1}^{n}(Q_i - Q'_i)^2}{\sum_{i=1}^{n}(Q_i - \bar{Q})^2}, \tag{5}$$

where Q_i is the observed daily discharge recorded by the station $(\mathrm{m^3/s})$, Q'_i is the simulated daily discharge ($\mathrm{m^3/s}$), $\bar{Q}$ is the mean observed discharge ($\mathrm{m^3/s}$) and n is the number of daily records.

The root mean square error ($RMSE$) is estimated through the square root of the difference between the observed and simulated values over the total number of data (Equation (6)) [52].

$$RMSE = \sqrt{\frac{\sum_{i=1}^{n}(Q_i - Q'_i)^2}{n}}. \tag{6}$$

The coefficient of determination (R^2) is defined as the squared Pearson's correlation coefficient (Equation (7)) [53].

$$R^2 = \left[\frac{\sum_{i=1}^{n} (Q_i - \bar{Q})(Q'_i - \bar{Q}')}{\sqrt{\sum_{i=1}^{n} (Q_i - \bar{Q})^2} \cdot \sqrt{\sum_{i=1}^{n} (Q'_i - \bar{Q}')^2}} \right]^2, \tag{7}$$

where $\bar{Q}'$ is the simulated average discharge (m^3/s).

The Percent Bias ($PBIAS$) assesses the average tendency of the simulated values to over or under estimate the observations (Equation (8)) [54].

$$PBIAS = \frac{\sum_{i=1}^{n} (Q_i - Q'_i) \cdot 100}{\sum_{i=1}^{n} Q_i}. \tag{8}$$

The Kling-Gupta Efficiency (KGE) is an improvement on the Nash-Sutcliffe Efficiency proposed by Gupta et al. [55],in which the correlation, deviation and variability components are weighted equally, solving systematic problems of maximum value underestimation [56].

$$KGE = 1 - \sqrt{(1-\alpha)^2 + (1-\beta)^2 + (1-p)^2}, \tag{9a}$$

$$\alpha = \frac{\sigma_{\text{obs}}}{\sigma_{\text{sim}}}, \qquad \beta = \frac{\mu_{\text{obs}}}{\mu_{\text{sim}}}. \tag{9b}$$

where σ_{obs} is the standard deviation of the observed daily discharge (m^3/s), σ_{sim} is the standard deviation of the simulated daily discharge (m^3/s), μ_{obs} is the mean of the observed daily discharge (m^3/s), μ_{sim} is the mean of the simulated daily discharge (m^3/s) and p is the Pearson correlation coefficient.

The Mean Absolute Error (MAE) is used in determining the global goodness of fit of simulated error (the difference between the observed data and the model predicted output). MAE values of 0 indicate a perfect fit.

$$MAE = \frac{\sum_{i=1}^{n} |Q_i - Q'_i|}{n}. \tag{10}$$

The logarithmic version of the NSE ($\log NSE$) was proposed by Oudin et al. [57] to increase the sensitivity of model evaluation to low-flow simulations. Unlike the conventional NSE, which tends to be dominated by high flows, the $\log NSE$ applies a logarithmic transformation to both observed and simulated discharges, giving greater weight to low flows. This makes it particularly useful when low-flow accuracy is important in hydrological applications.

$$\log NSE = 1 - \frac{\sum_{i=1}^{n} (\log(Q_i) - \log(Q'_i))^2}{\sum_{i=1}^{n} (\log(Q_i) - \log(\bar{Q}))^2}. \tag{11}$$

The Annual Peak Flow Bias ($APFB$) is an application specific metric that focuses on the annual peak flows. It measures the percent bias between the simulated and observed annual maximum flows (Mizukami et al. [58]).

$$APFB = \sqrt{\left(\frac{\mu_{peakQ_S}}{\mu_{peakQ_O}} - 1 \right)^2}, \tag{12}$$

where μ_{peakQ_S} is the mean of simulated annual peak flow series and is the mean of observed annual peak low series.

3. Results

3.1. Snowmelt-Runoff Model

Figures 4 and 5 compare simulated and observed streamflow. The model captures the overall trend across years; however, it does not adequately reproduce short-term variability, particularly abrupt changes. During the calibration period, streamflow tends to be overestimated, especially at low flows, while in the validation period a general underestimation is observed (Figure 5), with simulated values clustering around and below the 1:1 line. This bias is most evident in the final year of the record. In addition, the model consistently fails to reproduce extreme events throughout the simulation.

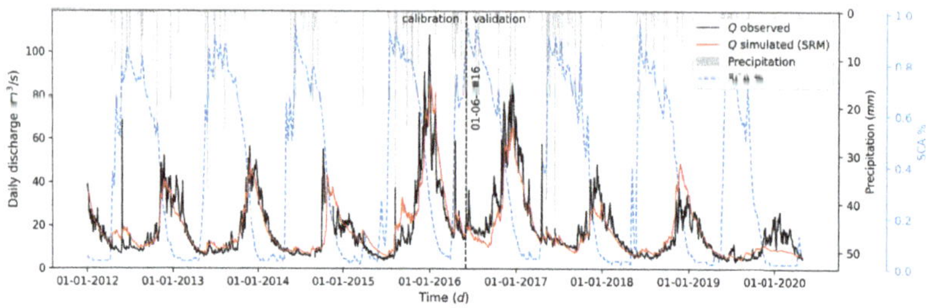

Figure 4. Simulated discharge using the SRM model for the Aconcagua River at Chacabuquito station (See parameters in Appendix A.1).

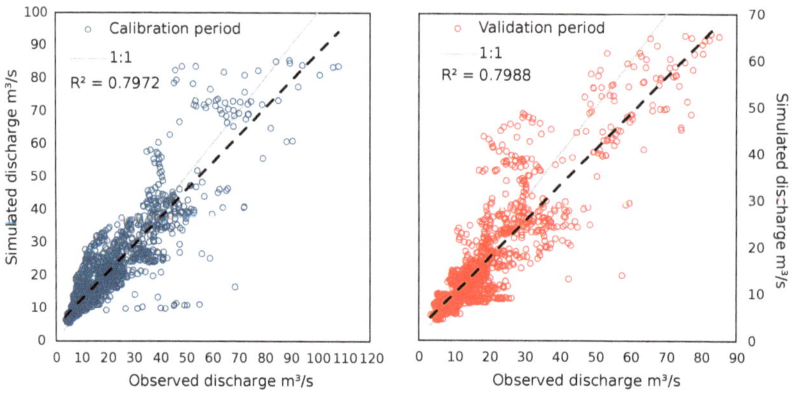

Figure 5. Comparison between observed and simulated streamflow using the SRM model for the Aconcagua River at Chacabuquito station (dashed line represents the trend line).

Additionally, a weak correlation is observed between precipitation events and streamflow response, suggesting that the hydrological regime of the basin is primarily governed by snow accumulation and melt processes rather than direct rainfall-runoff. This is further supported by Figure 4, which shows that streamflow reaches its minimum when the Snow-Covered Area (SCA) is at its maximum, emphasizing the role of snow accumulation in subsequent melt-driven flow. This behavior is consistent with the characteristics of the Aconcagua region.

The model simulation during the calibration period (1 January 2012 to 1 June 2016) yielded a Nash-Sutcliffe Efficiency (NSE) of 0.697, a Root Mean Square Error ($RMSE$) of 8.607 m^3/s, a coefficient of determination (R^2) of 0.797, a Percent Bias ($PBIAS$) of −9.763%, and a Kling-Gupta Efficiency (KGE) of 0.858, a MAE of 5.152 m^3/s, a log NSE of 0.785 and $APFB$ of 0.228. In the validation period, the model achieved an NSE of 0.783, $RMSE$ of 6.735 m^3/s, R^2 of 0.799, $PBIAS$ of 9.636%, and KGE of 0.783, MAE of 4.462 m^3/s, log NSE of 0.769 and $APFB$ of 0.317 (Table 1).

Table 1. Statistical performance indicators of hydrological simulation using the SRM model for the Aconcagua River at the Chacabuquito station.

	SRM	
Metric	**Calibration**	**Validation**
NSE	0.697	0.783
$RMSE$ (m^3/s)	8.607	6.735
R^2	0.797	0.799
$PBIAS(\%)$	−9.763	9.636
KGE	0.858	0.783
MAE	5.152	4.462
$\log NSE$	0.785	0.769
$APFB$	0.228	0.317

Figures 6 and 7 demonstrate a high-quality simulation, with the model successfully reproducing both the general behavior and the variability of streamflow, including extreme events. During both the calibration and validation periods, the model performance is notable. However, abrupt drops in simulated streamflow compared to the observed values are evident in certain instances. This behavior is more clearly seen in Figure 7, where some points are distinctly clustered relative to the 1:1 line.

In contrast to the Aconcagua basin, the Duqueco basin exhibits a strong correlation between precipitation events and streamflow response (see Figure 6), indicating a precipitationdominated hydrological regime. However, not all extreme precipitation events lead to peak streamflow events. This suggests that peak flows are not solely determined by the intensity of individual rainfall events, but rather by the frequency and accumulation of hydrological processes over time. For instance, when base flows are already elevated, an extreme rainfall event is more likely to trigger a high-flow response compared to the same event occurring under low-flow conditions.

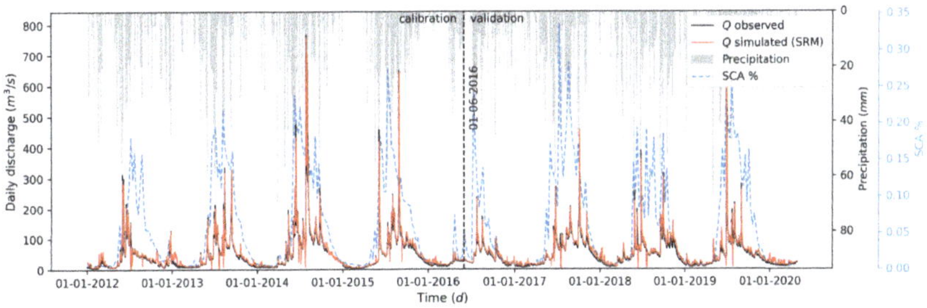

Figure 6. Simulated discharge using the SRM model for the Duqueco River at Cerrillos station (See parameters in Appendix A.2).

Moreover, considering that Figure 6 shows streamflow increases even during periods of snow accumulation, it is reinforced that, unlike the Aconcagua basin, snow accumulation is not the dominant process controlling streamflow in the Duqueco basin, indicating a pluvio-nival or predominantly precipitation-driven hydrological regime.

The model simulation during the calibration period (1 January 2012 to 1 June 2016) yielded a Nash-Sutcliffe Efficiency (NSE) of 0.943, a Root Mean Square Error ($RMSE$) of 15.709 m^3/s, a coefficient of determination (R^2) of 0.964, a Percent Bias ($PBIAS$) of −11.547%, and a Kling-Gupta Efficiency (KGE) of 0.936, a MAE of 8.465 m^3/s, a $\log NSE$ of 0.853 and $APFB$ of 0.024. In the validation period, the model achieved an NSE of 0.933,

$RMSE$ of 12.979 m^3/s, R^2 of 0.945, $PBIAS$ of −10.820%, and KGE of 0.907, a MAE of 7.644 m^3/s, a log NSE of 0.793 and $APFB$ of 0.233 (see Table 2).

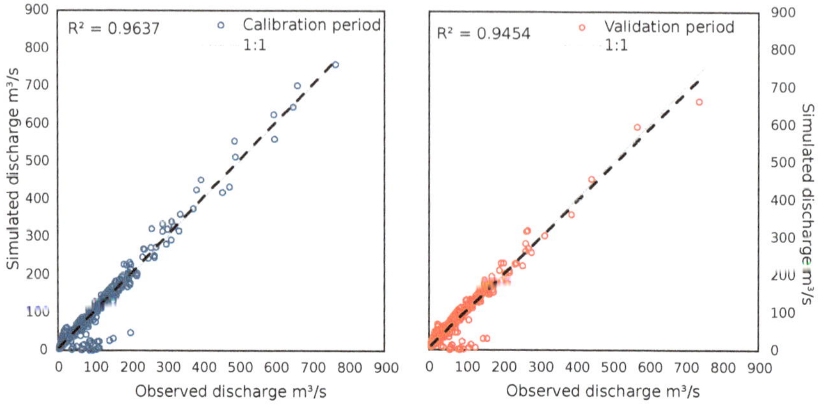

Figure 7. Comparison between observed and simulated streamflow using the SRM model for the Duqueco River at Cerrillos (dashed line represents the trend line).

Table 2. Statistical performance indicators of hydrological simulation using the SRM model for the Duqueco River at the Cerrillos station.

	SRM	
Metric	**Calibration**	**Validation**
NSE	0.943	0.933
$RMSE$ (m^3/s)	15.709	12.979
R^2	0.964	0.945
$PBIAS$(%)	−11.547	−10.820
KGE	0.936	0.907
MAE	8.465	7.644
log NSE	0.853	0.793
$APFB$	0.024	0.233

3.2. GRxJ Models (GR4J, GR5J, GR6J) with and Without the CemaNeige Module

3.2.1. Selection of the Evapotranspiration Model That Optimizes Model Performance

The evaluation of different potential evapotranspiration (PET) formulations, combined with the three GRxJ model variants (GR4J, GR5J, GR6J) with and without the CemaNeige snow module, revealed basin-dependent differences (Table 3). In the Aconcagua River basin at Chacabuquito, the CAMELS-CL (C) formulation yielded the best results with GR4J, regardless of the inclusion of CemaNeige. For GR5J, no differences were detected among PET models, whereas in GR6J, performance was slightly better with CAMELS-CL and Oudin's formulation (E_o). In contrast, in the Duqueco River basin at Cerrillos, Oudin's model (E_o) consistently outperformed all other formulations across GRxJ variants and CemaNeige configurations. Consequently, CAMELS-CL (C) was selected for the Aconcagua basin and Oudin's formulation (E_o) for the Duqueco basin.

Table 3. Performance of GRxJ hydrological models with and without the CemaNeige module under different PET estimation methods, using NSE as the objective function.

Basin	PET Model	CemaNeige	GR4J	GR5J	GR6J
Aconcagua River at the Chacabuquito station	C	No	0.112	0.005	0.005
		Yes	0.119	0.005	0.006
	E_o	No	0.074	0.005	0.004
		Yes	0.074	0.005	0.004
	E_h	No	−0.022	0.005	0.003
		Yes	−0.021	0.005	0.003
	E_p	No	−0.022	0.005	0.003
		Yes	−0.039	0.005	0.003
	E_{pa}	No	−0.43	0.005	0.003
		Yes	−0.022	0.005	0.003
Duqueco River at the Cerrillos station	C	No	0.826	0.819	0.829
		Yes	0.845	0.853	0.851
	E_o	No	0.834	0.832	0.835
		Yes	0.854	0.860	0.853
	E_h	No	0.807	0.751	0.810
		Yes	0.831	0.796	0.840
	E_p	No	0.810	0.744	0.810
		Yes	0.835	0.789	0.839
	E_{pa}	No	0.811	0.771	0.820
		Yes	0.836	0.814	0.844

3.2.2. Model Results

The simulations generated using the GRxJ models, both with and without the CemaNeige module, are presented in Figures 8 and 9. In both cases, the models fail to adequately reproduce the observed streamflow behavior. The simulations tend to represent the mean flow as a relatively linear trend throughout the period, without capturing temporal variability or extreme events. However, it is worth noting that the models exhibit some sensitivity to precipitation events, particularly in the GR4J and GR6J configurations. In these cases, increases in simulated streamflow are observed following rainfall events, although this response is limited and does not reflect the observed streamflow values. Consequently, the overall performance of the models for the Aconcagua River basin at the Chacabuquito station cannot be considered satisfactory.

In Table 4, the GR4J, GR5J, and GR6J models showed very poor performance during both the calibration and validation periods, with or without the inclusion of the CemaNeige module. The NSE values were close to zero or even negative in all cases, indicating a very limited ability of the models to reproduce the observed streamflow. GR4J showed the highest NSE value (0.119) during calibration, but it dropped to −0.028 in validation, both being poor. GR5J and GR6J exhibited NSE values between 0.005 and −0.002, with no improvements after incorporating CemaNeige.

The coefficient of determination (R^2) was also low in all cases, with values below 0.140, and the percent bias ($PBIAS$) was moderate, with some cases showing overestimation (e.g., GR4J with −1.902%) and others underestimation (e.g., GR4J with 8.244% in validation). The Kling-Gupta Efficiency (KGE) was closer zero or negative for all models and periods, further confirming the poor agreement between simulated and observed streamflow. The addition of the CemaNeige module resulted in only marginal changes in the metrics, with no clear evidence of improvement.

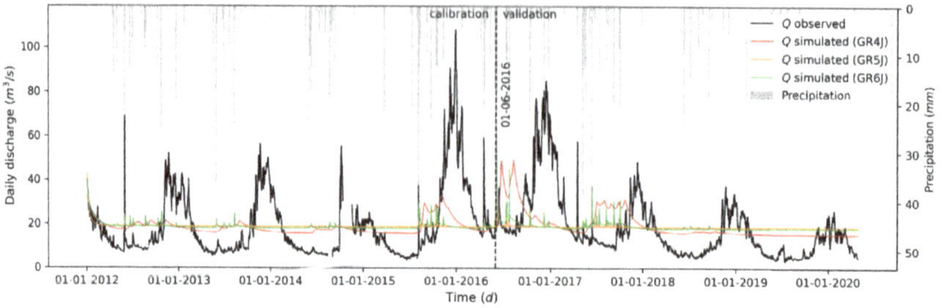

Figure 8. Simulated discharge using the GRxJ models for the Aconcagua River at the Chacabuquito station (See parameters in Appendix A.3).

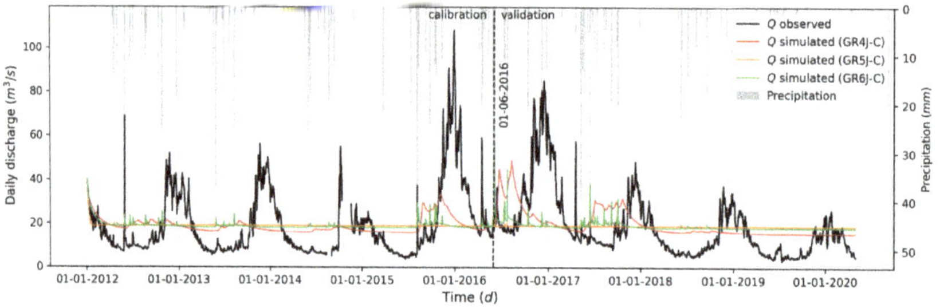

Figure 9. Simulated discharge using the GRxJ models with the CemaNeige module for the Aconcagua River at the Chacabuquito station (See parameters in Appendix A.3).

Table 4. Statistical performance indicators of hydrological simulations using the GRxJ models with/without the CemaNeige module for the Aconcagua River at the Chacabuquito station.

CemaNeige	Metric	GR4J		GR5J		GR6J	
		Calib.	Valid.	Calib.	Valid.	Calib.	Valid.
Without	NSE	0.112	−0.052	0.005	0.000	0.005	−0.002
	$RMSE$ (m^3/s)	14.362	14.833	15.203	14.467	15.200	14.474
	R^2	0.134	0.031	0.005	0.002	0.006	0.007
	$PBIAS$	−1.902	8.196	0.076	3.301	−0.295	3.904
	KGE	−0.007	0.003	−0.309	−0.359	−0.275	−0.242
	MAE	10.977	10.971	11.566	10.367	11.575	10.493
	log NSE	0.025	−0.065	−0.127	−0.158	−0.113	−0.134
	$APFB$	0.651	0.464	0.647	0.496	0.615	0.383
With	NSE	0.119	−0.028	0.005	−0.002	0.006	0.001
	$RMSE$ (m^3/s)	14.302	14.663	15.205	14.470	15.192	14.450
	R^2	0.140	0.038	0.005	0.000	0.007	0.008
	$PBIAS$	−1.057	8.244	−0.160	3.136	−0.297	3.701
	KGE	0.008	0.011	−0.323	−0.399	−0.271	−0.244
	MAE	10.998	10.861	11.596	10.353	11.568	10.463
	log NSE	0.027	−0.053	−0.130	−0.161	−0.111	−0.131
	$APFB$	0.644	0.463	0.658	0.510	0.616	0.394

To complement the evaluation of the GRxJ models' performance across different aspects of hydrological variability, the log NSE, $APFB$, and MAE metrics were considered. log NSE values were close to zero or negative (−0.161 to 0.027), highlighting the limited ability of the models to simulate low flows. The $APFB$ index showed notable deviations in the shape of the hydrographs, particularly for peak flows, with values ranging from 0.38 to 0.65. Finally, MAE ranged from 10.35 to 11.60 m^3/s, reflecting consistent discrepancies between simulated and observed streamflows throughout the hydrograph. The log NSE, $APFB$, and MAE values did not show significant improvements following the inclusion of the CemaNeige module, indicating that this module produced only marginal changes in the performance of $GR4J$, $GR5J$, and $GR6J$ in the studied basin.

The simulations using the GRxJ models coupled with the CemaNeige snow module demonstrate an improved ability to capture both the variability and general trend of streamflow. In contrast, the standalone GRxJ models tend to underestimate peak flows and exhibit reduced performance in reproducing hydrological extremes (Figure 10). Visual inspection of the hydrographs shows a more accurate fit during the validation period for all models, especially when CemaNeige is included (Figure 11). Overall, while all GRxJ configurations deliver acceptable performance, the inclusion of the snow module enhances model performance.

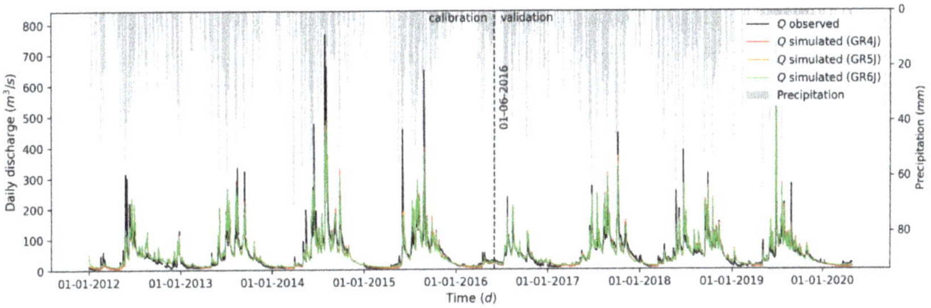

Figure 10. Simulated discharge using the GRxJ models for the Duqueco River at the Cerrillos station (See parameters in Appendix A.4).

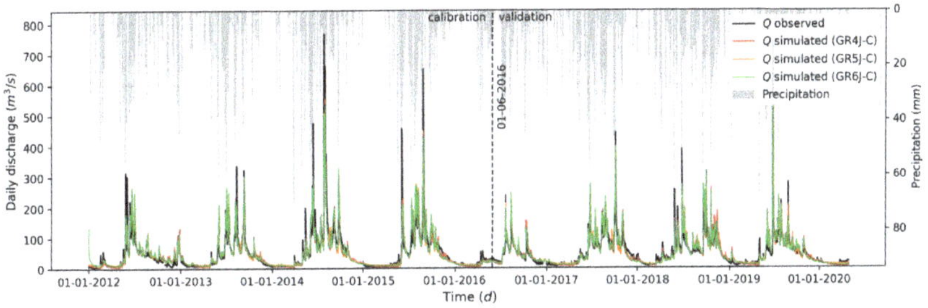

Figure 11. Simulated discharge using the GRxJ models with the CemaNeige module for the Duqueco River at the Cerrillos station (See parameters in Appendix A.4).

As previously observed in the Aconcagua basin, the GRxJ models exhibit a certain sensitivity to precipitation events. This characteristic likely explains their improved performance in the Duqueco basin, where streamflow generation is strongly correlated with precipitation. The direct relationship between rainfall and runoff in Duqueco provides more favorable conditions for the GRxJ models to adequately simulate streamflow dynamics, an effect that is further enhanced by the inclusion of the CemaNeige module, which accounts for processes related to snow accumulation and melt.

The performance of the hydrological models GR4J, GR5J, and GR6J was evaluated with and without the CemaNeige snow accumulation and melt module during the calibration and validation periods (Table 5). Incorporating CemaNeige generally led to improved or comparable performance across most evaluated metrics. For the GR4J model, the NSE increased from 0.834 to 0.854 during calibration and increased slightly from 0.837 to 0.844

during validation, while the KGE improved from 0.864 to 0.882 in calibration and from 0.873 to 0.901 in validation. The GR5J model showed consistent improvements with CemaNeige, with the NSE rising from 0.832 to 0.860 in calibration and from 0.840 to 0.865 in validation, and the KGE increasing from 0.844 to 0.868 in calibration and from 0.888 to 0.928 in validation, presenting the best overall performance. Meanwhile, the GR6J model also benefited from the inclusion of CemaNeige, achieving higher NSE values (from 0.835 to 0.853 in calibration and from 0.846 to 0.855 in validation) and higher KGE scores (from 0.837 to 0.876 in calibration and from 0.902 to 0.906 in validation), although its performance was slightly lower than that of GR5J. Overall, the use of CemaNeige improved streamflow representation, especially in the GR5J and GR6J models.

To complement the evaluation of the GRxJ models' performance across different aspects of hydrological variability, the log NSE, $APFB$, and MAE metrics were considered. log NSE values were close to 0.84 (0.777 to 0.872), highlighting the models' good ability to simulate low flows. The $APFB$ index showed moderate deviations for extreme flows, with values ranging from 0.15 to 0.30. Finally, MAE ranged from 11.01 to 14.09 m^3/s, reflecting a low error relative to the magnitude of streamflow. The log NSE, $APFB$, and MAE values showed improvements following the inclusion of the CemaNeige module, especially for GR6J in the studied basin.

Table 5. Statistical performance indicators of hydrological simulations using the GRxJ models with/without the CemaNeige module for the Duqueco River at the Cerrillos station.

CemaNeige	Metric	GR4J		GR5J		GR6J	
		Calib.	**Valid.**	**Calib.**	**Valid.**	**Calib.**	**Valid.**
Without	NSE	0.834	0.837	0.832	0.840	0.835	0.846
	$RMSE$ (m^3/s)	26.568	20.238	26.704	20.019	26.528	19.673
	R^2	0.834	0.865	0.834	0.861	0.838	0.862
	$PBIAS$	−0.441	5.801	1.160	5.999	1.096	4.802
	KGE	0.864	0.873	0.844	0.888	0.837	0.902
	MAE	13.931	11.864	14.087	11.605	14.064	11.366
	log NSE	0.860	0.807	0.872	0.848	0.853	0.854
	$APFB$	0.304	0.182	0.297	0.169	0.304	0.177
With	NSE	0.854	0.844	0.860	0.865	0.853	0.855
	$RMSE$ (m^3/s)	24.921	19.748	24.446	18.372	25.013	19.092
	R^2	0.854	0.862	0.861	0.872	0.853	0.871
	$PBIAS$	−1.165	3.705	0.412	2.612	1.070	3.060
	KGE	0.882	0.901	0.868	0.928	0.876	0.906
	MAE	14.001	11.638	13.366	11.175	13.538	11.065
	log NSE	0.827	0.777	0.859	0.864	0.863	0.865
	$APFB$	0.290	0.149	0.289	0.187	0.265	0.166

4. Discussion

4.1. Performance of Both Models

4.1.1. Aconcagua River in Chacabuquito

In the snow-dominated basin, the SRM model achieved a performance classified as good to very good, according to the performance ranges proposed by Moriasi et al. [59], during both calibration and validation periods. In contrast, the GRxJ family models, coupled with and without the CemaNeige module, showed unsatisfactory performance in the snow-dominated basin, with low and negative NSE values and poor capacity to adequately represent streamflow.

This difference can be partly explained by the structural characteristics of both models. As Martinec [43] pointed out, SRM allows extrapolation of input variables (precipitation and temperature) and considers parameters that vary in space (altitudinal zones or bands) and time (Martinec [43] notes can be parameterized every 15 days), in basins with a high correlation between precipitation and runoff, this allows for capturing greater precipitation inputs and adjusting runoff coefficients to simulate extreme events. However, in this basin, where such correlation is low, this extrapolation capability provides less advantage, as it does not significantly improve the estimation of extreme events. On the other hand, GRxJ models are designed with fixed parameters (space and time) and input variables constant in space, limiting their representativeness across the entire basin [25,45,46]. Although Valéry et al. [27] introduced the CemaNeige module to improve snow representation by extrapolating its input variables, its parameters remain fixed spatially and temporally, meaning it cannot simulate extreme events with the same performance as SRM.

Furthermore, the most significant difference in model performance in this study area lies in the approach each uses to represent snow processes. The SRM model directly incorporates the snow cover area (SCA) variable, mainly obtained from satellite images, allowing it to estimate snowmelt contributions even in the absence of precipitation. In contrast, the CemaNeige module simulates snow dynamics through an accumulation and melt routine (SAR), primarily based on daily temperature and precipitation data, which are extrapolated by altitudinal zones [27]. This difference is notable in areas like the one studied, where precipitation is infrequent. Under such conditions, CemaNeige faces a significant limitation since its ability to generate snow, and thus runoff, depends directly on the occurrence of precipitation events. If no precipitation is recorded, the model generates neither SCA nor runoff. In contrast, the SRM model is not affected by the absence of precipitation, as its snow cover estimation is primarily based on satellite observations. This allows it to more accurately represent hydrological dynamics in snow-dominated basins with scarce precipitation.

It is important to note that these results are specific to the Aconcagua River catchment and should not be interpreted as universally applicable to all snow-dominated basins.

4.1.2. Duqueco River in Cerrillos

In the rain–snow regime basin, both models showed performance ranging from good to very good, with the SRM model again achieving the best results. This is due to the high correlation between rainfall and runoff, which allows SRM, through the extrapolation of variables and spatially distributed parameterization, to capture greater precipitation inputs and convert them into runoff for extreme events. The GRxJ models, both with and without the CemaNeige module, also showed a significant improvement in performance compared to the snow-dominated basin. This improvement is primarily explained by the basin's hydrological regime, characterized by abundant and frequent precipitation. Since the GRxJ models heavily rely on precipitation to simulate both snow accumulation processes (in the case of CemaNeige) and runoff, their ability to reproduce streamflow is substantially enhanced in contexts with frequent rainfall. These results highlight the high sensitivity of the GRxJ models to available precipitation, which favors their performance in basins like Duqueco. However, this same characteristic limits their applicability in basins with lower precipitation regimes, as observed in Aconcagua.

4.2. Influence of the PET Source on GRxJ Models' Performance

Various methods to estimate potential evapotranspiration (PET) were evaluated as input variables for the GRxJ models, observing that an appropriate choice of PET source can partially improve their performance. For example, changing the source or estimation method of PET resulted in the most noticeable improvement in the GR5J model's performance, from an NSE of 0.744 to 0.832, for GR4J, from 0.807 to 0.834, and for GR6J, from 0.810 to 0.835. These improvements show that the appropriate selection of the PET estimation method positively influences the accuracy of the hydrological models evaluated.

In this study, the Oudin equation (E_o) was identified as the PET source that provided the best results, which aligns with reports from other authors who have demonstrated its reliability as an input variable for the GRxJ models [29,60]. Particularly, although the GRxJ models' performance in the Aconcagua River basin was poor, the best PET source corresponded to the CAMELS-CL (C) dataset, which is also based on the Oudin formula, reinforcing the evidence that among the evaluated options, this method is most suitable for GRxJ models.

These findings highlight the importance of adapting potential evapotranspiration estimation to the specific hydrometeorological context and suggest that, in data-scarce regions, the use of simple and robust methods such as Oudin's, which bases its calculations on mean daily temperature and estimated extraterrestrial radiation, can significantly optimize the calibration and performance of hydrological models.

4.3. Comparison of GR4J, GR5J, and GR6J Performance with and Without the CemaNeige Module

Performance evaluation among the GRxJ family models in the mixed regime basin shows consistent results with those obtained by Flores et al. [29] and Park et al. [61]. In those studies, GR5J and GR6J demonstrated better performance compared to GR4J, with GR6J being the best model, although the improvement was partially relative to the others.

When incorporating the CemaNeige module, improvement in performance was evident for all models, which align with Sezen et al. [62], who reported an improvement in GR6J simulations with CemaNeige compared to the GR4J and GR6J models without it. However, in the present study, the model with the best performance was GR5J + CemaNeige, even outperforming GR6J with and without the module.

In the snow regime basin, although overall model performance was poor, the incorporation of the CemaNeige module also helped improve results. In this case, the GR4J + CemaNeige model achieved the best performance.

These results demonstrate that incorporating the snow module improves simulation performance in rain–snow regime basins. Although GR5J + CemaNeige showed improvement in this study compared to other GRxJ models, the works of Flores et al. [29], Park et al. [61], and Sezen et al. [62] did not perform a complete comparison of all GRxJ models with and without the CemaNeige module. This limitation prevents generalizing that GR5J + CemaNeige is the optimal model in different contexts.

4.4. Performance of Both Models in Previous Studies

Previous studies comparing the GR4J and SRM models support the findings of this research. For example, in the Dudh Koshi River basin in Nepal, the SRM model achieved an NSE of 0.86, while GR4J obtained a value of 0.63, representing a significant difference in the performance of the two models. While both are characterized by a robust and simple structure, SRM demonstrates a greater responsiveness to variations in temperature and precipitation, whereas GR4J shows notable limitations, especially during the winter season. Specifically, Pokhrel et al. [63] noted that, in the GR4J model, a temperature increase between 1 °C and 4 °C generated an increase in winter runoff of only 0.46% to 1.49%.

In contrast, under the same conditions, SRM reported a much more significant increase, ranging from 10.2% to 70.1%. These results demonstrate that SRM has greater sensitivity and capacity to simulate runoff generated by snowmelt, a key aspect in basins dominated by this regime.

In a related study, Muñoz Castro et al. [47] evaluated the GRxJ model family, both with and without the CemaNeige snow module, across 95 catchments distributed throughout Chile. A notable aspect of that study is that individual results by catchment are not reported; instead, catchments are grouped into quartiles based on parameter agreement indices, linking these groups to hydroclimatic and physiographic characteristics. The authors found that wetter catchments, with higher precipitation and runoff ratios, tend to exhibit better performance and more stable parameters, whereas semiarid and snow-influenced catchments showed lower agreement and less consistent results. Consequently, the GRxJ models demonstrate limitations in catchments where runoff is primarily driven by snow accumulation and melt under low precipitation conditions. This evaluation approach, based on grouping, contrasts with the present study, in which explicit results are reported for each catchment. In particular, the findings for the Aconcagua basin show that the GRxJ models (with or without CemaNeige) perform poorly in basins with a dominant snow regime and reduced precipitation, thus reinforcing the limitations identified by Muñoz Castro et al. [47].

It is also worth noting that Perrin et al. [46] highlighted structural limitations of the GR4J model in its application under climate change scenarios, particularly due to the lack of an adequate snow module. This further supports the findings of the present research.

In addition, Calizaya et al. [64] successfully applied the SRM model in the Santa River subbasin (Peru) to assess climate change impacts on runoff. The model exhibited strong performance during calibration and validation, with high efficiency metrics. Hydrological projections extended to 2080 under future climate scenarios further demonstrated SRM's robustness in simulating watershed responses to changing climatic conditions. These findings affirm the model's value as a reliable tool for climate impact assessment at the basin scale.

Although the GRxJ model family has shown limitations in basins with a dominant snow regime, mainly associated with hydroclimatic conditions, both models analyzed in this study (SRM and GRxJ) have proven to be applicable across a wide range of spatial scales from a physiographic perspective. For instance, Martinec [43] reported that the SRM model has been successfully implemented in over 120 catchments worldwide, with basin areas ranging from 2.65 km^2 to 917,444 km^2, achieving good efficiency indicators in most cases. In Chile, Flores et al. [65] evaluated the GRxJ model family in 15 catchments in the Biobío Region, characterized by a strong correlation between rainfall and runoff, with areas ranging from 28.5 km^2 to 24,269 km^2. Overall, good performance was obtained; however, in cases where the NSE was below 0.4, the causes were attributed to factors such as input data quality, small catchment size, and potential anthropogenic influences.

Along these lines, the study by Flores et al. [65] highlights the importance of considering scale effects in parameter calibration and transfer, proposing a regionalization approach based on morphometric variables such as basin area, average slope, and form factor. This methodology allowed calibrated parameters to be transferred from monitored to ungauged catchments, achieving simulations with coefficients of determination (R^2) ranging from 0.68 to 0.99. These results suggest that, although the GRxJ models present limitations mainly related to hydrological regimes, their applicability can extend to various spatial scales, provided that catchment morphometric characteristics are properly accounted for. In this way, it is recognized that basin area, as a direct expression of scale effects, does not inherently limit the application of the model, but it does influence the stability and transferability of its parameters, which, to date, has not been studied in the case of SRM.

To date, no other studies have been identified that directly compare SRM with models from the GRxJ family, nor studies that apply GRxJ (with or without the CemaNeige module) in basins with such a dominant snow regime. At least in Andean catchments in South America, this remains an unexplored area of research. In this context, the results presented support the conclusion that the SRM model shows strong performance under various hydrological conditions, while GRxJ models achieve good results mainly in basins where precipitation is the primary driver of runoff.

5. Conclusions

This study evaluated the performance of the SRM and GRxJ (GR4J, GR5J, GR6J) hydrological models, with and without the CemaNeige snow module, in two Chilean catchments with contrasting hydrological regimes. The main findings and their implications lead to the following conclusions:

The results of this study indicate that model selection should be aligned with the dominant hydrological regime of each catchment. In particular, SRM demonstrated superior performance in the snow-dominated basin, likely due to its ability to incorporate satellite-derived snow cover information into the hydrological representation. In contrast, the GRxJ + CemaNeige models performed better in mixed (rain–snow) regimes where precipitation is more frequent. These results suggest that model selection should be guided by the prevailing hydrological regime.

The structural limitations of the GRxJ models, particularly their inability to adequately represent snow dynamics in the absence of precipitation, reduce their applicability in mountainous catchments of central–northern Chile. In this regard, the integration of satellite observations (as in SRM) represents a significant advantage for modeling in such regions.

The proper selection of the potential evapotranspiration (*PET*) source had a direct impact on the performance of the GRxJ models. The Oudin formula stood out as the most suitable option, reinforcing the need to evaluate multiple input options during data preparation and model calibration.

Given that many Andean catchments share similar characteristics with the Aconcagua basin, the use of models like SRM may be more reliable for hydrological planning, meltwater runoff estimation, and climate change scenario evaluation. However, in catchments with frequent precipitation, the conceptual GRxJ + CemaNeige models can offer efficient and transferable solutions to ungauged basins, provided they are properly calibrated.

The calibration process used in this study for the SRM model involves high parameter dimensionality, especially due to its spatial and temporal variability, resulting in higher computational costs compared to the GRxJ models. The latter, with a simpler structure and fewer parameters, are more efficient for rapid simulations or in contexts with limited computational resources.

An important limitation of this study is that the superior performance observed for SRM in snow-dominated basins is based solely on the analysis of two specific catchments, and therefore cannot be generalized to other basins without further investigation. Additionally, the findings highlight that the behavior of the GRxJ models, both with and without CemaNeige, is strongly conditioned by precipitation, which should be taken into account when interpreting the results.

The availability of hydrometeorological data is limited, particularly regarding snow and soil moisture observations, which restricts the validation of intermediate hydrological processes and contributes to model uncertainty. Concerning soil representation, the GRxJ models include a conceptual structure that accounts for soil water storage and retention, although in a simplified form, whereas the SRM model does not explicitly represent these processes. This difference may influence each model's ability to simulate the hydrological response in catchments where soil dynamics play a relevant role. Furthermore, the analysis was conducted in only two catchments with specific hydrological characteristics, which limits the generalization of the findings to other Andean basins. Finally, the SRM model involves a high level of parameter dimensionality, while the GRxJ models rely strongly on the frequency of precipitation events, which conditions their performance according to the dominant hydrological regime.

The fixed structure of the GRxJ model is a disadvantage in snow-influenced catchments or for capturing extreme events. Therefore, it is recommended that future research aimed at improving the GRxJ models should incorporate spatially and temporally variable parameterization, as well as the inclusion of snow cover area (SCA) as an input variable, to enhance their performance in mountainous regions and under infrequent precipitation scenarios. Additionally, we suggest evaluating the capacity of the SRM model to transfer its parameters to ungauged catchments, as well as explore semi-distributed models or models with dynamic vegetation cover to more fully represent hydrological processes in Andean catchments.

Author Contributions: Conceptualization, D.G. and S.Y.; methodology, B.R., V.O., D.G. and S.Y.; software, B.R. and V.O.; validation, B.R.; formal analysis, B.R.; writing—original draft preparation, B.R., V.O. and S.Y.; writing—review and editing, B.R., V.O., D.G., C.G. and S.Y.; visualization, B.R. and V.O.; supervision, C.G. and S.Y. All authors have read and agreed to the published version of the manuscript.

Funding: V.O is partially funded by UCM-IN-24200 internal grant, Universidad Católica de Maule.

Data Availability Statement: We provide open access to the dataset used in this study through the Zenodo repository. The dataset includes daily hydroclimatic time series (temperature, precipitation, evapotranspiration, and streamflow) as well as catchment physiographic information required for the GRxJ and CemaNeige models. It can be freely accessed and downloaded at: https://doi.org/10.5281/zenodo.17246190, accessed on 27 November 2025. This ensures full reproducibility and facilitates further research and validation.

Acknowledgments: The authors acknowledge the use of MODIS products distributed by NASA's Earth Science Data Systems (ESDS) Program, which were accessed and processed through the Google Earth Engine (GEE) platform. We also thank the Center for Climate and Resilience Research (CR2) for providing access to the CAMELS-CL dataset, as well as Chile's Dirección General de Aguas (DGA) for supplying hydrometric data for both watersheds in this study. These datasets were fundamental for the development of this study.

Conflicts of Interest: The authors declare no conflict of interest.

Abbreviations

The following abbreviations are used in this manuscript:

a.s.l.	above sea level
airGR	R package for GR models
BNA	Banco Nacional de Aforos (Chile's National Hydrometric Database)
CAMELS-CL	Catchment Attributes and Meteorology for Large-sample Studies—Chile
CemaNeige	Snow accounting routine (degree–day snow module for GR models)
CR2	Center for Climate and Resilience Research (Chile)
DGA	Dirección General de Aguas (Chilean Water Authority)
ESDS	(NASA) Earth Science Data Systems
GA	Genetic algorithm
GEE	Google Earth Engine

GR4J	Génie Rural daily rainfall–runoff model (4 parameters)
GR5J	Génie Rural daily rainfall–runoff model (5 parameters)
GR6J	Génie Rural daily rainfall–runoff model (6 parameters)
GRxJ	Family of GR daily rainfall–runoff models ($x \in \{4, 5, 6\}$)
KGE	Kling–Gupta Efficiency
MOD09A1	MODIS Terra Surface Reflectance (8-day, 500 m) product
MODIS	Moderate Resolution Imaging Spectroradiometer
NDSI	Normalized Difference Snow Index
NSE	Nash–Sutcliffe Efficiency
PBIAS	Percent Bias
PET	Potential Evapotranspiration
R	The R programming language
RMSE	Root Mean Square Error
SAR	Snow Accounting Routine (in CemaNeige)
SCA	Snow Cover Area
SRM	Snowmelt-Runoff Model
SWIR	Shortwave Infrared
UH	Unit hydrograph

Appendix A

Appendix A.1

Table A1. Monthly average values of SRM model parameters in the Aconcagua River at Chacabuquito, by elevation zone.

Parameter (Monthly Average)	Zone								
	1	2	3	4	5	6	7	8	9
a	0.8971	0.3096	0.3102	0.4670	0.8406	0.6885	0.7906	0.8684	0.4537
y	0.6390	0.6357	0.6389	0.6903	0.6979	0.4066	0.5149	0.5237	0.5932
C_s	0.3760	0.1881	0.2798	0.2574	0.6699	0.2191	0.6361	0.9985	0.1273
C_R	0.2461	0.0347	0.0086	0.0659	0.0562	0.0216	0.8936	0.6047	0.3107
T_{crit}	3.7922	1.5207	3.5385	2.6962	2.7044	2.4480	2.0601	0.5653	3.9698
X					1.1339				
Y					0.0480				

Appendix A.2

Table A2. Parameter values for SRM in the Duqueco River at Cerrillos, by elevation zone.

Parameter	Zone					
	1	2	3	4	5	6
a	0.4858	0.9890	0.2228	0.5453	0.3100	0.1577
y	0.6578	0.8510	0.9487	0.9006	0.8720	0.8256
C_s	0.5953	0.0045	0.0264	0.1228	0.8093	0.0250
C_R	0.0136	0.0016	0.0073	0.0172	0.4009	0.6573
T_{crit}	0.9747	3.0715	2.3894	3.4019	1.6414	2.8075
X			0.5750			
Y			0.1810			

Appendix A.3

Table A3. Set of parameter values obtained for the Aconcagua River at Chacabuquito that maximize the objective function.

Model	CemaNeige	GRxJ Parameters						CemaNeige Parameters	
		X_1	X_2	X_3	X_4	X_5	X_6	θ_{G1}	θ_{G2}
GR4J	Without	323.0413	6.4915	92.4645	20.0000	–	–	–	–
GR5J	Without	614.8404	−4.3659	233.2280	1.1625	0.4538	–	–	–
GR6J	Without	486.7449	0.5211	114.6054	1.2073	−0.0888	0.3751	–	–
GR4J	With	298.0756	6.9339	101.6963	20.0000	–	–	0.0015	0.0878
GR5J	With	1548.6440	−6.2439	218.5204	1.2664	0.4327	–	0.0015	0.0177
GR6J	With	455.1925	0.5211	114.8887	1.1971	−0.0891	0.3814	0.7052	0.0857

Appendix A.4

Table A4. Set of parameter values obtained for the Duqueco River at Cerrillos that maximize the objective function.

Model	CemaNeige	GRxJ Parameters						CemaNeige Parameters	
		X_1	X_2	X_3	X_4	X_5	X_6	θ_{G1}	θ_{G2}
GR4J	Without	1032.7702	2.0369	112.1683	2.0030	–	–	–	–
GR5J	Without	885.8531	0.6728	102.3547	1.5445	0.0991	–	–	–
GR6J	Without	796.8922	0.1472	118.9345	2.1298	−0.8336	3.2755	–	–
GR4J	With	665.1416	2.0597	117.9192	1.9932	–	–	0.0105	2.0786
GR5J	With	589.5081	−0.5694	119.4266	1.5082	1.0000	–	0.0425	20.7294
GR6J	With	543.5959	0.1482	51.9194	2.0492	−0.5749	11.5934	0.0913	7.4759

References

1. Basso, S.; Botter, G. Streamflow variability and optimal capacity of run-of-river hydropower plants. *Water Resour. Res.* **2012**, *48*, 1–13. [CrossRef]
2. Liu, Z.; Zhou, P.; Chen, X.; Guan, Y. A multivariate conditional model for streamflow prediction and spatial precipitation refinement. *J. Geophys. Res. Atmos.* **2015**, *120*, 10116–10129. [CrossRef]
3. Ketchum, D.; Hoylman, Z.H.; Huntington, J.; Brinkerhoff, D.; Jencso, K.G. Irrigation intensification impacts sustainability of streamflow in the Western United States. *Commun. Earth Environ.* **2023**, *4*, 479. [CrossRef]
4. Kumar, V.; Kedam, N.; Sharma, K.V.; Mehta, D.J.; Caloiero, T. Advanced Machine Learning Techniques to Improve Hydrological Prediction: A Comparative Analysis of Streamflow Prediction Models. *Water* **2023**, *15*, 2572. [CrossRef]
5. Bhasme, P.; Bhatia, U. Improving the interpretability and predictive power of hydrological models: Applications for daily streamflow in managed and unmanaged catchments. *J. Hydrol.* **2024**, *628*, 130421. [CrossRef]
6. Huntington, J.L.; Niswonger, R.G. Role of surface-water and groundwater interactions on projected summertime streamflow in snow dominated regions: An integrated modeling approach. *Water Resour. Res.* **2012**, *48*, 1–20. [CrossRef]
7. Barthel, R.; Banzhaf, S. Groundwater and Surface Water Interaction at the Regional-scale—A Review with Focus on Regional Integrated Models. *Water Resour. Manag.* **2015**, *30*, 1–32. [CrossRef]
8. Zhang, G.; Cui, P.; Gualtieri, C.; Zhang, J.; Ahmed Bazai, N.; Zhang, Z.; Wang, J.; Tang, J.; Chen, R.; Lei, M. Stormflow generation in a humid forest watershed controlled by antecedent wetness and rainfall amounts. *J. Hydrol.* **2021**, *603*, 127107. [CrossRef]
9. Zhang, G.; Cui, P.; Gualtieri, C.; Bazai, N.A.; Zhang, X.; Zhang, Z. Increased nonstationarity of stormflow threshold behaviors in a forested watershed due to abrupt earthquake disturbance. *Hydrol. Earth Syst. Sci.* **2023**, *27*, 3005–3020. [CrossRef]
10. Sofi, M.S.; Bhat, S.U.; Rashid, I.; Kuniyal, J.C. Socioeconomic Aspect of Disaster Risk in Kashmir: Contextualizing Village Vulnerability in Sindh Basin. In *Climate Change Adaptation, Risk Management and Sustainable Practices in the Himalaya*; Sharma, S., Kuniyal, J.C., Chand, P., Singh, P., Eds.; Springer International Publishing: Cham, Switzerland , 2023; pp. 199–212. [CrossRef]
11. Thongley, T.; Thakur, P.K.; Khajuria, V.; Bhattarai, N.; Choden, K.; Chokila, C. Estimation of glacier dynamics of the major glaciers of Bhutan using geospatial techniques. *J. Water Clim. Change* **2023**, *14*, 2825–2841. [CrossRef]

12. Rockström, J.; Gupta, J.; Qin, D.; Lade, S.J.; Abrams, J.F.; Andersen, L.S.; Armstrong McKay, D.I.; Bai, X.; Bala, G.; Bunn, S.E.; et al. Safe and just Earth system boundaries. *Nature* **2023**, *619*, 102–111. [CrossRef] [PubMed]
13. Migliavacca, F.; Confortola, G.; Soncini, A.; Senese, A.; Diolaiuti, G.A.; Smiraglia, C.; Barcaza, G.; Bocchiola, D. Hydrology and potential climate changes in the Rio Maipo (Chile). *Geogr. Fis. E Din. Quat.* **2015**, *38*, 155–168. [CrossRef]
14. Huh, K.I.; Baraër, M.; Mark, B.G.; Ahn, Y. Evaluating Glacier Volume Changes since the Little Ice Age Maximum and Consequences for Stream Flow by Integrating Models of Glacier Flow and Hydrology in the Cordillera Blanca, Peruvian Andes. *Water* **2018**, *10*, 1732. [CrossRef]
15. Koneti, S.; Sunkara, S.; Roy, P. Hydrological Modeling with Respect to Impact of Land-Use and Land-Cover Change on the Runoff Dynamics in Godavari River Basin Using the HEC-HMS Model. *ISPRS Int. J. Geo-Inf.* **2018**, *7*, 206. [CrossRef]
16. Krysanova, V.; Donnelly, C.; Gelfan, A.; Gerten, D.; Arheimer, B.; Hattermann, F.; Kundzewicz, Z.W. How the performance of hydrological models relates to credibility of projections under climate change. *Hydrol. Sci. J.* **2018**, *63*, 696–720. [CrossRef]
17. Duan, Q.; Pappenberger, F.; Wood, A.; Cloke, H.L.; Schaake, J.C. (Eds.) *Handbook of Hydrometeorological Ensemble Forecasting*, 1st ed.; Springer: Berlin/Heidelberg, Germany, 2019. [CrossRef]
18. Hrachowitz, M.; Clark, M.P. HESS Opinions: The complementary merits of competing modelling philosophies in hydrology. *Hydrol. Earth Syst. Sci.* **2017**, *21*, 3953–3973. [CrossRef]
19. Nemri, S.; Kinnard, C. Comparing calibration strategies of a conceptual snow hydrology model and their impact on model performance and parameter identifiability. *J. Hydrol.* **2020**, *582*, 124474. [CrossRef]
20. Ficklin, D.L.; Barnhart, B.L. SWAT hydrologic model parameter uncertainty and its implications for hydroclimatic projections in snowmelt-dependent watersheds. *J. Hydrol.* **2014**, *519*, 2081–2090. [CrossRef]
21. Arsenault, R.; Brissette, F.P. Continuous streamflow prediction in ungauged basins: The effects of equifinality and parameter set selection on uncertainty in regionalization approaches. *Water Resour. Res.* **2014**, *50*, 6135–6153. [CrossRef]
22. Rango, A.; Martinec, J.; Roberts, R. Relative importance of glacier contributions to streamflow in a changing climate. In Proceedings of the Second IASTED International Conference on Water Resource Management, Honolulu, HI, USA, 20–22 August 2007; ISBN 978-0-88986-680-5.
23. Panday, P.K.; Williams, C.A.; Frey, K.E.; Brown, M.E. Application and evaluation of a snowmelt runoff model in the Tamor River basin, Eastern Himalaya using a Markov Chain Monte Carlo (MCMC) data assimilation approach. *Hydrol. Process.* **2013**, *28*, 5337–5353. [CrossRef]
24. Tahir, A.A.; Chevallier, P.; Arnaud, Y.; Neppel, L.; Ahmad, B. Modeling snowmelt-runoff under climate scenarios in the Hunza River basin, Karakoram Range, Northern Pakistan. *J. Hydrol.* **2011**, *409*, 104–117. [CrossRef]
25. Pushpalatha, R.; Perrin, C.; Le Moine, N.; Mathevet, T.; Andréassian, V. A downward structural sensitivity analysis of hydrological models to improve low-flow simulation. *J. Hydrol.* **2011**, *411*, 66–76. [CrossRef]
26. Valéry, A.; Andréassian, V.; Perrin, C. 'As simple as possible but not simpler': What is useful in a temperature-based snow-accounting routine? Part 1—Comparison of six snow accounting routines on 380 catchments. *J. Hydrol.* **2014**, *517*, 1166–1175. [CrossRef]
27. Valéry, A.; Andréassian, V.; Perrin, C. 'As simple as possible but not simpler': What is useful in a temperature-based snow-accounting routine? Part 2—Sensitivity analysis of the Cemaneige snow accounting routine on 380 catchments. *J. Hydrol.* **2014**, *517*, 1176–1187. [CrossRef]
28. Escanilla-Minchel, R.; Alcayaga, H.; Soto-Alvarez, M.; Kinnard, C.; Urrutia, R. Evaluation of the Impact of Climate Change on Runoff Generation in an Andean Glacier Watershed. *Water* **2020**, *12*, 3547. [CrossRef]
29. Flores, N.; Rodríguez, R.; Yépez, S.; Osores, V.; Rau, P.; Rivera, D.; Balocchi, F. Comparison of Three Daily Rainfall-Runoff Hydrological Models Using Four Evapotranspiration Models in Four Small Forested Watersheds with Different Land Cover in South-Central Chile. *Water* **2021**, *13*, 3191. [CrossRef]
30. Webb, M.J.; Winter, J.M.; Spera, S.A.; Chipman, J.W.; Osterberg, E.C. Water, agriculture, and climate dynamics in central Chile's Aconcagua River Basin. *Phys. Geogr.* **2020**, *42*, 395–415. [CrossRef]
31. Bown, F.; Rivera, A.; Acuña, C. Recent glacier variations at the Aconcagua basin, central Chilean Andes. *Ann. Glaciol.* **2008**, *48*, 43–48. [CrossRef]
32. Martínez, C.; Fernández, A.; Rubio, P. Caudales y variabilidad climática en una cuenca de latitudes medias en Sudamérica: Río Aconcagua, Chile Central (33°S). *Bol. Asoc. Geógr. Esp.* **2012**, *58*, 227–248. [CrossRef]
33. Yépez, S.; Salas, F.; Nardini, A.; Valenzuela, N.; Osores, V.; Vargas, J.; Rodríguez, R.; Piégay, H. Semi-automated morphological characterization using South Rivers Toolbox. *Proc. IAHS* **2024**, *385*, 189–196. [CrossRef]
34. Dirección General de Aguas. *Balance hídrico de Chile*; Recursos Hídricos DGA, Ministerio de Obras Públicas (MOP): Santiago, Chile, 1987.
35. Alvarez-Garreton, C.; Mendoza, P.A.; Boisier, J.P.; Addor, N.; Galleguillos, M.; Zambrano-Bigiarini, M.; Lara, A.; Puelma, C.; Cortes, G.; Garreaud, R.; et al. The CAMELS-CL dataset: Catchment attributes and meteorology for large sample studies—Chile dataset. *Hydrol. Earth Syst. Sci.* **2018**, *22*, 5817–5846. [CrossRef]

36. Oudin, L.; Hervieu, F.; Michel, C.; Perrin, C.; Andréassian, V.; Anctil, F.; Loumagne, C. Which potential evapotranspiration input for a lumped rainfall–runoff model? *J. Hydrol.* **2005**, *303*, 290–306. [CrossRef]
37. Hargreaves, G.H.; Samani, Z.A. Estimating Potential Evapotranspiration. *J. Irrig. Drain. Div.* **1982**, *108*, 225–230. [CrossRef]
38. Priestley, C.H.B.; Taylor, R.J. On the Assessment of Surface Heat Flux and Evaporation Using Large-Scale Parameters. *Mon. Weather Rev.* **1972**, *100*, 81–92. [CrossRef]
39. Lhomme, J.P. A theoretical basis for The Priestley—Taylor Coefficient. *Bound.-Layer Meteorol.* **1997**, *82*, 179–191. [CrossRef]
40. Vermote, E. MODIS/Terra Surface Reflectance 8-Day L3 Global 500m SIN Grid V061. 2021. Available online: https://www.earthdata.nasa.gov/data/catalog/lpcloud-mod09a1-061 (accessed on 27 May 2025).
41. Riggs, G.; Barton, J.; Casey, K.; Hall, D.; Salomonson, V. MODIS Snow Products User Guide. 2006. Available online: https://modis-snow-ice.gsfc.nasa.gov/uploads/sug_c5.pdf (accessed on 27 May 2025) .
42. Härer, S.; Bernhardt, M.; Siebers, M.; Schulz, K. On the need for a time- and location-dependent estimation of the NDSI threshold value for reducing existing uncertainties in snow cover maps at different scales. *Cryosphere* **2018**, *12*, 1629–1642. [CrossRef]
43. Martinec, J. Snowmelt—Runoff Model for stream flow forecasts. *Hydrol. Res.* **1975**, *6*, 145–154. [CrossRef]
44. Martinec, J.; Rango, A.; Major, E. *The Snowmelt-Runoff Model (SRM) User's Manual*; NASA Reference Publication NASA-RP-1100; National Aeronautics and Space Administration (NASA): Washington, DC, USA, 1983.
45. Le Moine, N. Le Bassin Versant de Surface vu par le Souterrain: Une voie D'amélioration des Performances et du Réalisme des Modèles Pluie-Débit ? Ph.D. Thesis, Université Pierre-et-Marie-Curie, Paris, France, 2008.
46. Perrin, C.; Michel, C.; Andréassian, V. Improvement of a parsimonious model for streamflow simulation. *J. Hydrol.* **2003**, *279*, 275–289. [CrossRef]
47. Muñoz Castro, E.; Mendoza, P.A.; Vásquez, N.; Vargas, X. Exploring parameter (dis)agreement due to calibration metric selection in conceptual rainfall–runoff models. *Hydrol. Sci. J.* **2023**, *68*, 1754–1768. [CrossRef]
48. Nash, J.; Sutcliffe, J. River flow forecasting through conceptual models part I—A discussion of principles. *J. Hydrol.* **1970**, *10*, 282–290. [CrossRef]
49. Scrucca, L. GA: A Package for Genetic Algorithms in R. *J. Stat. Softw.* **2013**, *53*, 1–37. [CrossRef]
50. Bhagwat, N.; Kumar, R.; Qureshi, M.; Nagisetty, R.M.; Zhou, X. The Status, Applications, and Modifications of the Snowmelt Runoff Model (SRM): A Comprehensive Review. *Hydrology* **2025**, *12*, 156. [CrossRef]
51. Coron, L.; Thirel, G.; Delaigue, O.; Perrin, C.; Andréassian, V. The suite of lumped GR hydrological models in an R package. *Environ. Model. Softw.* **2017**, *94*, 166–171. [CrossRef]
52. Ritter, A.; Muñoz Carpena, R. Performance evaluation of hydrological models: Statistical significance for reducing subjectivity in goodness-of-fit assessments. *J. Hydrol.* **2013**, *480*, 33–45. [CrossRef]
53. Krause, P.; Boyle, D.P.; Bäse, F. Comparison of different efficiency criteria for hydrological model assessment. *Adv. Geosci.* **2005**, *5*, 89–97. [CrossRef]
54. Gupta, H.V.; Sorooshian, S.; Yapo, P.O. Status of Automatic Calibration for Hydrologic Models: Comparison with Multilevel Expert Calibration. *J. Hydrol. Eng.* **1999**, *4*, 135–143. [CrossRef]
55. Gupta, H.V.; Kling, H.; Yilmaz, K.K.; Martinez, G.F. Decomposition of the mean squared error and NSE performance criteria: Implications for improving hydrological modelling. *J. Hydrol.* **2009**, *377*, 80–91. [CrossRef]
56. Thiemig, V.; Rojas, R.; Zambrano-Bigiarini, M.; De Roo, A. Hydrological evaluation of satellite-based rainfall estimates over the Volta and Baro-Akobo Basin. *J. Hydrol.* **2013**, *499*, 324–338. [CrossRef]
57. Oudin, L.; Andréassian, V.; Mathevet, T.; Perrin, C.; Michel, C. Dynamic averaging of rainfall-runoff model simulations from complementary model parameterizations. *Water Resour. Res.* **2006**, *42*, 1–10. [CrossRef]
58. Mizukami, N.; Rakovec, O.; Newman, A.J.; Clark, M.P.; Wood, A.W.; Gupta, H.V.; Kumar, R. On the choice of calibration metrics for "high-flow" estimation using hydrologic models. *Hydrol. Earth Syst. Sci.* **2019**, *23*, 2601–2614. [CrossRef]
59. Moriasi, D.N.; Arnold, J.G.; Liew, M.W.V.; Bingner, R.L.; Harmel, R.D.; Veith, T.L. Model Evaluation Guidelines for Systematic Quantification of Accuracy in Watershed Simulations. *Trans. ASABE* **2007**, *50*, 885–900. [CrossRef]
60. Gharbia, S.S ; Smullen, T.; Gill, L.; Johnston, P.; Pilla, F. Spatially distributed potential evapotranspiration modeling and climate projections. *Sci. Total Environ.* **2018**, *633*, 571–592. [CrossRef] [PubMed]
61. Park, J.; Sung, J.H.; Seo, Y.H.; Kim, B.S. Applicability of the daily hydrological models (GR4J, GR5J, and GR6J) in the South Korean basins. *Sci. Rep.* **2025**, *15*, 31912. [CrossRef] [PubMed]
62. Sezen, C.; Bezak, N.; Šraj, M. Hydrological modelling of the karst Ljubljanica River catchment using lumped conceptual model. *Acta Hydrotech.* **2018**, *31* , 87–100. [CrossRef]
63. Pokhrel, B.K.; Chevallier, P.; Andréassian, V.; Tahir, A.A.; Arnaud, Y.; Neppel, L.; Bajracharya, O.R.; Budhathoki, K.P. Comparison of two snowmelt modelling approaches in the Dudh Koshi basin (eastern Himalayas, Nepal). *Hydrol. Sci. J.* **2014**, *59*, 1507–1518. [CrossRef]

64. Calizaya, E.; Mejía, A.; Barboza, E.; Calizaya, F.; Corroto, F.; Salas, R.; Vásquez, H.; Turpo, E. Modelling Snowmelt Runoff from Tropical Andean Glaciers under Climate Change Scenarios in the Santa River Sub-Basin (Peru). *Water* **2021**, *13*, 3535. [CrossRef]
65. Flores, N., Rodríguez, R.; Alzamora, R.; Gavilán, E.; Acuña, E.; Balocchi, F. Comparison of three daily rainfall-runoff models with multiple regression analysis for catchment parameters in Biobío, Chile. *Hydrol. Sci. J.* **2025**, 1–15 . [CrossRef]

water

MDPI

Article

Spatiotemporal Evolution of Compound Dry–Hot Events and Their Impacts on Vegetation Net Primary Productivity in the Yangtze River Basin

Hongqi Xi, Gengxi Zhang * and Hongkai Wang

College of Hydraulic Science and Engineering, Yangzhou University, Yangzhou 225009, China; 233202115@stu.yzu.edu.cn (H.X.); hongkaiwang2023@163.com (H.W.)
* Correspondence: gengxizhang@yzu.edu.cn; Tel.: +86-18792493190

Abstract

Compound dry–hot events increasingly threaten ecosystem productivity under global warming. Using ERA5-Land and MODIS NPP (2002–2024) for the Yangtze River Basin, we built climate indices and developed a Copula-based standardized compound dry–hot index (SCDHI) to detect events and examine spatiotemporal patterns. Trend and correlation analyses quantified NPP sensitivity and lag, and an NPP–SCDHI coupling framework assessed resistance and resilience across major vegetation types. Basin-wide monthly NPP increased slightly, while SCDHI decreased, indicating a warmer and drier tendency. Under dry–hot conditions, NPP was mainly negatively related to event intensity in the upper basin but positively related across much of the middle–lower plains. The mean NPP response time was approximately 2 months, with forests and croplands typically lagging 2–3 months. Under extreme stress, forests showed high resistance but limited recovery, whereas shrublands showed moderate resistance and low resilience. Cultivated vegetation exhibited the lowest resistance and weak resilience, grasslands had low resistance but relatively rapid recovery, and alpine vegetation showed moderate resistance and the highest resilience. Cultivated vegetation and grasslands may therefore represent high-risk types for ecological management.

Keywords: drought–heat interaction; ecological stability; resilience; resistance; standardized composite dry–hot index (SCDHI)

1. Introduction

Under global warming, extreme climate events exhibit significant increases in frequency, intensity, and duration, and drought and heat events not only occur in isolation but more often co-occur or occur in succession in a compound form, forming so-called compound dry–hot events [1,2]. Such events are often characterized by "high temperatures superimposed on drought" and "short-duration, high-intensity, concentrated outbreaks", and can markedly weaken vegetation photosynthesis and carbon sequestration capacity over short timescales, triggering concurrent degradation of ecosystem structure and function at the regional scale and posing severe challenges to agricultural production, water-resource security, and ecological security [3,4]. Therefore, under the context of global change and increasingly frequent extreme events, systematically characterizing the spatiotemporal evolution of compound dry–hot events and quantitatively assessing their impacts on terrestrial ecosystem productivity have become research frontiers in the fields of climate change and ecohydrology [3,5].

Net primary productivity (NPP) is a key indicator of the balance between vegetation photosynthesis and respiration, directly reflecting both the capacity of ecosystems to fix atmospheric CO_2 and their level of energy input, and constituting a fundamental variable for assessing regional carbon budgets and ecosystem functioning [6,7]. Numerous studies have demonstrated that climatic factors such as precipitation, temperature, and radiation are among the dominant drivers of the spatiotemporal variability of NPP, whereas drought and high-temperature extremes can substantially suppress NPP through mechanisms including water stress, stomatal closure, and leaf scorching, and may even trigger large-scale yield losses and vegetation degradation [8–10]. Most existing studies have examined the effects of either drought alone or heatwaves alone on vegetation growth and NPP, often relying on traditional drought indices and temperature anomaly metrics for correlation analyses. Although these efforts have yielded abundant insights, relatively limited attention has been paid to the synergistic mechanisms of "heat–drought" interactions, making it difficult to fully capture the actual impacts of compound dry–hot events on vegetation productivity [11–13].

From the perspective of monitoring and quantitative assessment, traditional drought and heat indices are primarily constructed from single variables and thus have limited ability to simultaneously characterize the joint distribution of moisture deficits and temperature anomalies [14]. In recent years, multivariate drought indices based on Copula theory and the standardized composite dry–hot index (SCDHI) have been progressively developed and applied, enabling the integration of precipitation, potential evapotranspiration, temperature, and other information on a unified probabilistic scale while accounting for the dependence structure among variables, thereby providing a new technical pathway for identifying and quantifying compound dry–hot events [15,16]. However, most existing studies have focused on issues such as drought risk assessment, the evolution of hydrological drought, or agricultural yield losses, and research that couples SCDHI with ecosystem productivity metrics such as NPP to systematically evaluate vegetation responses and recovery capacity to compound dry–hot events remains relatively limited, particularly with respect to large-basin scales and systematic comparisons across multiple vegetation types [15,17,18].

The Yangtze River Basin is a densely populated and economically important region, and it is also critical for China's ecological conservation and energy, food, and water security [19,20]. Influenced by both the East Asian monsoon and the Tibetan Plateau, the basin has a highly variable climate. Extreme droughts and high temperatures occur frequently. In recent years, droughts and heatwaves have increasingly co-occurred, leading to more compound dry–hot events [20–22]. However, existing studies on the Yangtze River Basin have largely focused on precipitation–runoff changes, the spatiotemporal evolution of single drought indices, and overall trends in vegetation productivity, and there is still a lack of systematic understanding of the occurrence patterns and intensity changes in compound dry–hot events and their impact mechanisms on NPP, particularly with respect to the quantitative assessment of vegetation resistance and resilience [23–25].

Accordingly, this study aims to examine the spatiotemporal evolution of compound dry–hot events and their impacts on vegetation net primary productivity in the Yangtze River Basin during 2002–2024. Three objectives are addressed in this study. The first objective is to characterize the spatiotemporal patterns and long-term changes in compound dry–hot conditions and to identify regions with intensified dry–hot stress. The second objective is to quantify the spatial variation in the direction, magnitude, and timing of vegetation NPP responses to compound dry–hot variability. The third objective is to assess differences in resistance and resilience among major vegetation types under extreme compound dry–hot events, thereby evaluating ecosystem stability and vulnerability. Together,

https://doi.org/10.3390/w18020276

these analyses provide an integrated assessment of compound dry–hot dynamics and ecosystem productivity responses, with implications for ecological security assessment and adaptive management in the Yangtze River Basin.

2. Materials and Methods

2.1. Study Area

The Yangtze River Basin is located between 90°33′–122°25′ E and 24°30′–35°45′ N, with a drainage area of about 1.8×10^6 km^2, accounting for roughly one-fifth of China's land area. Its main stream is approximately 6300 km long, with a total elevation drop of more than 5400 m from the source to the estuary, and flows eastward across the Tibetan Plateau, the Hengduan Mountains, the Sichuan Basin, and the middle–lower reaches plains, making it one of China's most water-abundant and ecologically crucial river basins.

Controlled by China's three-step topographic pattern, the basin exhibits an overall west–east elevation gradient, with higher terrain in the west and lower terrain in the east. The upper reaches are dominated by plateaus and deeply incised valleys, the middle reaches by mountains, hills, and basins, and the lower reaches by extensive alluvial plains and lake–wetland systems. This topographic transition produces pronounced relief and strong spatial heterogeneity across the basin.

Climatically, the Yangtze River Basin is generally situated in the subtropical monsoon zone, with a multi-year mean temperature of about 12.9 °C and a mean annual precipitation of approximately 1126.7 mm, exhibiting pronounced monsoonal and seasonal characteristics. Precipitation decreases roughly from the southeast to the northwest, and shows a "coincidence of rain and heat" in time, with flood-season rainfall (May–September) accounting for more than 60% of the annual total.

Marked interannual and intra-seasonal climatic variability, together with frequent extreme high-temperature and precipitation anomaly events, readily triggers regional droughts, floods, and dry–hot anomalies that may occur concurrently or in close succession at the monthly scale throughout the year, providing a favorable background for the occurrence of compound dry–hot events.

Ecologically, the basin hosts diverse vegetation types. Alpine meadows, shrublands, and coniferous forests dominate the upper reaches and northwestern areas, whereas evergreen broadleaved forests, mixed conifer–broadleaf forests, and shrublands prevail in the middle reaches and parts of the southern margins. while croplands, economic forests, and peri-urban artificial vegetation systems are predominant in the lower plains and riparian zones. Forest resources and cropland ecosystems jointly constitute the main body of terrestrial ecosystem productivity in the basin, playing a crucial role in maintaining regional carbon sink functions, soil and water conservation, and biodiversity. However, under the background of global warming, drought and high-temperature events in the Yangtze River Basin show increasing frequency, intensifying severity, and expanding spatial extent, and the potential impacts of compound dry–hot events on vegetation net primary productivity (NPP) and ecosystem stability are becoming increasingly prominent. Therefore, taking the Yangtze River Basin as the study area and systematically analyzing the spatiotemporal evolution of compound dry–hot events during 2002–2024 and their impacts on vegetation NPP is of considerable scientific and practical significance for elucidating the mechanisms of ecosystem responses to extreme climate in typical monsoonal basins in China and for assessing regional ecological security (see Figure 1).

https://doi.org/10.3390/w18020276

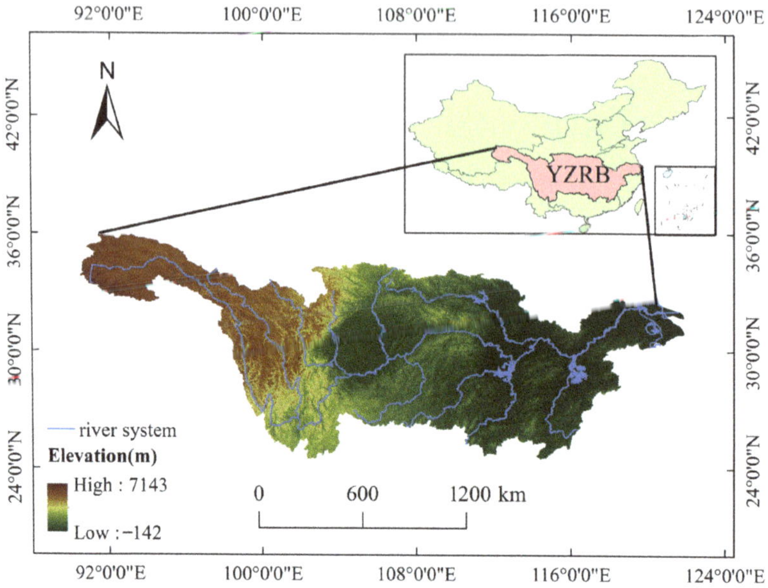

Figure 1. Geographic location of the Yangtze River Basin.

2.2. Data Sources

The primary datasets used in this study include 276 months (2002–2024) of meteorological variables such as potential evapotranspiration, temperature, and precipitation, together with NPP and GPP data. The meteorological data are all derived from the ERA5-Land Monthly Aggregates land reanalysis dataset (dataset ID: ECMWF/ERA5_LAND/MONTHLY_AGGR) provided by the European Centre for Medium-Range Weather Forecasts (ECMWF), with a spatial resolution of 0.1° and a monthly temporal resolution. The GPP data are obtained from the MOD17A3HGF V6.1 dataset (https://appeears.earthdatacloud.nasa.gov/ (accessed on 2 September 2025)), which has a spatial resolution of 500 m and a temporal resolution of 8 days. The NPP data are obtained from the MOD17A2HGF V6.1 dataset (https://appeears.earthdatacloud.nasa.gov/ (accessed on 5 September 2025)), which has a spatial resolution of 500 m and an annual temporal resolution (1 y). In this study, meteorological, GPP, and NPP data were processed using Python 3.10.6 for spatial resampling, and their spatial resolution was uniformly adjusted to 0.25° by applying a mean-resampling method. given that NPP is available only at an annual scale, whereas GPP has an 8-day temporal resolution, the annual NPP was allocated to the monthly scale according to the intra-annual relative contribution of GPP. Specifically, 8-day GPP values were first aggregated to each month, the proportion of monthly GPP to the annual total GPP was then calculated, and monthly NPP was obtained by scaling annual NPP according to these proportions [26,27], which has been widely applied to represent the intra-annual variability of vegetation productivity when monthly NPP products are unavailable.

The vegetation data used in this study are derived from the "1:1,000,000 Vegetation Atlas of China", downloaded from the Resource and Environment Science and Data Center of the Chinese Academy of Sciences (https://www.resdc.cn/ (accessed on 2 October 2025)). During preprocessing in ArcGIS 10.8, the main steps included projection, alignment, mosaicking, resampling, and clipping of the maps to ensure seamless spatial matching and consistency with the study area. Subsequently, vegetation types were reclassified according to the research objectives. The original vegetation classes were grouped into seven major

https://doi.org/10.3390/w18020276

categories, including cultivated vegetation, grassland, coniferous forest, broadleaved forest, shrubland, mixed conifer–broadleaf forest, and alpine vegetation. In addition, the source shapefile includes an "Others" class, which represents non-vegetated areas (i.e., areas without vegetation cover). This class is displayed in Figure 2 as background and is not counted as a vegetation type.

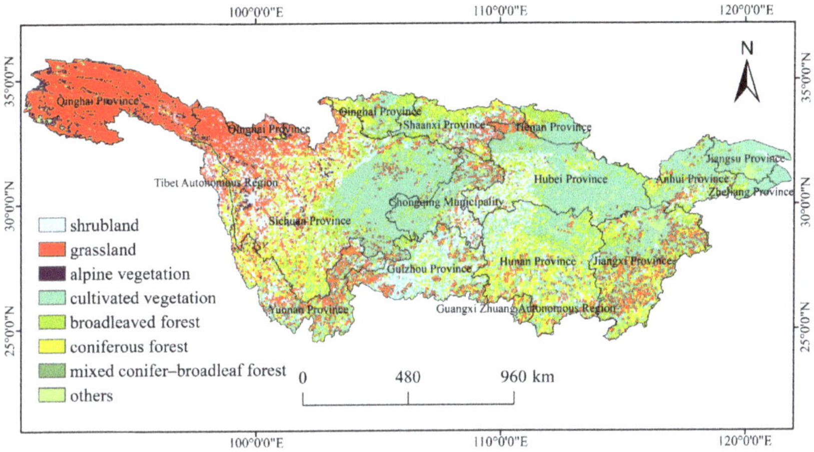

Figure 2. Spatial distribution of vegetation types in the Yangtze River Basin.

2.3. Methods

2.3.1. Standardized Temperature Index

The standardized temperature index (STI) is an indicator used to quantify and compare temperature conditions across different regions [28]. STI is commonly used to investigate the effects of climate change on temperature, the frequency and intensity of extreme temperature events, and to evaluate the impacts of temperature anomalies on ecosystems and human activities [29]. The index is first derived by computing the cumulative probability of monthly temperature using a Gamma distribution function, followed by a normal standardization of the results to obtain the final STI values [30]. The STI is calculated as follows

$$STI = \varphi^{-1}(F_{Gamma}(x)) \quad (1)$$

where x is the temperature value, F_{Gamma} denotes the cumulative distribution function of the Gamma distribution, and φ^{-1} is the inverse standard normal distribution function.

2.3.2. Standardized Precipitation Evapotranspiration Index

The standardized precipitation evapotranspiration index (SPEI) is a drought monitoring index that jointly considers precipitation and potential evapotranspiration. By accounting for these two components, SPEI provides a more comprehensive assessment of drought conditions [31]. The SPEI is calculated as follows

$$D_i = P_i - E_{p,i} \quad (2)$$

$$F(D_i - \gamma) = \frac{1}{1 + \left(\frac{\alpha}{D_i - \gamma}\right)^{\beta}} \quad (3)$$

$$SPEI = \phi^{-1}(F(D_i)) \quad (4)$$

https://doi.org/10.3390/w18020276

where D_i is the difference between precipitation and potential evapotranspiration, P_i is precipitation (mm), $E_{p,i}$ is potential evapotranspiration (mm), α is the scale parameter, β is the shape parameter, γ is the location parameter, $F(D_i)$ is the normalized cumulative probability, and $SPEI$ denotes the standardized precipitation evapotranspiration index. SPEI values greater than -0.5 indicate normal conditions, whereas SPEI values less than -0.5 indicate drought conditions [32].

2.3.3. Standardized Compound Dry–Hot Index

In this study, a pixel-wise standardized compound dry–hot index (SCDHI) is constructed within a Copula-based framework using the multitemporal gridded indices SPEI and STI [15]. Considering that moisture processes exhibit pronounced cumulative effects, whereas high-temperature processes are typically manifested as short-term anomalies at the monthly scale, this study adopts 3-month SPEI as the drought indicator and 1-month STI as the heat indicator. Compared with the 1-month scale, 3-month SPEI better integrates antecedent precipitation and evapotranspiration conditions, characterizes root-zone soil moisture and persistent moisture deficits, and is therefore more representative of the substantive stress on vegetation productivity [28,33,34]. A Copula function is a multidimensional joint distribution function defined on the 0–1 domain, which integrates the marginal distributions of multiple random variables and captures their dependence structure without relying on the specific form of the marginal distributions [35]. The joint probability of compound dry–hot events, P, can be expressed as

$$P(X \leq x, Y \geq y) = u - c(u, v) \tag{5}$$

where X is the climatic water balance (i.e., precipitation minus potential evapotranspiration) and Y is the maximum temperature. The values x and y are the corresponding drought and heatwave thresholds. A compound dry–hot condition is recorded when $X \leq x$ and $Y \geq y$. To characterize the joint behavior of drought and heat components on a unified probabilistic scale while allowing flexible dependence structures, copula-based multivariate standardized indices have been widely used in compound-event studies [15,35]. Accordingly, we adopted an Archimedean copula framework for pixel-wise implementation. Considering the large number of pixels and the need for robust parameter estimation, we selected the one-parameter Frank copula, which provides a parsimonious yet flexible representation of dependence and is computationally efficient for large-scale applications [35–37]. For each pixel, the copula parameter θ was estimated using maximum likelihood, and the copula-based joint cumulative distribution function was then computed to obtain the joint probability P. Uncertainty in P was quantified via a nonparametric bootstrap on 200 spatially stratified representative pixels (B = 200 resamples per pixel). For each pixel, P was recomputed after refitting the copula to resampled paired monthly (NPP, SCDHI) observations; the 95% CI was defined by the 2.5–97.5% percentiles of the bootstrap distribution, yielding a median CI width of 0.0025. Subsequently, P is transformed to a standard normal distribution using the inverse standard normal distribution function, yielding SCDHI as the indicator of compound dry–hot events, which is calculated as follows

$$SCDHI = \varphi^{-1}(P) \tag{6}$$

where φ^{-1} is the inverse standard normal distribution function. Similarly to STI and SPEI, SCDHI can also be classified into different categories, with lower SCDHI values indicating more severe compound dry–hot conditions. When $SCDHI < -0.5$ denotes a compound dry–hot event, whereas $SCDHI < -1.3$ denotes an extreme compound dry–hot event [38].

https://doi.org/10.3390/w18020276

2.3.4. Correlation Analysis and Response Timescale

To examine ecosystem responses to compound dry–hot events in the Yangtze River Basin, we computed pixel-wise Pearson correlation coefficients between monthly NPP and SCDHI over 2002–2024, as follows

$$R = \frac{\sum_{i=1}^{n}(x_i - \overline{x})(y_i - \overline{y})}{\sqrt{\sum_{i=1}^{n}(x_i - \overline{x})^2 \sum_{i=1}^{n}(y_i - \overline{y})^2}} \quad (7)$$

where R is the Pearson correlation coefficient between NPP and SCDHI. x_i and y_i are the NPP and SCDHI values in month i. $\overline{x}$ and $\overline{y}$ are the corresponding means of NPP and SCDHI over 2002–2024.

Vegetation growth typically responds to climatic factors with a time lag, and the response timescale varies among regions and vegetation types [39]. Previous studies have shown that the dominant lagged response of vegetation to climate variability is mostly within 3 months [39,40]. Accordingly, this study computes, at the pixel scale, the Pearson correlation coefficients between NPP and SCDHI in the current month (R_0), and with 1-month (R_1), 2-month (R_2), and 3-month (R_3) lags. For each pixel, the maximum absolute value among these four correlation coefficients is defined as the maximum correlation coefficient $R_{\max}$ between NPP and SCDHI, and the corresponding lag month is defined as the response timescale of NPP to dry–hot variability.

$$R_{\max} = \max(|R_i|), i = 0, 1, 2, 3 \quad (8)$$

$$T = t(R_{\max}) \quad (9)$$

where R_i is the correlation coefficient between NPP and SCDHI with an i-month lag. $R_{\max}$ represents the maximum correlation strength of NPP with dry–hot variability. T is the corresponding lag in months (0–3 months), indicating the dominant response timescale of vegetation to dry–hot variability.

2.3.5. Theil–Sen Median Slope Estimator and Mann–Kendall Trend Analysis

The Theil–Sen estimator is a nonparametric method for calculating slopes that can effectively characterize the magnitude of trends in a time series and serves as a useful complement to the Mann–Kendall trend test [41]. When the trend value is greater than 0, it indicates an increasing trend. Conversely, a negative value indicates a decreasing trend, and a value equal to 0 suggests no obvious trend in the region. The calculation formula is as follows

$$\beta = Median\left(\frac{x_i - x_j}{i - j}\right), \forall i > j \quad (10)$$

where x_i and x_j are the sample values at times i and j, respectively. *Median* denotes the median operator. $\beta > 0$ indicates an increasing trend in the SCDHI and NPP time series, whereas $\beta < 0$ indicates a decreasing trend.

The Mann–Kendall test is a nonparametric method used to detect whether a trend in a time series is statistically significant [42]. When the standardized test statistic $|Z| \geq 1.96$, it indicates passing the significance trend test at a 95% confidence level [41].

2.3.6. Assessment of Vegetation Resistance and Resilience Based on the NPP–SCDHI Framework

Building on the "resistance–resilience" framework proposed by Lloret et al. [40], this study quantifies vegetation resistance and resilience to extreme compound dry–hot events using the ratio of NPP during or after dry–hot conditions to NPP before dry–hot conditions (normal period), and extends this approach to the NPP–SCDHI framework at the pixel

https://doi.org/10.3390/w18020276

scale [43]. To quantitatively assess the response capacity of vegetation NPP to extreme compound dry–hot events in the Yangtze River Basin, vegetation responses to dry–hot events are divided into two dimensions, namely resistance during the event and recovery after the event. Based on SCDHI, extreme dry–hot months are identified using a threshold of −1.3. Months with $SCDHI < -1.3$ are defined as extreme dry–hot months, whereas other months are treated as non-extreme dry–hot months. At the pixel scale, the multi-year mean NPP of all non-extreme dry–hot months is calculated as the baseline productivity under normal climatic conditions and denoted as G.

$$\overline{G} = \frac{1}{N_{nor}} \sum_{t \in \Omega nor} G_t \tag{11}$$

where G_t is the NPP in month t. Ω_{nor} is the set of non-extreme dry–hot months. N_{nor} is the number of non-extreme dry–hot months, and G represents the multi-year mean NPP under normal conditions.

Considering that extreme dry–hot processes often persist for several consecutive months [44,45], all months with $SCDHI < -1.3$ are merged into individual extreme dry–hot events according to temporal continuity. Let the k-th extreme dry–hot event start in month s_k and end in month $(e_k(s_k \le e_k))$. The resistance of this event is defined as the ratio of the minimum NPP during the entire event to the baseline level.

$$R_k = \frac{G_{d_k}}{\overline{G}} \tag{12}$$

For each pixel, the mean resistance over all extreme dry–hot events during the study period is then calculated as

$$R = \frac{1}{N_{ext}} \sum_{k=1}^{N_{ext}} R_k \tag{13}$$

Here, N_{ext} is the number of extreme dry–hot events experienced by the pixel. If $R \approx 1$, NPP under extreme dry–hot conditions is close to the normal level, indicating strong resistance. $R < 1$ means that NPP during extreme dry–hot periods is lower than normal, and smaller R values indicate stronger suppression by dry–hot stress. $R > 1$ suggests that NPP during extreme dry–hot periods exceeds the normal level, reflecting strong maintenance capacity or pronounced effects of human management.

To characterize post-event recovery, this study adopts a one-year window after the end of each extreme compound dry–hot event as the recovery timescale. On the one hand, productivity at the annual (or full growing-season) scale can comprehensively reflect the overall recovery status of vegetation after a complete growing season [46]. On the other hand, many tree-ring- and remote-sensing-based studies commonly construct resistance–recovery–resilience metrics using "pre-drought year–drought year–first post-drought year (or the first 1–2 years)" and have found that the negative legacy effects of extreme drought are most pronounced in the first year after drought cessation [47–49]. For the k-th extreme dry–hot event, let $G_{\min,k}$ denote the minimum NPP during the event, $\overline{G}$ the multi-year mean NPP in non-event (normal) months, and G_{e_k+12} the NPP in the 12th month after the event ends. The resilience (relative recovery) of this event is defined as

$$\mathrm{Re}_k = \frac{G_{e_k+12} - G_{\min,k}}{\overline{G} - G_{\min,k}} \tag{14}$$

Similarly, by averaging Re_k over all extreme dry–hot events at each pixel, we obtain the pixel-scale mean resilience.

$$Re = \frac{1}{N_{ext}} \sum_{k=1}^{N_{ext}} Re_k \quad (15)$$

This index measures the fraction of the drought-induced productivity loss that has been regained one year after the event. When $Re \approx 1$, NPP has essentially recovered to the normal level, indicating strong resilience. When $Re < 1$ implies that only part of the loss has been compensated, indicating incomplete or delayed recovery. When $Re > 1$ indicates that NPP one year after the event exceeds the normal level, suggesting an over-recovery or compensatory growth effect.

3. Results

3.1. Spatiotemporal Variations in NPP

3.1.1. Temporal Variations in NPP

From 2002 to 2024, the annual mean of monthly NPP in the Yangtze River Basin showed a significant increasing trend (Figure 3), with a multi-year mean of approximately 46.12 $g{\cdot}m^{-2}{\cdot}month^{-1}$ and a Sen's slope of 0.255 $g{\cdot}m^{-2}{\cdot}month^{-1}{\cdot}yr^{-1}$ ($p < 0.001$). The Mann–Kendall test statistic (Kendall's *tau*) was 0.670. Basin-wide mean NPP ranged from 43.01 to 50.04 $g{\cdot}m^{-2}{\cdot}month^{-1}$, reaching a maximum in 2023 and a minimum in 2005.

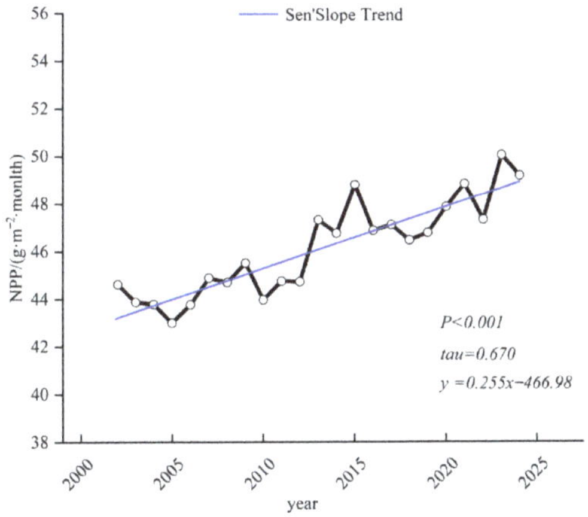

Figure 3. Interannual variations in annual mean of monthly NPP in the study area from 2002 to 2024.

3.1.2. Spatial Variations in NPP

The annual mean of monthly NPP in the Yangtze River Basin exhibits an overall spatial gradient, increasing from the northwest to the southeast, with low NPP values mainly occurring in the northwestern basin and high-elevation upstream areas, and high values concentrated in the middle–lower reaches plains and southeastern regions (Figure 4). This is because the middle–lower reaches and southeastern parts of the basin are characterized by low elevations, warm and humid climates, abundant precipitation, and favorable thermal conditions. These areas are dominated by forest and crops, thereby supporting relatively high NPP levels. In contrast, the upstream and northwestern basin are mostly high-cold or relatively arid areas with higher elevations, lower temperatures, and insufficient water supply in some regions, leading to shorter growing seasons, lower vegetation cover, and consequently generally lower NPP values.

https://doi.org/10.3390/w18020276

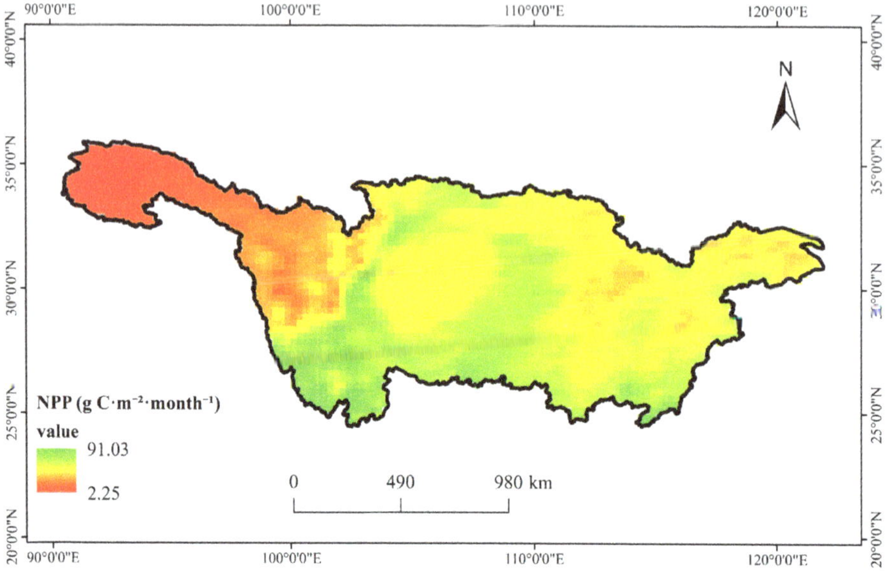

Figure 4. Spatial distribution of mean NPP in the Yangtze River Basin during 2002–2024.

The spatial patterns of NPP trends and their significance levels in the Yangtze River Basin are shown in Figures 5 and 6. During 2002–2024, NPP in most parts of the Yangtze River Basin shows an increasing trend, with approximately 85.8% of the area exhibiting NPP growth and about 16.7% identified as significantly increasing regions, mainly concentrated in the central basin and the middle–lower reaches plains. In contrast, NPP in the western basin and parts of the upper reaches is relatively stable, showing largely unchanged or only slight fluctuations. Meanwhile, about 2.5% of the area shows a decreasing NPP trend, sporadically distributed along the southern margin of the basin and in some hilly and mountainous areas in the east. Overall, during the study period, NPP changes in the Yangtze River Basin are characterized by "a generally slow increase, more pronounced growth in the central basin, relative stability in the west, and slight declines in some local areas," indicating an overall improvement in vegetation productivity across the basin, albeit with pronounced spatial heterogeneity.

3.2. Spatiotemporal Variations in SCDHI

3.2.1. Temporal Variations in SCDHI

From 2002 to 2024, the mean SCDHI in the Yangtze River Basin showed a significant decreasing trend (Figure 7), with a Sen's slope of -0.029 yr^{-1} ($p < 0.005$). The Mann–Kendall test statistic (Kendall's *tau*) was -0.447. During 2002–2024, the annual mean SCDHI fluctuated between approximately -2.03 and -0.73, with the highest value in 2012 (SCDHI = -0.73), indicating the weakest dry–hot conditions within the study period, and the lowest value in 2023 (SCDHI = -2.03), suggesting that dry–hot conditions in the Yangtze River Basin were most severe in 2023. Notably, although a severe drought occurred in the Yangtze River Basin in 2022, the basin-wide annual-mean SCDHI does not identify 2022 as the year with the lowest value. Overall, despite pronounced interannual fluctuations, SCDHI has shown a persistent decline over the past two decades, indicating an intensification of the compound drought background in the basin.

https://doi.org/10.3390/w18020276

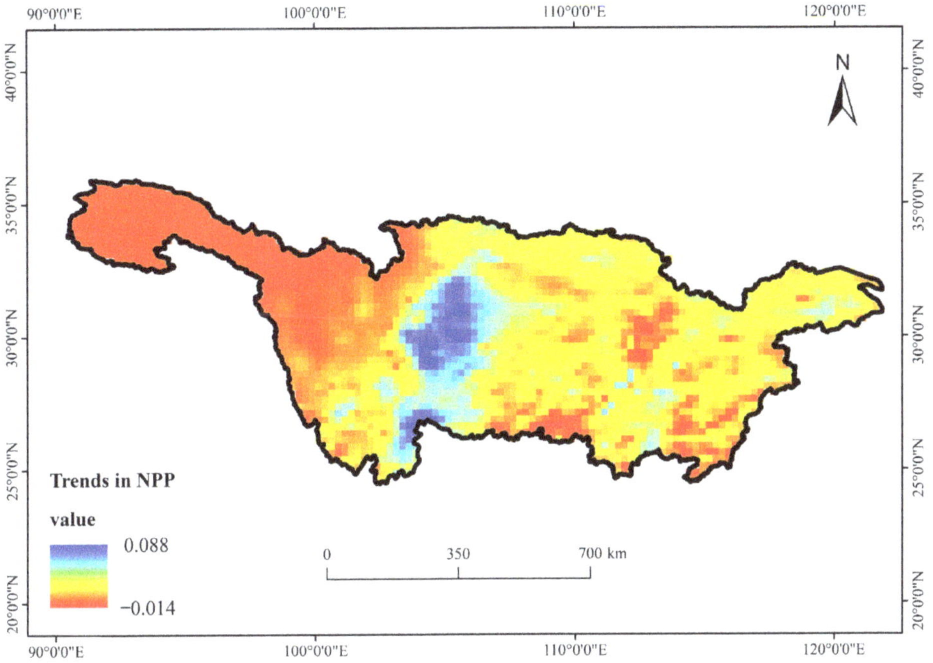

Figure 5. Spatial distribution of NPP trends in the Yangtze River Basin during 2002–2024.

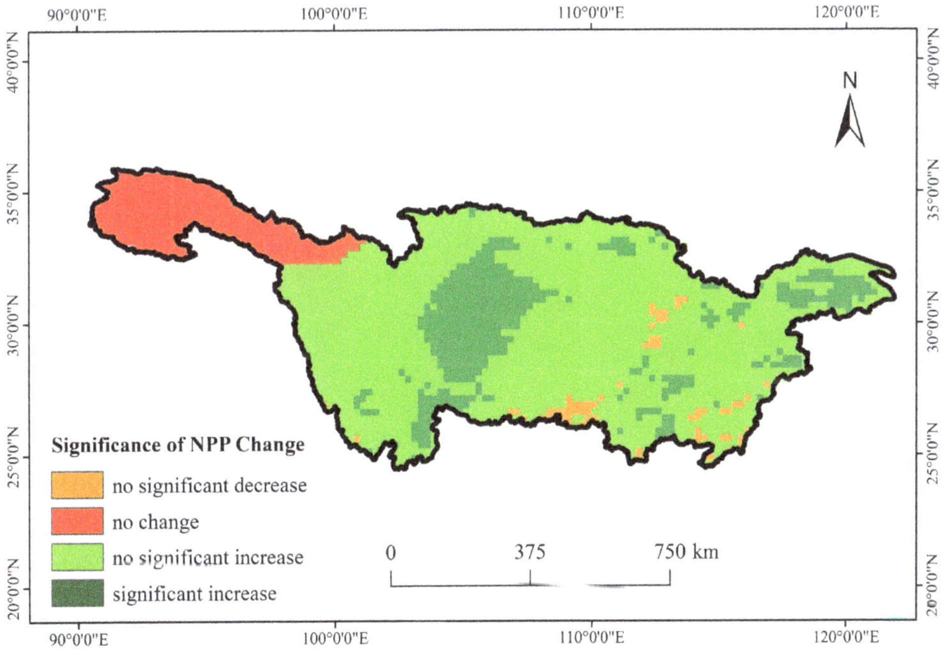

Figure 6. Spatial distribution of the significance levels of NPP trends in the Yangtze River Basin during 2002–2024.

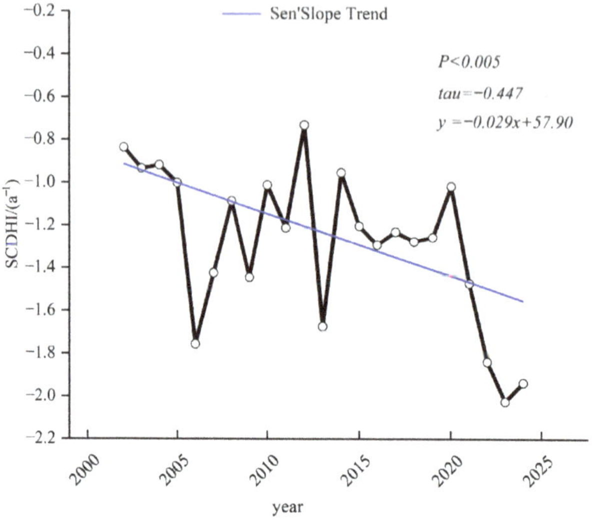

Figure 7. Interannual variations in SCDHI in the Yangtze River Basin from 2002 to 2024.

3.2.2. Spatial Variations in SCDHI

The spatial patterns of SCDHI trends and their significance levels in the Yangtze River Basin (Figures 8 and 9) show that during 2002–2024, SCDHI exhibits a decreasing trend over approximately 99.6% of the basin, with 70% of the area characterized by a significant decline, mainly distributed in the middle–lower reaches plains and the upper reaches around Sichuan–Chongqing and Yunnan–Guizhou, indicating that dry–hot conditions in these regions have been continuously intensifying.

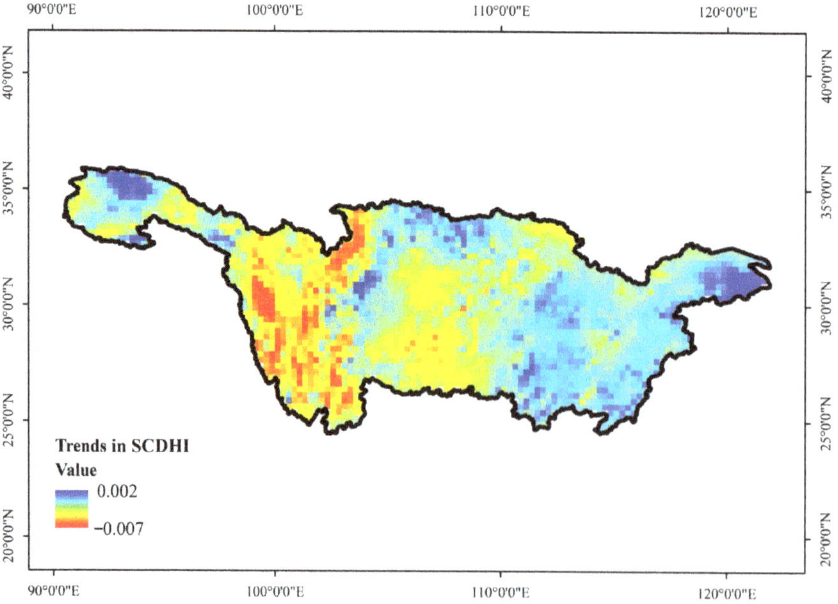

Figure 8. Spatial distribution of SCDHI trends in the Yangtze River Basin during 2002–2024.

https://doi.org/10.3390/w18020276

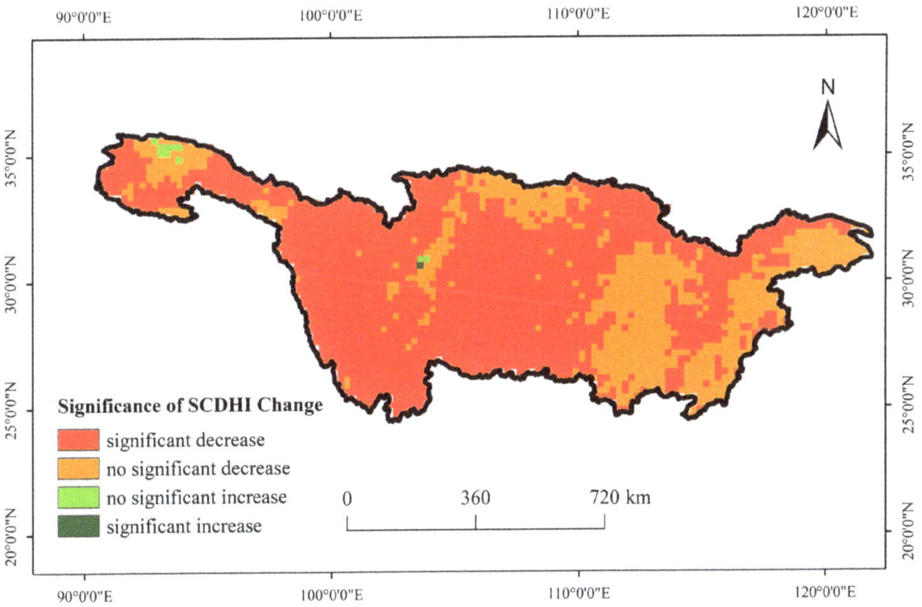

Figure 9. Spatial distribution of the significance levels of SCDHI trends in the Yangtze River Basin during 2002–2024.

Only about 0.4% of the area shows an increasing SCDHI trend, with significantly increasing regions sporadically distributed in the headwaters of the upper reaches and along the northwestern margin of the basin.

Overall, from 2002 to 2024, SCDHI trends in the Yangtze River Basin are dominated by declines, with a spatial pattern characterized by "slight local alleviation but a generally warmer and drier tendency," indicating a marked strengthening of the dry–hot background across the basin during the study period.

3.3. Response of NPP to Compound Dry–Hot Events

3.3.1. Relationship Between NPP and Compound Dry–Hot Events

During 2002–2024, NPP in approximately 47.00% of the Yangtze River Basin is positively correlated with compound dry–hot events (SCDHI < -0.5), among which about 39.72% of the area shows a significant positive correlation and 7.28% a non-significant positive correlation. These regions are mainly distributed in the middle–lower reaches plains of the Yangtze River and in the eastern part of the basin. In about 53.00% of the basin, NPP is negatively correlated with compound dry–hot events, with approximately 42.47% of the area exhibiting a significant negative correlation and 10.52% a non-significant negative correlation, mainly concentrated in the upper Yangtze and some hilly and mountainous areas of the middle reaches (Figures 10 and 11). Overall, NPP in the upper Yangtze tends to decline significantly under dry–hot conditions, whereas ecosystems in the middle–lower reaches are more likely to maintain or even increase NPP during compound dry–hot events, indicating marked differences among basin segments in both the direction and sensitivity of vegetation responses to such events.

https://doi.org/10.3390/w18020276

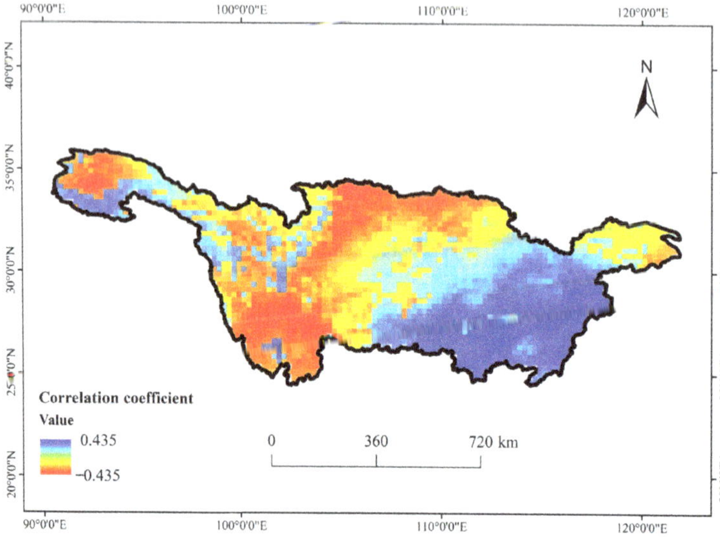

Figure 10. Spatial distribution of the maximum correlation coefficients between NPP and SCDHI in the Yangtze River Basin during 2002–2024.

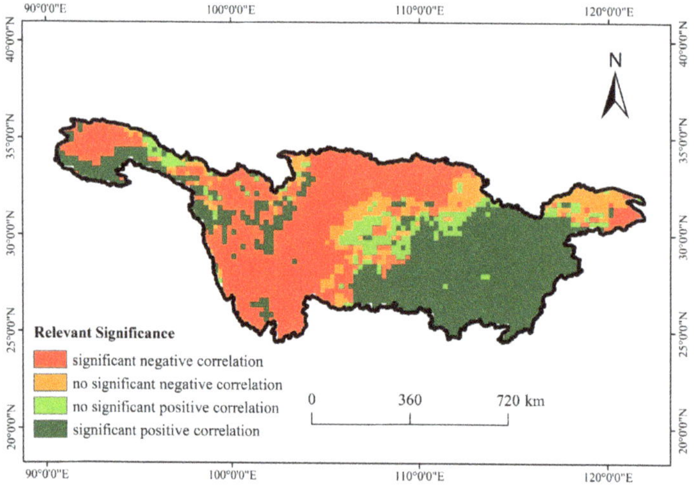

Figure 11. Spatial distribution of the significance levels of the maximum correlations between NPP and SCDHI in the Yangtze River Basin during 2002–2024.

3.3.2. Response Time of NPP to Compound Dry–Hot Events

From 2002 to 2024, the mean response time of NPP to compound dry–hot events in the Yangtze River Basin is about 2.03 months (Figure 12). Regions with a response time within the current month account for 21.50% of the basin and are mainly located on the Tibetan Plateau and in the alpine canyon areas of western Sichuan in the upper reaches, as well as in some headwater valleys. In these areas, vegetation is dominated by grasslands, shrublands, and alpine vegetation, with relatively shallow root systems and weak soil water-storage capacity, so that once a compound dry–hot event occurs, NPP responds rapidly within the same season. Same-season responses also occur in some tributary and main-stem valleys of the middle–lower reaches and around the Yangtze River Delta, but these areas are relatively small and mostly correspond to urban land, irrigated farmland,

and zones with dense hydraulic projects, where water and energy balance processes have been substantially modified, thereby amplifying and rapidly manifesting the impacts of compound dry–hot events on NPP. Areas with response times of 1 and 2 months account for 9.35% and 13.99%, respectively, and are patchily distributed in the transitional zones from the upper to the middle reaches and from the middle to the lower reaches, exhibiting a certain degree of lagged response to compound dry–hot variability. Regions with a response time of 3 months account for the largest proportion, reaching 55.15%, and are mainly distributed in the Jianghan Plain, the Dongting–Poyang Lake plains, and the surrounding hilly and mountainous areas in the middle–lower reaches. A similar 3-month lag-dominated response pattern also appears along the margins of the Sichuan Basin, in the mountains of western Sichuan, and in parts of the Yunnan–Guizhou Plateau in the southwestern basin. In these regions, vegetation is dominated by evergreen broadleaved forests, mixed conifer–broadleaf forests, and some cropland ecosystems, characterized by high vegetation cover, well-developed root systems, and generally deep soil layers that allow the use of deep soil water and groundwater to buffer short-term dry–hot stress. Together with the widespread lakes, wetlands, and irrigated agriculture in the middle–lower plains, and the strong soil water-storage and vegetation water-holding capacities in the southwestern mountains, these factors collectively enhance the "buffering–lag" effect of ecosystems in response to moisture fluctuations. Consequently, in these areas, anomalies in surface and shallow soil moisture require a certain time to propagate downward to deep soil layers and the root-active zone after the onset of a compound dry–hot event, so that the pronounced response of NPP to dry–hot stress often lags by more than one season.

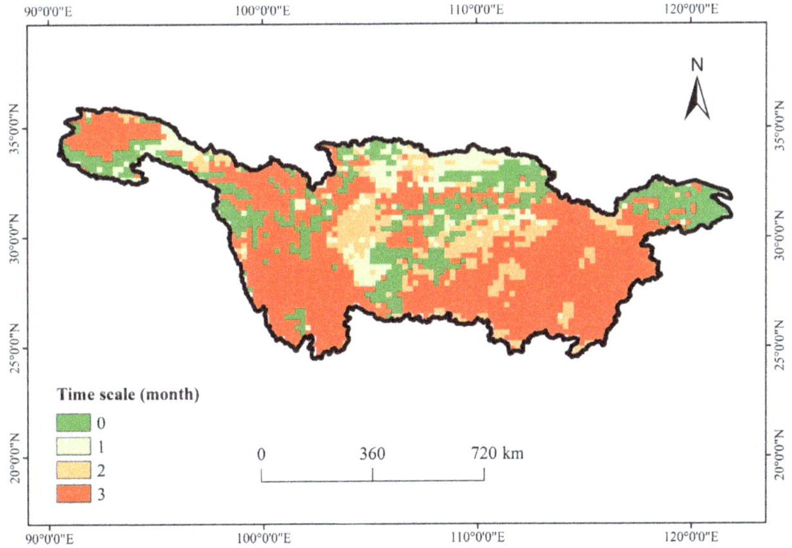

Figure 12. Spatial distribution of NPP responding time to compound dry–hot events in the Yangtze River Basin during 2002–2024.

3.4. Resistance and Resilience of Vegetation in the Yangtze River Basin to Extreme Compound Dry–Hot Events

At the basin scale, the mean resistance and resilience values of all vegetation types are less than 1 (Table 1). This indicates that during extreme compound dry–hot events, net primary productivity (NPP) across all vegetation types declines below the baseline under normal climatic conditions, and within one year after event termination only a fraction of the drought-induced productivity loss is recovered, precluding a full return to normal

https://doi.org/10.3390/w18020276

levels. Overall, extreme dry–hot events impose a pronounced and persistent suppressive effect on vegetation productivity in the Yangtze River Basin, although resistance during events and resilience after events differ substantially among vegetation types.

Table 1. Mean resistance and resilience of different vegetation types to extreme compound dry–hot events.

Vegetation Type	Mean Resistance	Mean Resilience
Coniferous forest	0.65	0.21
Mixed conifer–broadleaf forest	0.53	0.31
Broadleaved forest	0.65	0.37
Shrubland	0.57	0.18
Grassland	0.43	0.32
Alpine vegetation	0.60	0.46
Cultivated vegetation	0.25	0.17

Among vegetation types, coniferous and broadleaved forests both exhibit relatively high resistance, with mean values of approximately 0.65, indicating that NPP during extreme compound dry–hot events remains around 65% of the normal level. However, their resilience is generally low and differs markedly. Resilience is only 0.21 in coniferous forests and 0.37 in broadleaved forests, suggesting that forests—especially coniferous forests—can buffer immediate impacts but recover only a limited proportion of lost productivity within one year after the events. Mixed conifer–broadleaf forests show slightly lower resistance but relatively faster recovery, with resistance of 0.53 and resilience of 0.31, implying larger instantaneous losses yet somewhat quicker post-event rebound. Shrublands have moderate resistance (0.57) but low resilience (0.18), indicating pronounced NPP declines during extreme dry–hot events and recovery of less than one-fifth of the productivity loss within one year. Cultivated vegetation exhibits the strongest sensitivity to these events, with the lowest resistance among all vegetation types (0.25), reflecting the largest productivity reductions during extreme compound dry–hot events. Its resilience is also low (0.17), indicating that within one year after event termination, productivity losses are only partially compensated and productivity remains far below normal levels. Grasslands are characterized by relatively weak resistance (0.43) but comparatively rapid recovery (0.32), indicating substantial productivity losses during events yet recovery of nearly one-third of the drought-induced loss within one year. Alpine vegetation shows moderate resistance (0.60) and the highest resilience (0.46), recovering nearly half of the lost productivity within one year and demonstrating strong post-event recovery capacity. Taken together, these results reveal pronounced divergence in vegetation responses to extreme compound dry–hot events across the Yangtze River Basin. Forests exhibit higher resistance but limited recovery on annual timescales. Shrublands have moderate resistance but low resilience. Cultivated vegetation lies in a low-resistance/low-resilience regime. Grasslands have weak resistance but relatively fast recovery. Alpine vegetation combines moderate resistance with the highest resilience.

4. Discussion

The results of this study indicate that from 2002 to 2024, NPP in the Yangtze River Basin exhibited an overall slow increasing trend, whereas the compound dry–hot background (SCDHI) evolved markedly towards warmer and drier conditions. Consequently, the two variables display a clear pattern of divergent trends and regional differentiation in both time and space. Specifically, NPP increased in approximately 85.8% of the basin, while SCDHI decreased in roughly 99.6% of the area. This implies that despite ongoing global warming and intensifying compound dry–hot events (CDHEs), vegetation productivity in

https://doi.org/10.3390/w18020276

the basin remains, on the whole, on an upward trajectory. This spatial predominance of NPP increases is consistent with MODIS-based assessments reporting widespread productivity gains across the Yangtze River Basin since the early 2000s and attributing them to ecological restoration and land-use management in addition to climate effects [50]. At broader scales, satellite syntheses have similarly highlighted a strong greening signal since around 2000, in which land management and CO_2-driven physiological effects contribute alongside climate variability [51,52]. In contrast, the pronounced decline in SCDHI accords with the broader literature showing that compound drought–heat extremes have intensified in many regions under warming, and that their ecological impacts often exceed those of drought or heat alone because soil moisture deficits and high atmospheric demand co-occur [1,15]. Taken together, the divergence (increasing NPP but decreasing SCDHI) suggests that greening processes and the relaxation of energy limitation can partially offset the rising dry–hot risk at the basin scale, even though negative impacts remain evident in specific sub-regions.

Notably, in the basin-wide annual-mean SCDHI series, the widely reported 2022 Yangtze River Basin drought is not identified as the year with the lowest value. This is not inconsistent with evidence that 2022 constituted an exceptional drought–heat episode with extremely rare occurrence when both meteorological anomalies and impacts are considered [53]. Rather, the discrepancy likely reflects scale and metric effects. Short-lived extremes can be diluted by annual and spatial averaging, and the 3-month SPEI component embedded in SCDHI further smooths short-duration signals. Consistent with this interpretation, event-focused studies using sub-seasonal remote sensing indicators show that the 2022 summer event caused substantial suppression of vegetation activity and carbon uptake, including marked declines in photosynthesis-related proxies and GPP during July–August 2022 [54,55]. These comparisons highlight that basin-mean annual indices are well suited for characterizing background conditions, whereas event-scale diagnostics are necessary to capture short, intense compound extremes and their immediate ecosystem impacts.

Meanwhile, the NPP–SCDHI relationship is spatially heterogeneous. NPP is negatively correlated with SCDHI over approximately 53% of the basin and positively correlated over approximately 47%. This contrast likely reflects a shift in dominant controls along hydroclimatic gradients, from energy limitation in cold/high regions to water limitation in warm/humid lowlands [51]. In the upper reaches and some hilly middle-reach areas, NPP tends to increase or remain stable despite intensifying dry–hot conditions, suggesting that warming-induced relaxation of temperature/radiation constraints and ecological restoration can offset dry–hot stress [50–52]. In the humid middle–lower reaches and plains, however, stronger dry–hot conditions exacerbate soil moisture deficits and atmospheric demand (higher VPD), suppressing stomatal conductance and carbon assimilation and leading to more pronounced NPP declines [56]. Although water regulation, irrigation, improved management, and forestry practices can buffer impacts, they mainly reduce losses and accelerate recovery rather than reversing downward trajectories under moderate to severe dry–hot stress. In cropland-dominated areas of the mid–lower basin, irrigation and agricultural management may partially buffer dry–hot stress and thereby alter the apparent NPP response to dry–hot conditions. Similar management effects on CASA-based estimates of cropland NPP and carbon sequestration have been reported for maize in the Fen River Basin [57]. However, because spatially explicit irrigation intensity and management indicators are not included in our analysis, we discuss these anthropogenic influences as plausible mechanisms rather than providing a quantitative causal attribution. Future work incorporating irrigation intensity and management datasets would help quantify the contribution of anthropogenic regulation. Overall, these contrasting mechanisms jointly produce the coexisting positive and negative NPP responses across the basin.

https://doi.org/10.3390/w18020276

The distribution of response times reveals buffering and lag characteristics in hydro-ecological processes. The mean response time of NPP to CDHEs in the basin is approximately 2.03 months. Rapid responses (in the current month or with a lag of less than one month) dominate on the Tibetan Plateau and in the alpine canyons of western Sichuan. In these regions, shallow soils, limited water-storage capacity, and alpine vegetation types mean that soil moisture is rapidly depleted during dry–hot episodes, causing immediate vegetation stress. Conversely, lagged responses (2–3 months) prevail in the middle–lower plains. Here, thicker soil layers and vegetation with deeper root systems (e.g., forests) benefit from the groundwater, lakes, and wetlands. This buffers short-term drought, delaying the transmission of surface moisture anomalies to the root zone. Similar multi-month lags have been reported in satellite-based drought-impact studies, with response times modulated by rooting depth, soil water storage, and local hydroclimatic conditions [49]. Our results quantify, at the basin scale, the contrast between rapid responses in cold, high-elevation regions and delayed responses in humid regions under compound dry–hot stress.

From the perspective of ecosystem stability, differences in resistance and resilience among vegetation types delineate clear vulnerability patterns across the Yangtze River Basin. Vegetation productivity is broadly suppressed during extreme compound dry–hot events, and full recovery remains difficult within one year after event termination. Forest ecosystems generally exhibit strong resistance, but their post-event recovery is relatively constrained. This resistance–resilience trade-off is consistent with flux-based evidence showing that forests generally exhibit stronger resistance but weaker resilience than grasslands under extreme drought conditions [58]. Shrublands typically show moderate resistance yet limited resilience, whereas cultivated vegetation is characterized by both low resistance and low resilience, making it particularly susceptible to persistent productivity deficits. Grasslands often experience larger productivity reductions during events but tend to recover more rapidly afterward, while alpine vegetation shows comparatively strong recovery capacity. Global syntheses further indicate that regions with higher forest fractions are generally more resistant to extreme drought than croplands, and that irrigation can substantially enhance cropland drought resistance [59]. This helps explain why cultivated vegetation in the Yangtze River Basin remains vulnerable despite management buffering. Overall, these response characteristics reveal pronounced vulnerability contrasts among vegetation types and suggest that cultivated vegetation, together with some grassland systems, may represent especially sensitive components of basin ecosystems under intensifying compound dry–hot stress.

It should be noted that although this study provides a relatively systematic depiction of NPP responses to compound dry–hot events and their type-specific differences at the basin scale, it remains constrained by data and methodological limitations. Uncertainties exist in the spatial resolution and retrieval accuracy of ERA5-Land reanalysis data and MODIS NPP products, and after harmonizing multiple datasets to a 0.25° spatial grid, it is difficult to fully capture fine-scale differences in topography, soils, and land use at hillslope and small-basin scales. In addition, this study did not perform an independent accuracy evaluation or quantitative uncertainty propagation for the input datasets; therefore, the uncertainty of our results largely inherits the limitations documented for ERA5-Land and MODIS products as well as the uncertainties introduced by the harmonization workflow. SCDHI mainly integrates information on temperature and water balance and does not yet explicitly incorporate key factors such as soil moisture, radiation, wind speed, vegetation structural characteristics, and the intensity of human activities. Thus, its representation of compound dry–hot processes and their ecological effects remains relatively simplified, and it is not yet possible to quantitatively disentangle the relative contributions of climate change versus land-use/management changes to NPP. In the future, it will be necessary, under the

https://doi.org/10.3390/w18020276

support of higher-resolution meteorological and ecological datasets, to incorporate multi-source soil-moisture and vegetation-structure observations and to integrate ecosystem process models in order to assess, from the perspectives of individual event processes and different climate scenarios, the long-term impacts of compound dry–hot events on ecosystem structure and services in the Yangtze River Basin, thereby providing more refined and differentiated scientific evidence for basin-scale adaptive management and ecological restoration strategies.

5. Conclusions

This study used ERA5-Land reanalysis and MODIS NPP/GPP data (2002–2024) to construct a pixel-scale standardized compound dry–hot index (SCDHI) for the Yangtze River Basin and to quantify NPP responses (magnitude, direction, and lag), as well as vegetation resistance and resilience under compound dry–hot events. The main conclusions are:

(1) Basin-wide annual mean of monthly NPP increased slightly from 2002 to 2024 (43.01–50.04 g m^{-2} $month^{-1}$), peaking in 2023 and reaching a minimum in 2005. NPP increased from northwest to southeast and 85.8% of the basin exhibited an upward trend (16.7% significant), whereas only limited areas showed weak declines or remained stable.

(2) Annual mean SCDHI decreased significantly ($p < 0.005$) at approximately -0.029 yr^{-1}, indicating an overall drying and warming trend. The weakest dry–hot conditions occurred in 2012 and the most severe in 2023. Spatially, 99.6% of the basin showed decreasing SCDHI, with about 70% declining significantly.

(3) Under dry–hot conditions (SCDHI < -0.5), NPP was positively correlated with dry–hot intensity over approximately 47% of the basin (mainly the middle and lower reaches and eastern areas) and negatively correlated over 53% (primarily the upper reaches and parts of the middle-reach hills). The mean response time was about 2 months, with faster responses in the upper and high-elevation western regions (0–1 month) and longer lags in the humid middle–lower reaches (2–3 months).

(4) Extreme compound dry–hot events affected vegetation stability, with clear differences among vegetation types. Forests showed relatively high resistance but limited recovery, shrublands exhibited moderate resistance but low resilience, and cultivated vegetation had the lowest resistance and resilience and was therefore the most sensitive. Grasslands showed weak resistance but comparatively faster recovery, whereas alpine vegetation exhibited the highest resilience. Overall, these patterns suggest that cultivated vegetation and grasslands may represent high-risk vegetation types under extreme dry–hot stress.

Author Contributions: Conceptualization, G.Z.; methodology, H.X.; formal analysis, H.X.; investigation, H.X. and H.W.; data curation, H.X. and H.W.; writing—original draft preparation, H.X.; writing—review and editing, H.X., G.Z. and H.W.; supervision, G.Z.; funding acquisition, G.Z. All authors have read and agreed to the published version of the manuscript.

Funding: This study is supported by the China Postdoctoral Science Foundation (Grant No. 2024M752711), and the Natural Science Foundation of Jiangsu Province (Grant No. BK20220590).

Data Availability Statement: The datasets analyzed in this study are publicly available. ERA5 Land Monthly Aggregates (dataset ID: ECMWF/ERA5_LAND/MONTHLY_AGGR) are available from the European Centre for Medium-Range Weather Forecasts (ECMWF). MODIS MOD17A2HGF V6.1 and MODIS MOD17A3HGF V6.1 are available from the NASA Land Processes Distributed Active Archive Center (LP DAAC) (https://lpdaac.usgs.gov/). The vegetation data used in this study are derived from the 1:1,000,000 Vegetation Atlas of China, downloaded from the Resource and Environment Science and Data Center of the Chinese Academy of Sciences (https://www.resdc.cn/).

https://doi.org/10.3390/w18020276

Conflicts of Interest: The authors declare no conflicts of interest.

References

1. Mukherjee, S.; Mishra, A.K. Increase in Compound Drought and Heatwaves in a Warming World. *Geophys. Res. Lett.* **2021**, *48*, e2020GL090617. [CrossRef]
2. Liu, T.; Zhang, Y.; Guo, B.; Yin, Y.; Ge, J. Projected changes of compound droughts and heatwaves in China under 1.5 °C, 2 °C, and 3 °C of global warming. *Clim. Dyn.* **2024**, *62*, 6417–6431. [CrossRef]
3. Liu, M.; Yu, H.; Duan, W.; Wu, M. Differential impacts of compound dry- and humid-hot events on global vegetation productivity. *Front. Environ. Sci.* **2025**, *13*, 1597553. [CrossRef]
4. Wang, A.; Tao, H.; Ding, G.; Zhang, B.; Huang, J.; Wu, Q. Global cropland exposure to extreme compound drought heatwave events under future climate change. *Weather Clim. Extrem.* **2023**, *40*, 100559. [CrossRef]
5. Jiang, L.; Zhang, J.; Bai, L.; Han, J.; Meng, X.; Cao, D.; Al-Sakkaf, A.S. Increased frequency and severity of global compound dry and heat wave events in a daily scale. *J. Hydrol.* **2025**, *654*, 132857. [CrossRef]
6. Ma, Z.; Wu, J.; Yang, H.; Hong, Z.; Yang, J.; Gao, L. Assessment of vegetation net primary productivity variation and influencing factors in the Beijing-Tianjin-Hebei region. *J. Environ. Manag.* **2024**, *365*, 121490. [CrossRef]
7. Mehmood, K.; Anees, S.A.; Rehman, A.; Rehman, N.U.; Muhammad, S.; Shahzad, F.; Liu, Q.; Alharbi, S.A.; Alfarraj, S.; Ansari, M.J.; et al. Assessment of climatic influences on net primary productivity along elevation gradients in temperate ecoregions. *Trees For. People* **2024**, *18*, 100657. [CrossRef]
8. Yang, J.; Zhang, X.; Luo, Z.; Yu, X. Nonlinear Variations of Net Primary Productivity and Its Relationship with Climate and Vegetation Phenology, China. *Forests* **2017**, *8*, 361. [CrossRef]
9. Kamatchi, K.A.M.; Anitha, K.; Kumar, K.A.; Senthil, A.; Kalarani, M.K.; Djanaguiraman, M. Impacts of combined drought and high-temperature stress on growth, physiology, and yield of crops. *Plant Physiol. Rep.* **2023**, *29*, 28–36. [CrossRef]
10. Ciais, P.; Reichstein, M.; Viovy, N.; Granier, A.; Ogee, J.; Allard, V.; Aubinet, M.; Buchmann, N.; Bernhofer, C.; Carrara, A.; et al. Europe-wide reduction in primary productivity caused by the heat and drought in 2003. *Nature* **2005**, *437*, 529–533. [CrossRef]
11. Luo, M.; Meng, F.; Sa, C.; Bao, Y.; Liu, T.; De Maeyer, P. Detecting Drought-Related Temporal Effects on Global Net Primary Productivity. *Remote Sens.* **2024**, *16*, 3787. [CrossRef]
12. Li, Y.; Ma, S.; Zhang, X.; Gao, L. ASD and ADHD: Divergent activating patterns of prefrontal cortex in executive function tasks? *J. Psychiatr. Res.* **2024**, *172*, 187–196. [CrossRef] [PubMed]
13. Sun, H.; Wang, J.; Xiong, J.; Bian, J.; Jin, H.; Cheng, W.; Li, A.; García Mozo, H. Vegetation Change and Its Response to Climate Change in Yunnan Province, China. *Adv. Meteorol.* **2021**, *2021*, 8857589. [CrossRef]
14. Han, J.; Singh, V.P. A review of widely used drought indices and the challenges of drought assessment under climate change. *Environ. Monit. Assess.* **2023**, *195*, 1438. [CrossRef] [PubMed]
15. Li, J.; Wang, Z.; Wu, X.; Zscheischler, J.; Guo, S.; Chen, X. A standardized index for assessing sub-monthly compound dry and hot conditions with application in China. *Hydrol. Earth Syst. Sci.* **2021**, *25*, 1587–1601. [CrossRef]
16. Liu, Y.; Hu, T.; Zuo, Q.; Yu, L.; Yang, J. Understanding copula-based multivariate standardized drought indices for characterizing meteorological, hydrological and agricultural droughts across global land areas. *Agric. Water Manag.* **2025**, *320*, 109864. [CrossRef]
17. Zhang, Y.; Hao, Z.; Zhang, Y. Agricultural risk assessment of compound dry and hot events in China. *Agric. Water Manag.* **2023**, *277*, 108128. [CrossRef]
18. Faiz, M.A.; Zhang, Y.; Zhang, X.; Ma, N.; Aryal, S.K.; Ha, T.T.V.; Baig, F.; Naz, F. A composite drought index developed for detecting large-scale drought characteristics. *J. Hydrol.* **2022**, *605*, 127308. [CrossRef]
19. Zhu, M.; Zhang, Z.; Zhu, B.; Kong, R.; Zhang, F.; Tian, J.; Jiang, T. Population and Economic Projections in the Yangtze River Basin Based on Shared Socioeconomic Pathways. *Sustainability* **2020**, *12*, 127308. [CrossRef]
20. Xia, J.; Li, Z.; Zeng, S.; Zou, L.; She, D.; Cheng, D. Perspectives on eco-water security and sustainable development in the Yangtze River Basin. *Geosci. Lett.* **2021**, *8*, 18. [CrossRef]
21. Duan, A.; Wu, G.; Wang, B.; Turner, A.G.; Hu, J.; Hu, W.; Zhang, P.; Hu, D.; Tang, Y. Drivers of East Asian summer monsoon variability: Global oceans and the Tibetan Plateau. *Sci. Bull.* **2024**, *69*, 2487–2490. [CrossRef] [PubMed]
22. Zhu, J.; Guo, H.; Sun, Y. Responses of Terrestrial GPP to Extreme Compound Heatwave and Drought Events of Different Intensities in the Yangtze River Basin. *Remote Sens.* **2025**, *17*, 848. [CrossRef]
23. Bai, W.; Zhang, C.; Xiao, X.; Zou, Z.; Liu, Z.; Li, P.; Tang, J.; Li, T.; Zhou, X.; Peng, C. Characteristics of Meteorological Drought Evolution in the Yangtze River Basin. *Water* **2024**, *16*, 3391. [CrossRef]
24. Huang, L.; Zhou, P.; Cheng, L.; Liu, Z. Dynamic drought recovery patterns over the Yangtze River Basin. *Catena* **2021**, *201*, 105194. [CrossRef]
25. Zhang, Y.; Gong, J.; Yang, J.; Peng, J. Evaluation of Future Trends Based on the Characteristics of Net Primary Production (NPP) Changes over 21 Years in the Yangtze River Basin in China. *Sustainability* **2023**, *15*, 10606. [CrossRef]

https://doi.org/10.3390/w18020276

26. Song, S.; Niu, J.; Singh, S.K.; Du, T. Projection of net primary production under changing environment in Xinjiang using an improved wCASA model. *J. Hydrol.* **2023**, *620*, 129314. [CrossRef]
27. Du, K.; Huang, J.; Wang, W.; Zeng, Y.; Li, X.; Zhao, F. Monitoring Low-Temperature Stress in Winter Wheat Using TROPOMI Solar-Induced Chlorophyll Fluorescence. *IEEE Trans. Geosci. Remote Sens.* **2024**, *62*, 4402111. [CrossRef]
28. Zhang, Q.; Yu, X.; Qiu, R.; Liu, Z.; Yang, Z. Evolution, severity, and spatial extent of compound drought and heat events in north China based on copula model. *Agric. Water Manag.* **2022**, *273*, 107918. [CrossRef]
29. Monteleone, B.; Borzí, I. Drought in the Po Valley: Identification, Impacts and Strategies to Manage the Events. *Water* **2024**, *16*, 1187. [CrossRef]
30. Wang, H.; Zhang, G.; Zhang, S.; Shi, L.; Su, X.; Song, S.; Feng, K.; Zhang, T.; Fu, X. Development of a novel daily-scale compound dry and hot index and its application across seven climatic regions of China. *Atmos. Res.* **2023**, *287*, 106700. [CrossRef]
31. Vicente-Serrano, S.M.; Beguería, S.; López-Moreno, J.I. A Multiscalar Drought Index Sensitive to Global Warming: The Standardized Precipitation Evapotranspiration Index. *J. Clim.* **2010**, *23*, 1696–1718. [CrossRef]
32. Yang, Y.; Dai, E.; Yin, J.; Jia, L.; Zhang, P.; Sun, J. Spatial and Temporal Evolution Patterns of Droughts in China over the Past 61 Years Based on the Standardized Precipitation Evapotranspiration Index. *Water* **2024**, *16*, 1012. [CrossRef]
33. Kong, J.; Zan, M.; Chen, Z.; Xue, C.; Yang, S. Study on the Response of Vegetation Water Use Efficiency to Drought in the Manas River Basin, Xinjiang, China. *Forests* **2024**, *15*, 114. [CrossRef]
34. Zhan, C.; Liang, C.; Zhao, L.; Jiang, S.; Niu, K.; Zhang, Y. Drought-related cumulative and time-lag effects on vegetation dynamics across the Yellow River Basin, China. *Ecol. Indic.* **2022**, *143*, 109409. [CrossRef]
35. Hao, Z.; AghaKouchak, A. Multivariate Standardized Drought Index: A parametric multi-index model. *Adv. Water Resour.* **2013**, *57*, 12–18. [CrossRef]
36. Ayantobo, O.O.; Wei, J. Appraising regional multi-category and multi-scalar drought monitoring using standardized moisture anomaly index (SZI): A water-energy balance approach. *J. Hydrol.* **2019**, *579*, 124139. [CrossRef]
37. Feng, K.; Su, X.; Singh, V.P.; Ayantobo, O.O.; Zhang, G.; Wu, H.; Zhang, Z. Dynamic evolution and frequency analysis of hydrological drought from a three-dimensional perspective. *J. Hydrol.* **2021**, *600*, 126675. [CrossRef]
38. Wu, X.; Hao, Z.; Zhang, X.; Li, C.; Hao, F. Evaluation of severity changes of compound dry and hot events in China based on a multivariate multi-index approach. *J. Hydrol.* **2020**, *583*, 124580. [CrossRef]
39. Wu, D.; Zhao, X.; Liang, S.; Zhou, T.; Huang, K.; Tang, B.; Zhao, W. Time-lag effects of global vegetation responses to climate change. *Glob. Change Biol.* **2015**, *21*, 3520–3531. [CrossRef]
40. Wang, Y.; Dai, E.; Wu, C. Spatiotemporal heterogeneity of net primary productivity and response to climate change in the mountain regions of southwest China. *Ecol. Indic.* **2021**, *132*, 108273. [CrossRef]
41. Yan, F.; Zhang, Y.; Wang, X.; Xu, Z.; Liang, Y.; Wang, Z.; Wang, J.; Chen, Y.; Zhu, Z. Characteristics of spatial and temporal non-stationarity of groundwater storage in different basins of China and its driving mechanisms. *J. Hydrol.* **2025**, *655*, 132882. [CrossRef]
42. Cao, D.; Zhang, J.; Han, J.; Zhang, T.; Yang, S.; Wang, J.; Prodhan, F.A.; Yao, F. Projected Increases in Global Terrestrial Net Primary Productivity Loss Caused by Drought Under Climate Change. *Earth's Future* **2022**, *10*, e2022EF002681. [CrossRef]
43. Lloret, F.; Keeling, E.G.; Sala, A. Components of tree resilience: Effects of successive low-growth episodes in old ponderosa pine forests. *OIKOS* **2011**, *120*, 1909–1920. [CrossRef]
44. Feng, S.; Wu, X.; Hao, Z.; Hao, Y.; Zhang, X.; Hao, F. A database for characteristics and variations of global compound dry and hot events. *Weather Clim. Extrem.* **2020**, *30*, 100299. [CrossRef]
45. Yu, R.; Zhai, P. Changes in compound drought and hot extreme events in summer over populated eastern China. *Weather Clim. Extrem.* **2020**, *30*, 100295. [CrossRef]
46. Shao, X.; Zhang, Y.; Ma, N.; Zhang, X.; Tian, J.; Xu, Z.; Liu, C. Drought-induced ecosystem resistance and recovery observed at 118 flux tower stations across the globe. *Agric. For. Meteorol.* **2024**, *356*, 110170. [CrossRef]
47. Schwarz, J.; Skiadaresis, G.; Kohler, M.; Kunz, J.; Schnabel, F.; Vitali, V.; Bauhus, J. Quantifying Growth Responses of Trees to Drought—A Critique of Commonly Used Resilience Indices and Recommendations for Future Studies. *Curr. For. Rep.* **2020**, *6*, 185–200. [CrossRef]
48. Serra-Maluquer, X.; Mencuccini, M.; Martinez-Vilalta, J. Changes in tree resistance, recovery and resilience across three successive extreme droughts in the northeast Iberian Peninsula. *Oecologia* **2018**, *187*, 343–354. [CrossRef]
49. Dong, B.; Yu, Y.; Pereira, P. Non-growing season drought legacy effects on vegetation growth in southwestern China. *Sci. Total Environ.* **2022**, *846*, 157334. [CrossRef]
50. Liu, C.; Shi, S.; Wang, T.; Gong, W.; Xu, L.; Shi, Z.; Du, J.; Qu, F. Analysis of Net Primary Productivity Variation and Quantitative Assessment of Driving Forces-A Case Study of the Yangtze River Basin. *Plants* **2023**, *12*, 3412. [CrossRef]
51. Zhu, Z.; Piao, S.; Myneni, R.B.; Huang, M.; Zeng, Z.; Canadell, J.G.; Ciais, P.; Sitch, S.; Friedlingstein, P.; Arneth, A.; et al. Greening of the Earth and its drivers. *Nat. Clim. Change* **2016**, *6*, 791–795. [CrossRef]

52. Chen, C.; Park, T.; Wang, X.; Piao, S.; Xu, B.; Chaturvedi, R.K.; Fuchs, R.; Brovkin, V.; Ciais, P.; Fensholt, R.; et al. China and India lead in greening of the world through land-use management. *Nat. Sustain.* **2019**, *2*, 122–129. [CrossRef]
53. Yang, S.; Sun, H.; Zhao, R.; Xing, L.; Tan, Z.; Ning, Y.; Li, M. Was the 2022 drought in the Yangtze River Basin, China more severe than other typical drought events by considering the natural characteristics and the actual impacts? *Theor. Appl. Climatol.* **2023**, *155*, 5543–5556. [CrossRef]
54. Wang, J.; Yan, R.; Wu, G.; Liu, Y.; Wang, M.; Zeng, N.; Jiang, F.; Wang, H.; He, W.; Wu, M.; et al. Unprecedented decline in photosynthesis caused by summer 2022 record-breaking compound drought-heatwave over Yangtze River Basin. *Sci. Bull.* **2023**, *68*, 2160–2163. [CrossRef] [PubMed]
55. Chen, S.; Qiu, R.; Chen, Y.; Gong, W.; Han, G. Impacts of Drought and Heatwave on the Vegetation and Ecosystem in the Yangtze River Basin in 2022. *Remote Sens.* **2024**, *16*, 2889. [CrossRef]
56. Yuan, W.; Zheng, Y.; Piao, S.; Ciais, P.; Lombardozzi, D.; Wang, Y.; Ryu, Y.; Chen, G.; Dong, W.; Hu, Z.; et al. Increased atmospheric vapor pressure deficit reduces global vegetation growth. *Sci. Adv.* **2019**, *5*, eaax1396. [CrossRef]
57. Tian, S.; Van Dijk, A.; Tregoning, P.; Renzullo, L.J. Forecasting dryland vegetation condition months in advance through satellite data assimilation. *Nat. Commun.* **2019**, *10*, 469. [CrossRef]
58. Lu, J.; Yan, F. The Divergent Resistance and Resilience of Forest and Grassland Ecosystems to Extreme Summer Drought in Carbon Sequestration. *Land* **2023**, *12*, 1672. [CrossRef]
59. Xiao, C.; Zaehle, S.; Yang, H.; Wigneron, J.-P.; Schmullius, C.; Bastos, A. Land cover and management effects on ecosystem resistance to drought stress. *Earth Syst. Dyn.* **2023**, *14*, 1211–1237. [CrossRef]

water

MDPI

Article

Climate Change and Thermal Dynamics of the Lake Sevan Basin (Armenia): Observational Insights and Future Projections

Gor Khachatryan [1], Artur Gevorgyan [2], Ashok Vaseashta [3,4,*], Amalya Misakyan [2], Karsten Rinke [5], Artak Gevorgyan [6], Lilit Ghukasyan [7] and Gor Gevorgyan [1,*]

1 Scientific Center of Zoology and Hydroecology, National Academy of Sciences of the Republic of Armenia, Yerevan 0014, Armenia; gor_xachatryan_9898@mail.ru
2 Hydrometeorology and Monitoring Center, Ministry of Environment of the Republic of Armenia, Yerevan 0025, Armenia; agm86@yandex.ru (A.G.); miamalya@yandex.ru (A.M.)
3 Office of Strategic Research, International Clean Water Institute, Manassas, VA 20108, USA
4 Research Institute of the University of Bucharest, Bucharest 050663, Romania
5 Helmholtz-Centre for Environmental Research–UFZ, Magdeburg 39114, Germany; karsten.rinke@ufz.de
6 Institute of Mechanics, National Academy of Sciences of the Republic of Armenia, Yerevan 0019, Armenia; agevorgyanv@gmail.com
7 Faculty of Biology, Yerevan State University, Yerevan 0025, Armenia; lghukasyan@ysu.am
* Correspondence: prof.vaseashta@ieee.org (A.V.); gor.gevorgyan@sczhe.sci.am (G.G.)

Abstract

The Lake Sevan basin is particularly sensitive to climate change due to its continental climate and mountainous terrain, which collectively amplify climatic impacts. This study aimed to assess the influence of climate change on the thermal dynamics of the basin by analyzing both historical and projected temperature variations. Over the past three decades, the region has experienced a marked rise in air temperatures. Seasonal variability revealed distinct contrasts between winter and summer, with winter exhibiting greater fluctuations, ranging from 1.67 to 2.41 °C, compared to the more stable summer range of 0.81 to 1.41 °C. An analysis of heat inflow and outflow patterns demonstrated a moderating effect of Lake Sevan on temperature extremes. Stations, located near the lake, recorded lower levels of heat inflow and outflow, indicating that the lake's thermal inertia helps buffer seasonal temperature extremes. In contrast, stations situated farther from the lake exhibited more pronounced fluctuations, reflecting the absence of this stabilizing influence. These results underscore the lake's critical role in modulating the local climate by dampening extreme thermal variations. Additionally, comparative analysis of air and water temperature trends revealed that, while both exhibit warming, air temperatures show greater interannual variability. In contrast, water temperatures remained more stable, particularly during winter, due to the lake's thermal inertia. Future climate projections for the Lake Sevan region, based on CMIP6 (Coupled Model Intercomparison Project phase 6) ensemble outputs under four Shared Socioeconomic Pathways (SSP1–2.6, SSP2–4.5, SSP3–7.0, and SSP5–8.5), suggest a persistent warming trend throughout the 21st century. We project that the most significant increases are expected during summer months, with an anticipated mean annual temperature rise of up to 6 °C by the end of the century under the high-emission scenario (SSP5–8.5).

Keywords: climate change; thermal dynamics; Lake Sevan; seasonal cycles

1. Introduction

Climate change is one of the most pressing global challenges of the 21st century, affecting ecosystems, biodiversity, and human societies worldwide [1–3]. According to the Intergovernmental Panel on Climate Change (IPCC) Climate Change Synthesis Reports, the global surface temperature was, on average, 1.1 °C warmer in 2011–2020 than in the preindustrial period [4,5]. Among regions most vulnerable to these changes is the Lake Sevan basin in Armenia, a critical freshwater resource that supports both ecological and economic systems. As one of the world's largest high-altitude freshwater lakes, Lake Sevan holds immense ecological importance due to its biodiversity, role in regulating the local climate, and contributions to agriculture, energy, and tourism [6,7].

The Sevan basin is a large intermontane region bordered by the Geghama, Vardenis, Pambak, Areguni, Sevan, and Zangezur mountain ranges. More than half of the basin lies within the 2000–2600 m altitude zone, with the largest portion covering the slopes and foothills of the Geghama and Vardenis ranges [8]. The climatic conditions of Lake Sevan are strongly influenced by its high altitude, making the analysis of air temperature variability crucial for predicting seasonal changes in water temperature, ice cover, and lake stratification. The continental climate and mountainous topography further amplify Armenia's vulnerability to climate change, especially in the Lake Sevan basin [9]. Previous studies have also highlighted elevation-dependent warming in regions such as the south-central Himalayas, the Pyrenees, and the Tibetan Plateau [10–13]. However, most of these studies emphasize mean temperature trends, while largely overlooking seasonal extremes and basin-scale spatial variability, both of which are critical for understanding high-altitude lake systems such as Lake Sevan. Studying these variations over time offers valuable insights into how climate change is reshaping seasonal cycles [14].

The unique topography of Lake Sevan, along with its large surface area and high altitude, creates microclimatic zones where temperatures can vary substantially from one location to another. Understanding this spatial variability is essential for developing effective climate adaptation strategies, as different areas around the lake may experience climate impacts to varying degrees [15].

Decadal variations in air temperature provide key indicators of long-term climate shifts in the Lake Sevan basin. Understanding these trends helps identify periods of accelerated warming or cooling, which affect ecosystems, water resources, and socioeconomic activities [16–18]. Analyzing these patterns also establishes a foundation for assessing future climate impacts and informs the study's focus on how temperature extremes and heat circulation influence the lake's thermal regime.

Annual variations in air temperature, especially the timing of monthly extremes, reveal how the Lake Sevan basin's climate is evolving. Shifts in the hottest and coldest months, influenced by broader climate drivers, directly affect growing seasons, snowmelt, and water availability [19,20]. Understanding these patterns provides context for assessing the lake's thermal regime and potential climate impacts.

This study analyzes how temperature extremes and heat inflow/outflow have evolved over time in the Lake Sevan basin and considers their implications for the lake and the surrounding environment. Understanding these dynamics is essential for predicting future climate scenarios, assessing impacts on the lake's thermal regime, and informing adaptation strategies to maintain ecological balance [21,22].

Climate change may also result in less distinct transition periods between spring and autumn, leading to longer summers. Gevorgyan (2016) has reported that temperature increases in Armenia are most pronounced in summer [23]. These seasonal shifts are especially important for agriculture, as they influence the length of the growing season and the hydrological cycle, affecting snowmelt and water availability in spring and early summer.

https://doi.org/10.3390/w18030352

By examining the evolution of temperature patterns over recent decades, this study offers a deeper insight into the seasonal manifestations of climate change in the Lake Sevan basin. Climate change is projected to influence not only long-term average temperatures but also the frequency and intensity of extreme weather events. Therefore, analyzing both positive and negative anomalies in monthly mean air temperatures relative to long-term averages provides a more comprehensive understanding of the region's climatic dynamics and variability [24].

No specific studies have focused on air temperature changes and projections for the Lake Sevan basin. Although no studies have specifically examined air temperature variability or long-term climatic characteristics of the Lake Sevan basin, several existing studies are indirectly relevant to the region [15,25,26]. The absence of dedicated basin-scale climate analyses underscores a clear knowledge gap that the present study seeks to address. However, several related studies explored the impact of climate change on Lake Sevan. For instance, Shikhani et al. (2024) assessed climate change effects using a multi-lake model ensemble [27], while another key study examined the thermal dynamics of Lake Sevan and their role in regional climate processes [28]. Although these studies contributed to a broader understanding of regional climate dynamics, they primarily focus on lake thermal processes or employ multi-lake modeling approaches with relatively coarse spatial resolution. As a result, they do not provide a detailed assessment of basin-scale air temperature variability, seasonal extremes, or long-term trends specific to the Lake Sevan basin, nor do they explicitly link historical trends with future projections.

This study employed climate change projections to evaluate future temperature changes in the Lake Sevan basin under various climate scenarios. These models, which represent a spectrum of greenhouse gas emissions from low to high, provide important insights into the expected evolution of the climate under different global warming scenarios [5,29,30]. In response to the limitations of previous studies, the primary objective of this research is to provide a basin-scale, temporally detailed analysis of air temperature dynamics in the Lake Sevan basin, integrating historical trends with future climate projections and explicitly examining seasonal patterns, temperature extremes, and spatial variability. Integrating climate projections into management practices is crucial for ensuring sustainable agricultural systems and effective water resource management [31–33]. By comparing these projections with past temperature trends, this study provides a forward-looking assessment of potential future climate conditions in the Lake Sevan basin. The findings can inform policy decisions, adaptation strategies, and future research initiatives to mitigate the impacts of climate change.

This study presents a comprehensive analysis of air temperature dynamics in the Lake Sevan basin, incorporating both past trends and future projections. By examining decadal variations, seasonal and annual patterns, extreme temperature events, and spatial distribution, it provides a robust assessment of the region's climate and its projected trajectory under different global warming scenarios. The findings are expected to serve as a foundational resource for developing agricultural adaptation strategies and optimizing water resource management in the Lake Sevan basin. By establishing a clear analytical framework, this research aims to enhance resilience to climate variability and support sustainable resource use, both of which are increasingly critical as climate change continues to affect agricultural productivity and water availability.

2. Data and Methods

Climate projection data for the Lake Sevan basin were obtained from the Climate Change Knowledge Portal (CCKP) and the Coupled Model Intercomparison Project Phase 6 (CMIP6), developed under the framework of the World Climate Research Program [34]. The CMIP6 dataset, provided at a spatial resolution of 0.25° × 0.25° (approximately 25 km × 25 km), represents one of the most advanced global climate modeling systems to date. It incorporates

https://doi.org/10.3390/w18030352

substantial improvements in physical parameterizations, including enhanced cumulus convection schemes, refined radiation calculations, and more sophisticated land surface models. Additionally, the improved representation of aerosol–cloud interactions contributes to greater accuracy in simulating climate processes and projecting future climatic conditions [35,36]. CMIP6 outputs obtained from the CCKP were first resampled to a common 1° × 1° grid and subsequently bias-corrected. This method aligns modeled climate distributions with observations while explicitly preserving the original temperature trends of the global climate models. Spatial disaggregation was performed using harmonic analysis-based scaling, with ERA5 reanalysis data at 0.25° resolution serving as the reference dataset. As a result, the downscaled products inherit fine-scale spatial patterns from the reference data, although these patterns are not fully resolved by the original global climate models. All CMIP6 data used in this study are based on these bias-corrected and downscaled CCKP products. A key advantage of the CCKP-CMIP6 framework over earlier climate models lies in its enhanced spatial and temporal resolution, which enables more precise regional climate projections. This refinement allows for better representation of localized climate dynamics, particularly in complex terrains such as the Lake Sevan basin. Moreover, advances in cloud microphysics, atmospheric convection, and radiation schemes significantly enhance the accuracy of simulated atmospheric processes and extreme weather events [37].

Unlike the earlier RCP (Representative Concentration Pathway)-based framework, CMIP6 incorporates Shared Socioeconomic Pathways (SSPs), which integrate climate projections with socioeconomic development trajectories. The SSPs, developed through Integrated Assessment Models (IAMs), represent the most recent advancements in model structures, boundary conditions, and emission trends. They form the foundation of the ScenarioMIP component within CMIP6. Rather than offering deterministic forecasts, the SSPs explore five distinct socioeconomic pathways, illustrating how different policy decisions and development strategies can influence future climate outcomes [36,38,39].

The CMIP6 reduced daily surface fields available through the CCKP provide high-resolution (0.25°) daily climate data from 30 global models spanning the period 1950–2100. These datasets undergo bias-correction spatial disaggregation (BCSD) and are calibrated against a historical baseline (1961–2014) to maintain the integrity of long-term temperature trends. Spatial refinement is further supported through the application of Fast Fourier Transform methods and the ERA5 reanalysis dataset. The outputs are delivered in a globally gridded NetCDF format and include multiple temporal resolutions, climate indices, and overlays of administrative boundaries and watershed datasets [38,40,41]. The CCKP-CMIP6 ensemble dataset integrates outputs from up to 30 climate models (Table 1), each associated with a specific SSP.

Table 1. List of models used in CCKP CMIP6-x0.25 compilation [42,43].

Model Name	Modeling Center
hadgem3-gc31-ll	UK Met Office Hadley Centre, UK
ipsl-cm6a-lr	The Institute Pierre Simon Laplace, France
inm-cm4-8	Institute for Numerical Mathematics, Russia
kiost-esm	Korea Institute of Ocean Science and Technology, Republic of Korea
kace-1-0-g	National Institute of Meteorological Research, Republic of Korea
miroc6	Atmosphere and Ocean Research Institute, The University of Tokyo, Japan
access-cm2	CSIRO (Commonwealth Scientific and Industrial Research Organization, Australia), and ARCCS (Australian Research Council Centre of Excellence for Climate System Science, Australia

https://doi.org/10.3390/w18030352

Table 1. *Cont.*

Model Name	Modeling Center
miroc-es2l	Atmosphere and Ocean Research Institute, The University of Tokyo, Center for Climate System Research-National Institute for Environmental Studies, Japan
mpi-esm1-2-hr	Max Planck Institute for Meteorology (MPI-M), Germany
giss-e2-1-g	Goddard Institute of Space Studies, NASA, USA
access-esm1.5	CSIRO (Commonwealth Scientific and Industrial Research Organization, Australia), and ARCCS (Australian Research Council Centre of Excellence for Climate System Science, Australia
gfdl-esm4	Geophysical Fluid Dynamics Laboratory, NOAA, USA
cnrm-cm6-1	Centre National de Recherches Meteorologiques, France
hadgem3-gc31-mm	UK Met Office Hadley Centre, UK
gfdl-cm4	Geophysical Fluid Dynamics Laboratory, NOAA, USA
canesm5	Canadian Centre for Climate Modeling and Analysis, Canada
mpi-esm1-2-lr	Max Planck Institute for Meteorology (MPI-M), Germany
bcc-csm2-mr	Beijing Climate Center, China Meteorological Administration, China
mri-esm2-0	Meteorological Research Institute, Japan
taiesm1	Research Center for Environmental Changes, Academia Sinica, Taiwan
ec-earth3-veg-lr	EC-Earth-Consortium
ukesm1-0-ll	UK's Met Office and Natural Environment Research Council (NERC), UK
cmcc-esm2	Euro-Mediterranean Center on Climate Change
nesm3	Nanjing University of Information Science and Technology, China
noresm2-mm	Norwegian Climate Centre, Norway
ec-earth2	EC-Earth-Consortium
noresm2-lm	Norwegian Climate Centre, Norway
inm-cm5-0	Institute for Numerical Mathematics, Russia
cnrm-esm2-1	Centre National de Recherches Meteorologiques/Centre Européen de Recherche et Formation Avancées en Calcul Scientifique, France
fgoals-g3	China Academy of Sciences, China

This ensemble provides a broad range of climate projections for the Lake Sevan basin, delivering essential insights for climate research, adaptation planning, and risk management. Data processing was conducted using the Enhanced Climate Risk Management Engine (CRMe), with final outputs rendered in ArcGIS (version 10.3, Esri, Redlands, CA, USA) to ensure compatibility with geospatial data standards. This facilitates the integration of climate projections into spatial analyses and supports informed decision-making in climate-sensitive sectors [35].

The SSPs represent a crucial framework for understanding climate change impacts, outlining various socioeconomic futures and their corresponding implications for climate dynamics. These scenarios project how different societal and economic trends could influence future climate outcomes. The four distinct SSPs are as follows:

1. SSP1–2.6 (Sustainability—Taking the Green Road): This pathway envisions a sustainable future with high social equity, focusing on renewable energy and reduced emissions. It requires data on economic growth, energy consumption, and land use practices.

2. SSP2–4.5 (Middle of the Road): A balanced approach with moderate economic growth, blending continued fossil fuel use with sustainability efforts. It necessitates datasets on economic trends, energy production, and demographic changes.
3. SSP3–7.0 (Regional Rivalry—A Rocky Road): A fragmented world marked by nationalism and limited international cooperation, where high fossil fuel reliance and environmental degradation prevail. It requires data on geopolitical trends and energy systems.
4. SSP5–8.5 (Fossil-Fueled Development—Taking the Highway): This pathway is characterized by rapid economic growth fueled by fossil fuel consumption, projecting severe climate impacts. It requires data on fossil fuel investments and environmental degradation.

By aligning the output from these climate models with the SSP scenarios, projections of air temperature changes in the Lake Sevan basin have been generated. These projections provide a range of potential future climate conditions, offering valuable insights for researchers and policymakers. Understanding these scenarios is crucial for climate adaptation planning and policy formulation, as it provides a structured framework for managing future climate risks. The temperature projection analysis utilized the reference period 1991–2020 as a baseline to represent current climate conditions. The 1991–2020 period was selected as the baseline to maintain consistency between the observational trend analysis and the reference period used for future climate projections. Future projections were made for the periods 2020–2039, 2040–2059, 2060–2079, and 2080–2099.

This study utilized average monthly temperature data from six meteorological stations in the Lake Sevan basin, spanning the period 1991–2020. Air temperature was measured in full shade at a height of 2 m above the ground surface. All measurements were performed in accordance with the procedures established by the World Meteorological Organization (WMO) and relevant international standards [44,45]. Observations were recorded at eight standard synoptic times: 00:00, 03:00, 06:00, 09:00, 12:00, 15:00, 18:00, and 21:00 (UTC). Subsequently, the average monthly temperature was determined as the arithmetic mean of the daily averages for each month. These stations are strategically positioned at varying distances from the lake and at different geographical elevations, providing a diverse representation of local climate conditions. The selected stations offer high-quality data with extensive temporal coverage, ensuring reliable insights for climate analysis. Some stations are located near Lake Sevan, while others are situated at varying distances, contributing to a well-distributed spatial coverage across the basin (Figure 1). Detailed information for each station is provided in Table 2.

Water temperature observations near the shore of Lake Sevan were conducted at four hydrological observation points: Sevan Peninsula, Shorzha, Martuni, and Karchaghbyur. Measurements were taken at a depth of 0.1–0.5 m below the surface layer, under ice-free conditions. Water temperature was recorded twice daily at 08:00 and 20:00. The average daily water temperature was calculated as the arithmetic mean of the two daily measurements. Subsequently, the average monthly temperature was determined as the arithmetic mean of the daily averages for each month. The final dataset presented here shows the arithmetic mean of the average monthly water temperatures recorded at all four hydrological observation points. The Hydrometeorology and Monitoring Center of the Ministry of Environment of the Republic of Armenia carried out continuous measurements of air and water temperatures.

During the period 1991–2020, monthly air temperature data were missing for certain months at the following meteorological stations: Gavar in January 2006, Martuni in October 2020, Masrik in March 1992, and Semyonovka in March, August, and December 1997, as well as in April and May 1998, and from April to December in both 2001 and 2002. For average monthly water temperature, data were unavailable for February, March, and April of both 1992 and 1993, as well as for January, February, and March of 2008. The missing

https://doi.org/10.3390/w18030352

average monthly air temperature values for these periods were not estimated or interpolated; instead, decadal and long-term mean values were calculated using only available monthly observations, with months containing missing data excluded from the analysis.

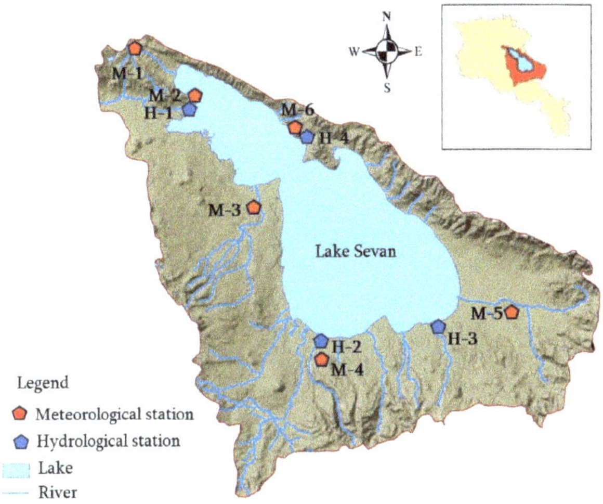

Figure 1. Spatial distribution of meteorological and hydrological stations across the Lake Sevan basin, referenced by their respective codes as detailed in Table 2.

Table 2. A summary of the selected meteorological and hydrological stations in the Lake Sevan basin.

N	Meteorological Station	Station Code	Geographic Coordinates		Altitude Above Sea Level (m)
			Latitude	Longitude	
1	Semyonovka	M-1	40.6597	44.8981	2104
2	Sevan	M-2	40.5653	45.0083	1917
3	Gavar	M-3	40.3486	45.1300	1961
4	Martuni	M-4	40.1369	45.2969	1943
5	Masrik	M-5	40.2075	45.7644	1940
6	Shorzha	M-6	40.5006	45.2717	1917
	Hydrological station				
1	Sevan Peninsula	H-1	40.5628	45.0084	1890
2	Martuni	H-2	40.1623	45.3076	1890
3	Karchaghbyur	H-3	40.1783	45.5644	1890
4	Shorzha	H-4	40.4972	45.2700	1890

To analyze the climatic characteristics of the Lake Sevan basin, a comprehensive assessment of air and surface water temperature variability was conducted using observational data from 1991 to 2020. The study applied several methods to evaluate temperature trends and patterns. First, average annual air temperatures were calculated for each year and grouped into three decades (1991–2000, 2001–2010, 2011–2020), with decadal means determined by arithmetic averaging to identify long-term warming trends. Second, long-term average monthly air temperatures were computed by averaging daily temperature values over the 30-year period, serving as a baseline for characterizing climatic conditions at each station and across the basin. Third, seasonal average air temperatures were calculated for winter (December–February), spring (March–May), summer (June–August), and autumn (September–November) based on the corresponding long-term monthly values. Fourth, extreme positive and negative deviations from the 30-year monthly means were identified for each calendar month to assess the

https://doi.org/10.3390/w18030352

degree of anomalous warming or cooling. Fifth, long-term monthly heat inflow and outflow were inferred from variations in monthly temperatures, based on the direction and magnitude of changes between adjacent months and the seasonal cycle, reflecting heat accumulation in warmer months and loss in colder months. For this analysis, stations were classified into three categories based on their distance from Lake Sevan: (1) Lake stations, including Sevan (1 km) and Shorzha (1–2 km); (2) Coastal stations, including Gavar (6–7 km) and Martuni (3–4 km); and (3) Distant stations, including Semyonovka (8–9 km) and Masrik (10–11 km).

To ensure a robust statistical characterization of temperature variability, the analysis explicitly incorporates classical variability metrics, including standard deviation and temperature anomaly ranges, which are widely used in climatological studies to assess intra-seasonal, interannual, and spatial variability. Standard deviation was employed to quantify the dispersion of monthly and seasonal air temperature values around long-term means, thereby providing insight into the stability of temperature regimes and the magnitude of seasonal fluctuations across the Lake Sevan basin. In addition, positive and negative temperature anomalies relative to the 1991–2020 baseline were analyzed to identify extreme deviations and to evaluate the intensity of anomalous warming and cooling events. Together, these metrics allow for a clear interpretation of climate variability and long-term change, highlighting differences related to elevation, distance from the lake, and seasonal controls. The use of these well-established statistical measures ensures transparency, comparability with previous regional and global climate studies, and reliable interpretation of observed and projected temperature changes in the Lake Sevan basin.

Finally, seasonal and interannual variability of both air and water temperatures was examined to detect intra-annual patterns and interannual fluctuations across the study period. All data visualizations and graphical outputs were generated using Origin software (version 2018, OriginLab Corporation, Northampton, MA, USA). One-way statistical calculations were performed in Excel (version 2016, Microsoft Corporation, Redmond, WA, USA). Spatial data processing and map generation were conducted using ArcGIS (version 10.3, Esri, Redlands, CA, USA).

Figure 2 presents a schematic workflow illustrating the methodology used in this study. It depicts the sequential steps from observational data collection at meteorological and hydrological stations, through quality control, computation of mean values, and CMIP6-based climate projections under various SSP scenarios, to temperature analysis, statistical processing, and spatial visualization.

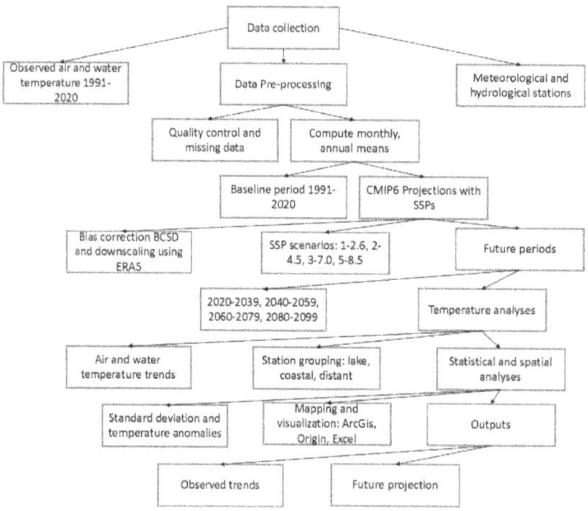

Figure 2. Flowchart of data collection, processing, and climate projections.

3. Results

3.1. Long-Term Decadal Variability of Air Temperature

Statistical analysis of average annual temperature data from six meteorological stations in the Lake Sevan basin for the decades 1991–2000, 2001–2010, and 2011–2020 reveals a clear and consistent warming trend across the region (Figure 3).

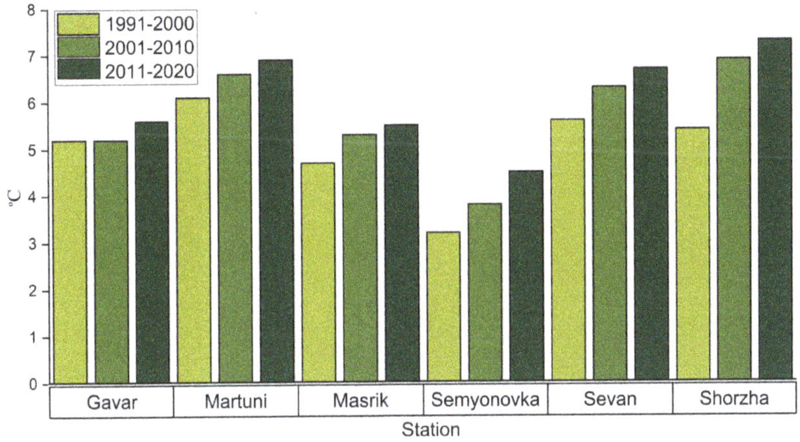

Figure 3. Average annual air temperatures recorded at six meteorological stations in the Lake Sevan basin during the decades 1991–2000, 2001–2010, and 2011–2020.

All stations exhibited positive temperature trends over the three decades with cumulative warming ranging from 0.4 to 1.9 °C (Figure 3). Gavar station showed the smallest increase, with mean annual temperature rising by only 0.4 °C (from 5.2 to 5.6 °C). This comparatively weak warming signal may be influenced by the station's concave topographic setting, which can promote cold-air pooling and dampen observed temperature increases. In contrast, Shorzha station experienced pronounced warming, with temperatures rising from 5.4 to 7.3 °C, a total increase of 1.9 °C, most of which occurred during the first two decades (Figure 3).

The remaining stations exhibited intermediate warming, with increases of 0.8 °C at Martuni and Masrik, 1.1 °C at Sevan, and 1.2 °C at Semyonovka. Despite being the coldest station due to its high elevation, Semyonovka showed a consistent warming rate of approximately 0.6 °C per decade, indicating that elevation does not necessarily constrain long-term regional warming trends (Figure 3).

Temperature increases generally accelerated between the first and second decades, followed by a weaker but continued rise in the third. Despite this slowdown, cumulative warming remains substantial, consistent with global and regional trends showing enhanced warming since the early 2000s, particularly in mountainous and continental interior regions [46,47]. According to the World Meteorological Organization (WMO), the decade 2011–2020 was the warmest on record globally. The WMO's High-Level Climate Report notes that the global average temperature during this period was 1.10 ± 0.12 °C above the 1850–1900 pre-industrial baseline. Furthermore, the six warmest years on record worldwide occurred between 2015 and 2020 [48].

3.2. Monthly Air Temperature Variations

The variability of monthly temperature values, particularly their deviations from long-term averages, provides valuable insights into the stability of mean monthly temperatures

https://doi.org/10.3390/w18030352

and the reliability of these averages for individual months. Figure 4 presents the long-term mean monthly air temperatures (1991–2020) for the six meteorological stations in the Lake Sevan basin, while seasonal variability is quantified using standard deviation as a measure of intra-seasonal dispersion.

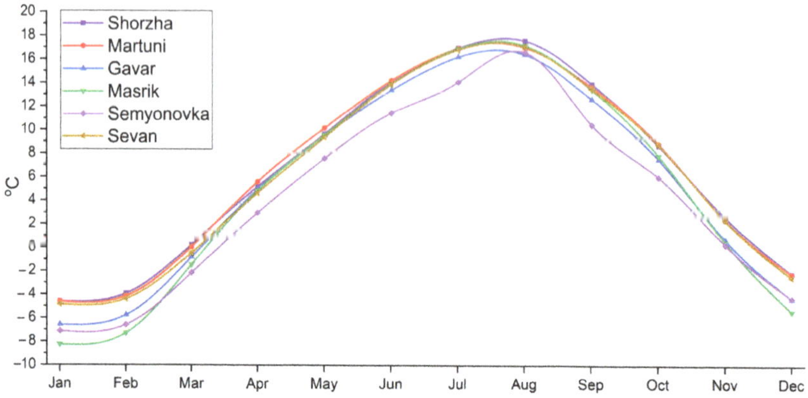

Figure 4. Long-term average monthly air temperatures (1991–2020) in the Lake Sevan basin.

To quantify this, the average variability for each season was calculated and is presented in Table 3.

Table 3. Long-term mean seasonal air temperature variability (STDEV, ±°C) at six meteorological stations in Lake Sevan basin for period 1991–2020.

Stations	Winter	Spring	Summer	Autumn
Shorzha	±1.77	±1.67	±1.41	±1.51
Martuni	±1.68	±1.30	±0.86	±1.01
Gavar	±1.69	±1.25	±0.81	±0.91
Masrik	±2.04	±1.15	±0.89	±1.06
Semyonovka	±2.41	±1.63	±1.10	±1.19
Sevan	±1.67	±1.35	±1.04	±1.07

January consistently exhibited the lowest and August the highest mean monthly temperatures across all stations (Figure 3). In January, the warmest conditions were observed at Sevan (−4.9 °C), Shorzha (−4.6 °C), and Martuni (−4.6 °C), whereas August maxima occurred at Shorzha (17.6 °C), Masrik (17.2 °C), and Sevan (17.2 °C). Stations located near Lake Sevan (Sevan and Shorzha) consistently recorded higher winter and summer temperatures, indicating a moderating lacustrine influence (Figure 4).

Analysis of STDEV values (Table 3) shows that temperature variability was lowest in summer (±0.81–1.41 °C) and highest in winter (±1.67–2.41 °C) across all stations. This seasonal contrast reflects stronger thermal regulation by Lake Sevan during summer, while winter conditions are characterized by greater atmospheric instability and continental influences.

Spatial patterns of variability reveal a clear gradient related to both elevation and distance from the lake (Table 3; Figure 4). Stations closest to Lake Sevan (Shorzha and Sevan) exhibited the most stable temperature regimes throughout the year, particularly in summer and autumn. Moderate variability was observed at Martuni and Gavar, where the lake's influence remains detectable but reduced. In contrast, the higher-elevation and more distant stations (Masrik and especially Semyonovka) experienced substantially

https://doi.org/10.3390/w18030352

greater variability, most pronounced during winter, consistent with stronger continental and altitudinal controls [49]. Overall, proximity to Lake Sevan and lower elevation act to dampen seasonal temperature fluctuations in the basin.

Figure 5 illustrates the largest positive and negative deviations of average monthly air temperatures from the long-term mean (1991–2020), highlighting the magnitude of thermal extremes.

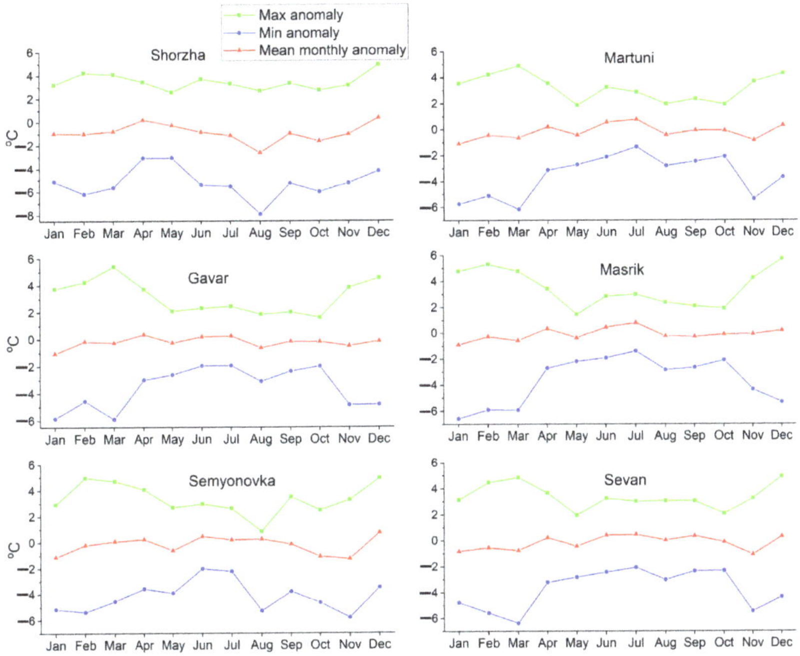

Figure 5. Extreme positive and negative deviations of average monthly air temperatures from the long-term mean (1991–2020) at six meteorological stations in the Lake Sevan basin.

These extreme anomalies provide quantitative indicators of climate variability and the occurrence of unusually warm or cold periods, serving as important indicators of climate variability and potential long-term climate change in the Lake Sevan basin. Meteorological stations in the Lake Sevan basin exhibited notable monthly variability in temperature anomalies, encompassing both maximum and minimum extremes. The spatial distribution of these stations, ranging from lower elevations near the lake to higher-altitude inland areas, played a significant role in shaping the observed differences in temperature patterns. To better understand the thermal characteristics and underlying dynamics of the regional climate, the largest positive and negative deviations from the monthly mean temperatures at each station were analyzed and compared (Figure 5).

Average monthly temperature anomalies at six meteorological stations in the Lake Sevan basin reflect a cold continental climate. The strongest negative anomalies occurred in March (Gavar, Martuni, Sevan), August (Shorzha), November (Semyonovka), and January (Masrik), with minimum deviations ranging from −5.8 °C to −7.9 °C (Figure 5). These extreme cold events, observed consistently over several decades, underscore the harsh climatic conditions in the Lake Sevan basin and carry important implications for the vulnerability of local ecosystems, water resources, and biodiversity to seasonal extremes.

The largest positive temperature anomalies occurred in March (Gavar, Martuni), December (Masrik, Shorzha, Sevan), and February (Semyonovka), with maximum deviations

https://doi.org/10.3390/w18030352

of 4.9 to 5.7 °C (Figure 5). These warm anomalies indicate pronounced winter warming in recent decades, consistent with global climate-change trends [50]. Overall, the results suggest a seasonal warming signal in the Lake Sevan basin influenced by regional topography, elevation, lake effects, and intra-seasonal variability.

Overall, the data indicates that the most pronounced warming in average monthly temperatures occurred during winter and early spring. This trend aligns with broader climate change patterns observed in mountainous regions, where minimum temperatures have increased more rapidly, particularly during the colder seasons [51,52]. Such seasonal warming is indicative of a shift in regional climate dynamics and underscores the heightened sensitivity of high-altitude regions to global warming.

3.3. *Intensity of Heat Inflow and Outflow: Heat Circulation*

The intensity of heat inflow and outflow, also referred to as heat circulation, is characterized by deviations of average monthly temperatures from the mean annual temperature. In this study, heat inflow and outflow are defined as temperature-based indicators derived from these deviations, rather than direct measurements of heat flux. Specifically, heat inflow/outflow for month i was calculated as the deviation of the monthly mean air temperature from the annual mean air temperature:

$$AT_i = T_{month,i} - \bar{T}_{year}$$

where AT_i is the air temperature deviation for month i, $T_{month,i}$ is the monthly mean temperature, and $\bar{T}_{year}$ is the annual mean temperature. Positive deviations ($AT_i > 0$) indicate heat inflows, which can increase surface temperatures and enhance stratification, while negative deviations ($AT_i < 0$) indicate heat outflows, which cool the surface and affect mixing patterns.

This circulation substantially impacts the lake's ecosystem by regulating water temperatures, which are essential for the survival, growth, and reproduction of aquatic species. Additionally, heat circulation affects water quality, as warmer water holds less oxygen, potentially leading to hypoxia in deeper layers, while cooler temperatures help maintain higher oxygen levels. The heat balance also influences the lake's hydrological processes, such as evaporation and water levels. In the context of climate change, heat circulation helps predict the lake's response to temperature variations [53,54]. Average values of heat inflow and outflow were then calculated for each station group, as shown in Figure 6.

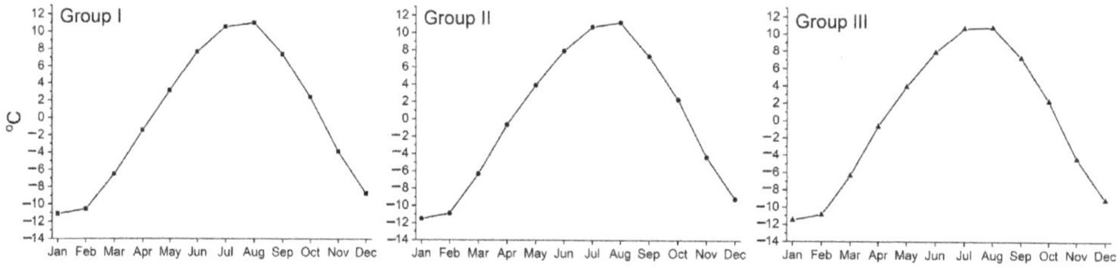

Figure 6. Long-term average monthly heat inflow and outflow values (1991–2020) in the Lake Sevan basin.

In the Sevan basin, negative deviations from the annual mean indicate heat outflow (loss of thermal energy), typically occurring between November and April, while positive deviations reflect heat inflow (heat accumulation), usually observed between May and October (Figure 6). The maximum heat inflow values were recorded during the summer months (June–August) (Figure 6), coinciding with increased solar radiation and higher

https://doi.org/10.3390/w18030352

temperatures, which contributed to surface warming and heat accumulation across all station groups. Conversely, the maximum heat outflow values were observed in the winter months (December–February), with the degree of heat outflow varying with distance from the lake (Figure 6). Group I (lake stations: Sevan and Shorzha), located closest to the lake, showed clear signs of thermal moderation due to the lake's high heat capacity (Figure 6). The cold months were relatively mild, and the warm months less extreme (Figure 6). For instance, heat outflow in January was −11.1 °C, and heat inflow in August was +11.0 °C (Figure 6), indicating more moderate deviations. This stabilizing effect of Lake Sevan likely reduced the intensity of both heat outflow and inflow. Large bodies of water tend to retain heat in winter and provide cooling in summer, helping buffer against extreme temperature changes [55]. Group II (coastal stations: Gavar and Martuni), located relatively close to the lake, exhibited characteristics transitional between lacustrine influence and a more terrestrial climate. Compared to the lake stations, coastal stations showed slightly more pronounced summer warming and winter cooling (Figure 6), reflecting a weaker moderating influence from the lake, though it still exerted some effect. Specifically, the heat outflow in January was −11.5 °C, and the heat influx in August was +11.2 °C (Figure 6). The influence of Lake Sevan was less pronounced here, allowing for more pronounced seasonal temperature variations than at the lakeside stations. Group III (distant stations: Semyonovka and Masrik), located farthest from the lake, exhibited the highest temperature extremes (Figure 6). January heat outflow reached −12.3 °C, while the August heat inflow reached +11.3 °C (Figure 6). These stations experienced the least thermal moderation from Lake Sevan, resulting in larger seasonal temperature fluctuations. The lack of the lake's stabilizing effect meant that temperatures in these areas were more directly influenced by atmospheric conditions, leading to a more pronounced seasonal cycle.

In summary, the analysis of long-term average monthly heat inflow and outflow confirms the significant role of Lake Sevan in regulating the basin's thermal regime. A clear spatial gradient emerged, with thermal moderation being strongest at stations closest to the lake and gradually decreasing with increasing distance. This pattern is crucial for understanding the local microclimate, informing agricultural planning, guiding hydrological modeling, and supporting the management of ecosystem dynamics within the Lake Sevan basin.

3.4. Seasonal and Interannual Patterns of Air and Water Temperature Variability

As shown in Figure 7, air temperatures in the Lake Sevan basin exhibited notable variability, while surface water temperatures in the lake followed a more stable trend. Pronounced seasonal variability was observed. In winter (December–February), water temperatures remained relatively stable (1–3 °C), whereas air temperatures showed greater variability, reaching minima of about −10 °C (Figure 7). Both air and water temperatures increased progressively throughout the year. During spring (March–May), rising air temperatures induced concurrent warming of surface waters, with the air–water temperature difference decreasing in April and May (Figure 7). Overall, both air and water temperatures showed a warming trend, with air temperatures displaying more pronounced interannual variability and water temperatures increasing more consistently, underscoring the potential impacts of climate change in the region (Figure 7).

Pronounced seasonal variability was observed. In winter (December–February), water temperatures remained relatively stable (1–3 °C), whereas air temperatures showed greater variability, reaching minima of about −10 °C (Figure 7). Both air and water temperatures increased progressively throughout the year. During spring (March–May), rising air temperatures induced concurrent warming of surface waters, with the air–water temperature difference decreasing in April and May (Figure 7).

https://doi.org/10.3390/w18030352

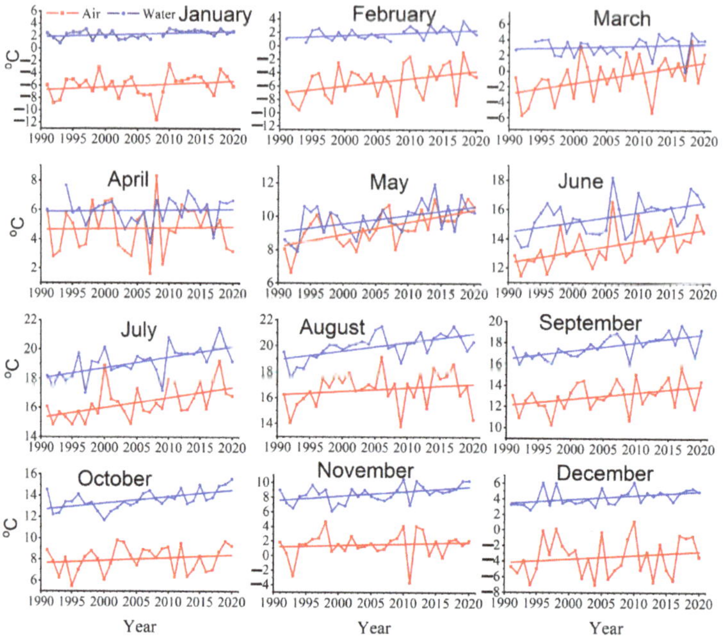

Figure 7. Seasonal and interannual variability of mean annual air and water temperatures in the Lake Sevan basin.

Maximum temperatures occurred in summer (June–August), with water temperatures occasionally exceeding 20 °C and air temperatures exhibiting strong interannual variability (Figure 7). In autumn (September–November), air temperatures declined more rapidly than water temperatures, highlighting the lake's heat retention capacity. Throughout the year, water temperature lagged behind air temperature, particularly during transitional seasons, reflecting the lake's thermal inertia. The analysis confirmed an overall warming trend in both air and water temperatures over the past three decades (Figure 7). The lake's thermal regime is increasingly influenced by climate change, which can have profound effects on its hydrology, biodiversity, and water quality.

Positive temperature increases were observed for both air and water throughout the entire study period (Figure 7), reflecting an overall warming trend. Lake water temperatures showed consistent increases, with the most significant rises in September (+2.31 °C) and April (+2.20 °C), while the smallest increases were recorded in winter, particularly in January (+0.82 °C) and February (+1.23 °C) (Figure 7). This suggests that warming was more pronounced in the fall and spring. These temperature changes may have significant implications for the lake's thermal stratification, water quality, and ecosystem health.

Air temperatures also exhibited a steady rise across all months, with notable increases in March (+4.08 °C) and February (+3.39 °C) (Figure 7). The smallest increases were observed in April (+0.18 °C) and November (+0.66 °C) (Figure 7), indicating that air temperatures are more sensitive to climate change in winter and spring, while summer temperatures showed a more stable increase (Figure 7). On average, water temperature increases (+1.77 °C) were slightly higher than air temperature increases (+1.69 °C) (Figure 7), highlighting a more significant warming of the lake's surface waters.

Overall, lake water temperature exhibited a more stable trend, ranging from +0.82 °C in January to +2.31 °C in September (Figure 7). In contrast, air temperature changes showed greater variability, ranging from +0.18 °C in November to +4.08 °C in March (Figure 7). These differences underscore the buffering role of Lake Sevan in mitigating temperature extremes. Air temperature changes were notably high, particularly in February (+3.39 °C) and March (+4.08 °C) (Figure 7), signaling significant warming trends in late winter and early spring. By contrast, water temperature increases were more moderate during this period, from +1.23 °C in February to +1.60 °C in March (Figure 7). This disparity reflects the lake's thermal inertia, in which water warms more gradually than the surrounding atmosphere.

In April, the change in air temperature (+0.18 °C) was notably lower than in the preceding months (Figure 7), likely reflecting atmospheric variability or cooling trends associated with transitional weather patterns. Meanwhile, lake water temperature increased steadily, with April showing a rise of +2.20 °C, one of the highest increases for water (Figure 7), indicating that solar radiation and calm atmospheric conditions played a more significant role in warming the water than air temperature alone during this period. In June, both air and water temperature changes approached similar levels (+2.32 °C for air and +2.04 °C for water) (Figure 7), marking the onset of the summer period of thermal synchronization. From July to September, lake water temperatures remained relatively stable and high (+2.01 to +2.31 °C), while air temperature changes began to decrease, from +2.09 °C in July to +0.83 °C in September (Figure 7). This reversal demonstrates how water retains heat longer than air, thus prolonging the warm period in the aquatic environment even as atmospheric temperatures begin to cool. Figure 8 shows the estimated average monthly air and lake water temperatures in the Lake Sevan basin for the period 1991–2020.

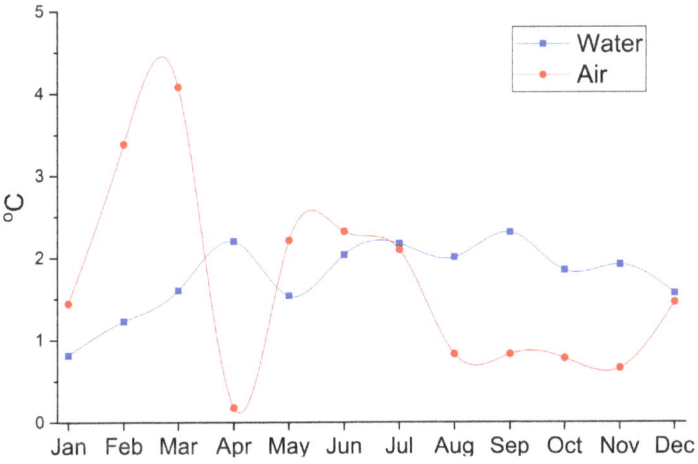

Figure 8. Estimated average monthly air and lake water temperatures (1991–2020).

Air temperature displayed more abrupt and pronounced fluctuations than lake water temperature, which remained relatively stable due to the lake's thermal inertia (Figure 8). Air temperatures peak in March (+4.1 °C) and reach a minimum (+0.1 °C) in April (Figure 8), reflecting high seasonal variability, possibly influenced by the transitional conditions of spring. In contrast, water temperatures peak in September (+2.3 °C) and reach a minimum in January (+0.8 °C) (Figure 8). Notably, the water remained warmer than the air in April, July, August, September, October, November, and December (Figure 8). This pattern highlights the lake's buffering capacity against abrupt atmospheric changes and has important implications for studies of limnological processes, aquatic ecosystems, and regional climate.

https://doi.org/10.3390/w18030352

3.5. Projected Temperature Changes in the Lake Sevan Basin

The temperature projections for the period 2020–2099, based on four SSPs (SSP1–2.6, SSP2–4.5, SSP3–7.0, and SSP5–8.5), reveal a marked departure from the long-term average for 1991–2020 (Figure 9).

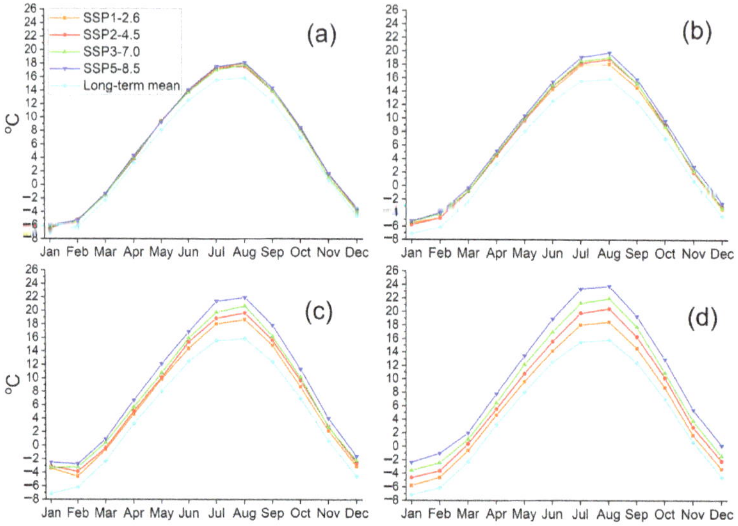

Figure 9. Monthly temperature variability in the Lake Sevan basin: (**a**) 2020–2039, (**b**) 2040–2059, (**c**) 2060–2079, and (**d**) 2080–2099.

The long-term average temperatures for 1991–2020 were used as a baseline for comparison, reflecting the region's typical climate with cold winters and mild summers (e.g., January: –7.1 °C; August: 15.9 °C; Figure 9).

Deviations from these baseline values indicate potential shifts in the Lake Sevan region's seasonal climate, with possible impacts on water resources, ecosystems, agriculture, and local communities.

Temperature projections for the Lake Sevan basin under four SSP scenarios indicate moderate warming during 2020–2039 relative to the 1991–2020 baseline, with variations depending on seasonal patterns and emission pathways (Figure 9).

Winter (December–February) temperatures in the Lake Sevan basin are projected to rise by 0.7–1.1 °C during 2020–2039, while summer months, particularly July and August, could increase by up to 2.3 °C under SSP5–8.5; lower warming is projected under SSP1–2.6 (Figure 9). Spring (April–May) and autumn (September–October) are expected to warm by 0.7–1.5 °C, contributing to an overall shift toward warmer conditions year-round and potentially impacting hydrology, ecosystems, and socioeconomic activities.

Mid-century projections (2040–2059) show more pronounced warming, with summer increases of 3.6–3.9 °C and winter rises of 1.1–2.1 °C under SSP5–8.5. Spring and autumn are projected to warm by 1.2–3.4 °C, indicating shifts in seasonal boundaries and ecological timing (Figure 9).

For 2060–2079, SSP5–8.5 projects extreme summer warming, with July and August reaching 21.4–22.0 °C (5.8–6.1 °C above baseline), while winter months warm by 1.4–4.6 °C. Transitional seasons are projected to increase by 1.3–5.5 °C (Figure 9).

By the end of the century (2080–2099), SSP5–8.5 indicates extreme annual warming, with July and August reaching 23.4–23.8 °C (≈8 °C above historical averages) and January rising from −7.1 °C to −2.3 °C. SSP3–7.0 also shows strong warming, with summer

https://doi.org/10.3390/w18030352

temperatures exceeding 21 °C and spring/autumn increases of 1.0–7.0 °C, highlighting significant implications for Lake Sevan's climate, water resources, ecosystems, and agriculture (Figure 9).

Projected temporal changes in mean annual air temperature in the Lake Sevan basin throughout the 21st century, based on CMIP6 climate model outputs under four SSPs, indicate a significant and progressive warming trend relative to the 1991–2020 baseline average of 4.7 °C (Figure 10).

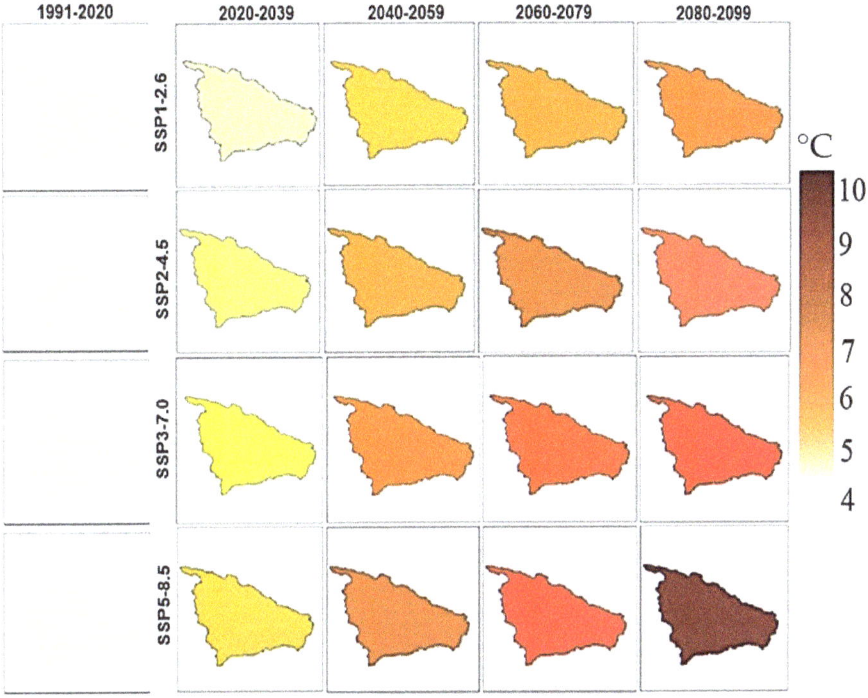

Figure 10. Temporal changes in mean annual air temperature according to scenarios.

Under the low-emission SSP1–2.6 scenario, mean annual temperatures in the Lake Sevan basin are projected to increase from 5.7 °C (2020–2039) to 6.7 °C (2080–2099), representing a total rise of ~2.0 °C above the historical baseline (Figure 10).

Under SSP2–4.5, mean annual temperatures are projected to rise from 5.8 °C (2020–2039) to 7.7 °C (2080–2099), an increase of ~3.0 °C above the baseline (Figure 10). For SSP3–7.0, temperatures are projected to increase from 5.9 °C to 8.8 °C over the same period, a total rise of ~4.1 °C (Figure 10).

Under SSP5–8.5, mean annual temperatures are projected to reach 10.4 °C by 2080–2099, an increase of 5.7 °C above the 1991–2020 baseline, representing the greatest warming among the scenarios and highlighting potential impacts on Lake Sevan's hydrology, ecosystems, and water resources (Figure 10). Overall, higher emissions scenarios lead to greater warming and more pronounced ecological and hydrological impacts on the Lake Sevan basin.

4. Discussion

The analysis of long-term temperature data in the Lake Sevan basin highlights clear trends and spatial patterns in both air and water temperatures. Observed warming occurred

https://doi.org/10.3390/w18030352

across all meteorological stations during 1991–2020, with cumulative increases ranging from 0.4 °C (Gavar) to 1.9 °C (Shorzha). The results confirm that proximity to the lake moderates seasonal temperature extremes, as stations near Lake Sevan exhibit lower variability in heat inflow and outflow than more distant, higher-elevation stations. The methodological approach, including calculation of decadal averages, seasonal means, and heat circulation (deviation of monthly temperatures from annual means), allowed us to quantify these patterns rigorously.

The data show that air temperatures fluctuate more strongly than lake water temperatures, especially in winter and transitional seasons, illustrating the buffering capacity of Lake Sevan due to its thermal inertia. Water temperatures increased more steadily, with maximum rises observed in spring (April +2.20 °C) and autumn (September +2.31 °C), while winter increases remained moderate (January +0.82 °C). These patterns have important implications for limnological processes, aquatic ecosystems, and regional climate adaptation strategies. The observed seasonal warming signal, particularly in late winter and early spring, aligns with global trends in mountainous continental regions, where minimum temperatures have increased more rapidly.

Analysis of historical data reveals several key aspects of Lake Sevan's role as a primary climatic regulator in the region. Notably, the rate of air temperature increase accelerated from the first to the second decade, followed by a more moderate but sustained rise in the third decade. Lake Sevan functions as a climatic buffer, moderating temperature extremes by keeping adjacent areas warmer during winter and hotter during late summer. This highlights the significant influence that large water bodies exert on local temperature regimes. Meteorological stations near Lake Sevan, such as Sevan and Shorzha, experience more stable temperatures due to the lake's moderating effect. Stations farther away, like Semyonovka and Masrik, show greater seasonal temperature swings, especially in winter, reflecting typical inland and high-elevation variability. Long-term analysis of heat inflow and outflow highlights Lake Sevan's key role in regulating the region's thermal regime. Thermal moderation is strongest near the lake and gradually decreases with distance. The lake's water temperature remains relatively stable, ranging from +0.82 °C in January to +2.31 °C in September, whereas air temperatures fluctuate more, from +0.18 °C in November to +4.08 °C in March, with notable warming in late winter and early spring.

This study analyzes long-term temperature changes, which are key climate variables, in the Lake Sevan basin through the end of the 21st century. The projections indicate a significant warming trend under all SSP scenarios, underscoring the substantial impact of cumulative greenhouse gas emissions on the regional climate. The results suggest a marked increase in temperature during the 21st century, aligning with earlier studies that also forecast substantial warming across Armenia [23]. Both historical observations and future projections based on CCSM6 model data show that the most intense warming in the Lake Sevan basin is expected during the summer months. Under the SSP5–8.5 scenario, a temperature rise of 4–6 °C is projected by mid-century and by 2100, consistent with projections for Armenia as a whole. These findings are in close agreement with previous climate assessments for both the Sevan basin and the broader region, which likewise anticipate a 4–6 °C increase by the end of the century [23,56].

Rising summer temperatures are expected to intensify evaporation from Lake Sevan, further exacerbating regional aridity. This process will reduce river inflows, which are closely tied to the lake's water level, thereby threatening the overall availability of freshwater resources. According to Armenia's Fourth National Communication on Climate Change, total river inflows into the Lake Sevan basin are projected to decline by approximately 34% (equivalent to 265 million m^3) by 2100, relative to the baseline period of 1961–1990. Moreover, climate models consistently indicate that declining lake levels will intensify

https://doi.org/10.3390/w18030352

evaporation losses, creating a feedback loop that accelerates water depletion [57]. Elevated temperatures are also likely to diminish snowmelt contributions, extend dry periods, and reduce water availability for human consumption, posing significant risks to both local communities and ecosystems.

Another significant consequence of rising temperatures in the Lake Sevan region is the increased risk of harmful algal blooms, particularly those caused by cyanobacteria. Elevated temperatures create favorable conditions for the proliferation of these organisms, a process further intensified by nutrient runoff from surrounding areas. These blooms pose serious ecological threats, including oxygen depletion, toxin production, and disruption of aquatic ecosystems [58–60]. Notably, the highest recorded temperatures in the Lake Sevan basin occurred in 2010 and 2018, both years coinciding with cyanobacterial blooms [61]. Gevorgyan et al. (2020) reported that the 2018 bloom reached unprecedented levels, hypothesizing that rising air and water temperatures contributed to the expansion of cyanobacteria [61]. Given this relationship, it is likely that cyanobacterial blooms in Lake Sevan will become increasingly frequent and severe in the coming years, particularly during the summer months, as regional temperatures continue to rise.

Despite advances in climate modeling, considerable uncertainty remains regarding the precise patterns of temperature change projected for the Lake Sevan basin throughout the 21st century. Key uncertainties include the representation of fine-scale local climate dynamics in complex terrain and the response of extreme temperature and precipitation events to climate change. To address these uncertainties, future studies could employ high-resolution dynamical downscaling, coupled lake–atmosphere modeling, and detailed analysis of CMIP6 precipitation and extreme indices. While much attention has been given to annual and seasonal temperature trends, it is equally important to examine changes in daily precipitation variability and the frequency and intensity of extreme events. Investigating these aspects will require substantial computational resources and methodological sophistication and will form a critical focus of future climate research in the region.

The observed winter variability aligns with broader continental climate patterns, particularly those characteristic of the Eurasian region. Lee et al. (2025) noted that winter months generally exhibit higher temperature fluctuations due to the influence of unstable synoptic weather conditions, varying air masses, snow cover, and fluctuations in atmospheric pressure systems [62]. Our findings are also consistent with those of Gevorgyan et al. (2015), who reported that temperature variability in Armenia is most pronounced during winter [63]. These results reinforce the broader understanding that winter months are typically characterized by greater temperature deviations than other seasons, especially in continental and high-altitude regions such as the Lake Sevan basin. These spatial and seasonal patterns emphasize Lake Sevan's importance in regulating local climate and impacting ecosystems, agriculture, hydrology, and sustainable water management.

5. Conclusions

Analysis of long-term temperature data in the Lake Sevan basin shows a clear warming trend at all meteorological stations between 1991 and 2020, with cumulative increases ranging from 0.4 °C to 1.9 °C. Proximity to Lake Sevan moderates seasonal temperature extremes, as stations near the lake show lower air temperature variability than those farther away or at higher elevations. Air temperatures fluctuate more strongly than water temperatures, particularly in winter, while water temperatures rise more steadily. All scenarios for the Lake Sevan basin project show significant warming over the 21st century, with summer temperatures increasing by 4–6 °C in the mid- to late-century, potentially increasing evaporation, reducing river inflows, and increasing the frequency of cyanobacterial blooms, thereby impacting water resources, aquatic ecosystems, and regional climate adaptation

https://doi.org/10.3390/w18030352

strategies. Despite these clear trends, there is still considerable uncertainty about local climate dynamics and responses to extreme events, highlighting the need for high-resolution modeling and detailed future analyses. Overall, Lake Sevan plays an important role in regulating local climate, mitigating temperature extremes, and supporting sustainable management of regional water and ecological resources.

Author Contributions: G.K., A.G. (Artur Gevorgyan), A.V., A.M., K.R., A.G. (Artak Gevorgyan) and G.G. Conceptualization, G.G., G.K. and A.V.; methodology, G.K., G.G., A.G. (Artur Gevorgyan) and A.G. (Artak Gevorgyan); software, G.K.; validation, G.G. and G.K.; formal analysis, G.K., A.G. (Artak Gevorgyan) and L.G.; investigation, G.K., G.G., A.G. (Artur Gevorgyan), A.M. and K.R.; resources, G.G., data curation, G.K. and A.M.; writing—original draft preparation, G.K., G.G. and A.V.; writing—review and editing, G.G., A.G. (Artur Gevorgyan), A.V., K.R. and L.G.; visualization, G.K.; supervision, G.G.; project administration, G.G. All authors have read and agreed to the published version of the manuscript.

Funding: The study was partly supported by the Higher Education and Science Committee of the Ministry of Education, Science, Culture, and Sports (MESCS) of the Republic of Armenia under Research Project No. 23LCG-1F005.

Data Availability Statement: The original contributions presented in this study are included in the article. Further inquiries can be directed to the corresponding authors.

Conflicts of Interest: The authors declare no conflicts of interest–financial or otherwise.

Abbreviations

The following abbreviations are used in this manuscript:

ARCCS	Australian Research Council Centre of Excellence for Climate System Science
BCSD	Bias-correction-spatial-disaggregation
CCKP	Climate Change Knowledge Portal
CMIP	Coupled Model Intercomparison Project
CSIRO	Commonwealth Scientific and Industrial Research Organization
IAM	Integrated Assessment Models
IPCC	Intergovernmental Panel on Climate Change
MESCS	Ministry of Education, Science, Culture, and Sports
NERC	Natural Environment Research Council
NOAA	National Oceanic and Atmospheric Administration
RCP	Representative Concentration Pathway
SSP	Shared Socioeconomic Pathways
WMO	World Meteorological Organization

References

1. Shivanna, K.R. Climate change and its impact on biodiversity and human welfare. *Proc. Indian Natl. Sci. Acad.* **2022**, *88*, 160–171. [CrossRef]
2. Muluneh, M.G. Impact of climate change on biodiversity and food security: A global perspective—A review article. *Agric. Food Secur.* **2021**, *10*, 36. [CrossRef]
3. Feulner, G. Global challenges: Climate change. *Glob. Chall.* **2015**, *1*, 5–6. [CrossRef] [PubMed]
4. Pörtner, H.-O.; Roberts, D.C.; Adams, H.; Adelekan, I.; Adler, C.; Adrian, R.; Aldunce, P.; Ali, E.; Ara Begum, R.; Bednar-Friedl, B.; et al. *Climate Change 2022: Impacts, Adaptation and Vulnerability*; Working Group II Contribution to the Sixth Assessment Report of the Intergovernmental Panel on Climate Change; Cambridge University Press: Cambridge, UK; New York, NY, USA, 2022. [CrossRef]
5. Méndez, C.; Simpson, N.; Johnson, F.; Birt, A. *Climate Change 2023: Synthesis Report (Full Volume)*; Contribution of Working Groups I, II and III to the Sixth Assessment Report of the Intergovernmental Panel on Climate Change; Intergovernmental Panel on Climate Change: Geneva, Switzerland, 2023. [CrossRef]
6. Gevorgyan, G.; Tümpling, W.; Shahnazaryan, G.; Friese, K.; Schultze, M. Lake-wide assessment of trace elements in surface sediments and water of Lake Sevan. *J. Limnol.* **2023**, *81*, 2096. [CrossRef]

https://doi.org/10.3390/w18030352

7. Hayrapetyan, A.; Gevorgyan, G.; Schultze, M.; Shikhani, M.; Khachikyan, T.; Krylov, A.; Rinke, K. Contemporary community composition, spatial distribution patterns, and biodiversity characteristics of zooplankton in large alpine Lake Sevan, Armenia. *J. Limnol.* **2023**, *81*, 2150. [CrossRef]
8. Mkrtchyan, S.S.; Gabrielyan, A.H.; Maghakyan, H.G.; Pafengolts, K.N.; Vardanyants, L.A. (Eds.) *Geology of the Armenian SSR. Volume I: Geomorphology*; Academy of Sciences Press: Yerevan, Armenia, 1962. (In Russian)
9. Climate Risk Country Profile: Armenia. The World Bank Group and the Asian Development Bank: Yerevan, Armenia. 2021. [Cited 10 December 2025]. Available online: https://www.adb.org/publications/climate-risk-country-profile-armenia (accessed on 1 December 2025).
10. Sudeep, T.; Dahal, S.; Shrestha, D.; Guyennon, N.; Romano, E.; Colombo, N.; Salerno, F. Elevation-dependent warming of maximum air temperature in Nepal during 1976–2015. *Atmos. Res.* **2019**, *228*, 261–269. [CrossRef]
11. Wang, Y.; Ding, Z.; Ma, Y. Spatial and temporal analysis of changes in temperature extremes in the non-monsoon region of China from 1961 to 2016. *Theor. Appl. Climatol.* **2019**, *137*, 2697–2713. [CrossRef]
12. Gao, L.; Deng, H.; Lei, X.; Wei, J.; Chen, Y.; Li, Z.; Ma, M.; Chen, X.; Chen, Y.; Liu, M.; et al. Evidence of elevation-dependent warming from the Chinese Tian Shan. *Cryosphere* **2021**, *15*, 5765–5783. [CrossRef]
13. Sigro, J.; Cisneros, M.; Perez-Luque, A.J.; Perez-Martinez, C.; Vegas-Vilarrubia, T. Trends in temperature and precipitation at high and low elevations in the main mountain ranges of the Iberian Peninsula (1894–2020): The Sierra Nevada and the Pyrenees. *Int. J. Climatol.* **2024**, *44*, 2897–2920. [CrossRef]
14. Adrian, R.; O'Reilly, C.M.; Zagarese, H.; Baines, S.B.; Hessen, D.O.; Keller, W.; Livingstone, D.M.; Sommaruga, R.; Straile, D.; Van Donk, E.; et al. Lakes as sentinels of climate change. *Limnol. Oceanogr.* **2009**, *54*, 2283–2297. [CrossRef]
15. Armenian Legal Information System (ARLIS). RA Government Resolution N. 1912-N Dated 08.12.2022, Sevan Water Basin Management Area Management Plan for 2022–2027. 2022; [Cited 10 December 2025]. Available online: http://arlis.am/DocumentView.aspx?DocID=171479 (accessed on 1 December 2025). (In Armenian)
16. Muller, R.; Curry, J.; Groom, D.; Jacobsen, R.; Perlmutter, S.; Rohde, R.; Rosenfeld, A.; Wickham, C.; Wurtele, J. Decadal variations in the global atmospheric land temperatures. *J. Geophys. Res. Atmos.* **2013**, *118*, e50458. [CrossRef]
17. Wang, T.; Yin, S.; Wei, H.; Wang, H.; Luo, F.; Miao, J.; Fu, Y. Decadal variability of extreme high temperature in mid- and high-latitude Asia and its associated North Atlantic air–sea interaction. *Clim. Dyn.* **2023**, *61*, 4587–4601. [CrossRef]
18. Neukom, R.; Barboza, L.; Erb, M.; Shi, F.; Emile-Geay, J.; Evans, M.; Franke, J.; Kaufman, D.; Luecke, L.; Rehfeld, K.; et al. Consistent multi-decadal variability in global temperature reconstructions and simulations over the Common Era. *Nat. Geosci.* **2019**, *12*, 643–649. [CrossRef] [PubMed]
19. González Hidalgo, J.C.; Beguería, S.; Peña Ángulo, D.; Sandonis, L. Variability of maximum and minimum monthly mean air temperatures over mainland Spain and their relationship with low variability atmospheric patterns for the period 1916–2015. *Int. J. Climatol.* **2022**, *42*, 1723–1741. [CrossRef]
20. Duhan, D.; Pandey, A.; Gahalaut, K.P.S.; Pandey, R.P. Spatial and temporal variability in maximum, minimum and mean air temperatures at Madhya Pradesh in central India. *Comptes Rendus Geosci.* **2013**, *345*, 3–21. [CrossRef]
21. Buytaert, W. Assessment of the impacts of climate change on mountain hydrology: Development of a methodology through a case study in the Andes of Peru. *Mt. Res. Dev.* **2012**, *32*, 385–386. [CrossRef]
22. Vergara, W.; Deeb, A.; Leino, I.; Kitoh, A.; Escobar, M. *Assessment of the Impacts of Climate Change on Mountain Hydrology: Development of a Methodology Through a Case Study in the Andes of Peru*; World Bank Publications, No. 2278; The World Bank Group: Washington, DC, USA, 2011.
23. Gevorgyan, A.; Melkonyan, H.; Aleksanyan, T.; Iritsyan, A.; Khalatyan, Y. An assessment of observed and projected temperature changes in Armenia. *Arab. J. Geosci.* **2016**, *9*, 27. [CrossRef]
24. Alexander, L.; Zhang, X.; Peterson, T.C.; Caesar, J.; BA, G.; Tank, A.; Haylock, M.; Collins, D.; Trewin, B.; Rahimzadeh, F.; et al. Global observed changes in daily climate extremes of temperature and precipitation. *J. Geophys. Res.* **2006**, *111*, D05109. [CrossRef]
25. Melkonyan, A. Climate change impact on water resources and crop production in Armenia. *Agric. Water Manag.* **2015**, *161*, 86–101. [CrossRef]
26. Robles, M.; Peyron, O.; Brugiapaglia, E.; Ménot, G.; Dugerdil, L.; Ollivier, V.; Ansanay-Alex, S.; Develle, A.-L.; Tozalakyan, P.; Meliksetian, K.; et al. Impact of climate changes on vegetation and human societies during the Holocene in the South Caucasus (Vanevan, Armenia): A multiproxy approach including pollen, NPPs and brGDGTs. *Quat. Sci. Rev.* **2022**, *277*, 107297. [CrossRef]
27. Shikhani, M.; Feldbauer, J.; Ladwig, R.; Mercado-Bettín, D.; Moore, T.N.; Gevorgyan, A.; Misakyan, A.; Mi, C.; Schultze, M.; Boehrer, B.; et al. Combining a multi-lake model ensemble and a multi-domain CORDEX climate data ensemble for assessing climate change impacts on Lake Sevan. *Water Resour. Res.* **2024**, *60*, e2023WR036511. [CrossRef] [PubMed]
28. Shikhani, M.; Mi, C.; Gevorgyan, A.; Gevorgyan, G.; Misakyan, A.; Azizyan, L.; Barfus, K.; Schulze, M.; Shatwell, T.; Rinke, K. Simulating thermal dynamics of the largest lake in the Caucasus region: The mountain Lake Sevan. *J. Limnol.* **2021**, *81*, 2024. [CrossRef]

https://doi.org/10.3390/w18030352

29. Siabi, E.K.; Awafo, E.A.; Kabo-bah, A.T.; Derkyi, N.S.A.; Akpoti, K.; Mortey, E.M.; Yazdanie, M. Assessment of Shared Socioeconomic Pathway (SSP) climate scenarios and its impacts on the Greater Accra region. *Urban Clim.* **2023**, *49*, 101432. [CrossRef]
30. Nazarenko, L.; Tausnev, N.; Russell, G.; Rind, D.; Miller, R.; Schmidt, G.; Bauer, S.; Kelley, M.; Ruedy, R.; Ackerman, A.; et al. Future climate change under SSP emission scenarios with GISS-E2.1. *J. Adv. Model. Earth Syst.* **2022**, *14*, e2021MS002871. [CrossRef]
31. Ciampittiello, M.; Marchetto, A.; Boggero, A. Water resources management under climate change: A review. *Sustainability* **2024**, *16*, 3590. [CrossRef]
32. Yuan, X.; Li, S.; Chen, J.; Yu, H.; Yang, T.; Wang, C.; Huang, S.; Chen, H.; Ao, X. Impacts of global climate change on agricultural production: A comprehensive review. *Agronomy* **2024**, *14*, 1360. [CrossRef]
33. Zhao, M.; Boll, J. Adaptation of water resources management under climate change. *Front. Water* **2022**, *4*, 983228. [CrossRef]
34. Earth System Model Evaluation Project. CMIP6—Coupled Model Intercomparison Project Phase 6. In *Program for Climate Model Diagnosis and Intercomparison (PCMDI) [Internet]*; Cited 10 December 2025; LLNL: Livermore, CA, USA, 2014. Available online: https://pcmdi.llnl.gov/CMIP6/ (accessed on 1 December 2025).
35. Eyring, V.; Bony, S.; Meehl, G.; Senior, C.; Stevens, B.; Stouffer, R.; Taylor, K. Overview of the Coupled Model Intercomparison Project Phase 6 (CMIP6) experimental design and organization. *Geosci. Model. Dev.* **2016**, *9*, 1937–1958. [CrossRef]
36. Riahi, K.; van Vuuren, D.P.; Kriegler, E.; Edmonds, J.; O'Neill, B.; Fujimori, S.; Bauer, N.; Calvin, K.; Dellink, R.; Fricko, O.; et al. The Shared Socioeconomic Pathways and their energy, land use, and greenhouse gas emissions implications: An overview. *Glob. Environ. Change* **2017**, *42*, 153–168. [CrossRef]
37. Bock, L.; Lauer, A. Cloud properties and their projected changes in CMIP models with low to high climate sensitivity. *Atmos. Chem. Phys.* **2024**, *24*, 1587–1605. [CrossRef]
38. O'Neill, B.; Kriegler, E.; Riahi, K.; Ebi, K.; Hallegatte, S.; Carter, T.; Mathur, R.; Vuuren, D. A new scenario framework for climate change research: The concept of Shared Socioeconomic Pathways. *Climatic Change* **2014**, *122*, 387–400. [CrossRef]
39. Vuuren, D.; Edmonds, J.; Kainuma, M.; Riahi, K.; Thomson, A.; Hibbard, K.; Hurtt, G.; Kram, T.; Krey, V.; Lamarque, J.-F.; et al. The representative concentration pathways: An overview. *Clim. Change* **2011**, *109*, 5–31. [CrossRef]
40. Soares, P.M.M.; Johannsen, F.; Lima, D.; Lemos, G.; Bento, V.; Bushenkova, A. High-resolution downscaling of CMIP6 Earth system and global climate models using deep learning for Iberia. *Geosci. Model. Dev.* **2024**, *17*, 229–259. [CrossRef]
41. Grose, M.; Narsey, S.; Trancoso, R.; Mackallah, C.; Delage, F.; Dowdy, A.; Virgilio, G.; Watterson, I.; Dobrohotoff, P.; Rashid, H.; et al. A CMIP6-based multi-model downscaling ensemble to underpin climate change services in Australia. *SSRN Prepr.* **2022**. [CrossRef]
42. Gebrechorkos, S.; Leyland, J.; Slater, L.; Wortmann, M.; Ashworth, P.J.; Bennett, G.L.; Boothroyd, R.; Cloke, H.; Delorme, P.; Griffith, H.; et al. A high-resolution daily global dataset of statistically downscaled CMIP6 models for climate impact analyses. *Sci. Data* **2023**, *10*, 611. [CrossRef]
43. World Bank. Climate Metadata. [Internet]. World Bank Climate Knowledge Portal. 2025. [Cited 10 December 2025]. Available online: https://climateknowledgeportal.worldbank.org/metadata (accessed on 3 December 2025).
44. Yang, Y.Q.; Hou, Q.; Zhou, C.H.; Liu, H.L.; Wang, Y.Q.; Niu, T. Sand/dust storm processes in Northeast Asia and associated large-scale circulations. *Atmos. Chem. Phys.* **2008**, *8*, 25–33. [CrossRef]
45. Data Buoy Cooperation Panel (DBCP). *Reference Guide to the GTS Sub-System of the Argos Processing System. Rev. 1.6*; DBCP Technical Document No. 2; WMO & IOC Data Buoy Cooperation Panel: Geneva, Switzerland, 2005; 102p. [CrossRef]
46. Hegerl, G.C.; Zwiers, F.W.; Braconnot, P.; Gillet, N.P.; Luo, Y.; Marengo, J.; Nicholls, N.; Penner, J.E.; Stott, P.A. Understanding and attributing climate change. In *Climate Change 2007: The Physical Science Basis*; Solomon, S., Qin, D., Manning, M., Chen, Z., Marquis, M., Averyt, K.B., Tignor, M., Mille, H.L., Eds.; Contribution of Working Group I to the Fourth Assessment Report of the Intergovernmental Panel on Climate Change; Cambridge University Press: Cambridge, UK, 2007.
47. Pepin, N.C.; Arnone, E.; Gobiet, A.; Haslinger, K.; Kotlarski, S.; Notarnicola, C.; Palazzi, E.; Seibert, P.; Serafin, S.; Schöner, W.; et al. Climate changes and their elevational patterns in the mountains of the world. *Rev. Geophys.* **2022**, *60*, e2020RG000730. [CrossRef]
48. World Meteorological Organization (WMO). *The Global Climate 2011–2020: A Decade of Accelerating Climate Change*; WMO-No. 1338; World Meteorological Organization: Geneva, Switzerland, 2023.
49. Subin, Z.M.; Murphy, L.N.; Li, F.; Bonfils, C.; Riley, W.J. Boreal lakes moderate seasonal and diurnal temperature variation and perturb atmospheric circulation: Analyses in the Community Earth System Model 1 (CESM1). *Tellus A* **2012**, *64*, 15639. [CrossRef]
50. Jiang, Y.; Chi, Y.; Wang, W.; Li, W.; Wang, H.; Sun, J. Responses of the East Asian winter climate to global warming in CMIP6 models. *Atmosphere* **2025**, *16*, 1143. [CrossRef]
51. Hu, S.; Hsu, P.-C. Drivers of elevation-dependent warming over the Tibetan Plateau. *Atmos. Ocean. Sci. Lett.* **2023**, *16*, 100289. [CrossRef]
52. Wang, H.; Wang, B.-B.; Cui, P.; Ma, Y.-M.; Wang, Y.; Hao, J.-S.; Wang, Y.; Li, Y.-M.; Sun, L.-J.; Wang, J.; et al. Disaster effects of climate change in High Mountain Asia: State of art and scientific challenges. *Adv. Clim. Change Res.* **2024**, *15*, 367–389. [CrossRef]

https://doi.org/10.3390/w18030352

53. Lenters, J.; Kratz, T.; Bowser, C. Effects of climate variability on lake evaporation: Results from a long-term energy budget study of Sparkling Lake, northern Wisconsin (USA). *J. Hydrol.* **2005**, *308*, 168–195. [CrossRef]
54. Kundzewicz, Z.; Mata, L.; Arnell, N.; Doll, P.; Jiménez, B.; Miller, K.; Oki, T.; Şen, Z.; Shiklomanov, I. The implications of projected climate change for freshwater resources and their management. *Hydrol. Sci. J.* **2008**, *53*, 3–10. [CrossRef]
55. Verma, A.; Agrawal, S. Evaluating the natural cooling potential of waterbodies in dense urban landscape: A case study of Bengaluru, India. *Urban Clim.* **2024**, *58*, 102200. [CrossRef]
56. Galstyan, H.; Sfîcă, L.; Ichim, P. Long-term variability of temperature in Armenia in the context of climate change. *Int. J. Environ. Chem. Ecol. Geol. Geophys. Eng.* **2016**, *10*, 19–25.
57. UNDP Armenia. *Fourth National Communication on Climate Change*; UNDP Armenia: Yerevan, Armenia, 2020; 213p, Available online: http://env.am/storage/files/fnc-eng.pdf (accessed on 4 December 2025).
58. Paerl, H.W.; Huisman, J. Climate change: A catalyst for global expansion of harmful cyanobacterial blooms. *Environ. Microbiol. Rep.* **2009**, *1*, 27–37. [CrossRef]
59. Li, X.; Hang, X.; Zhu, S.; Sun, L.; Li, Y. Response of cyanobacterial blooms to climate warming: Evidence from satellite observations and long-term trends in Lake Taihu in China. *Sci. Rep.* **2025**, *15*, 38820. [CrossRef]
60. Erratt, K.J.; Creed, I.F.; Lobb, D.A.; Smol, J.P.; Trick, C.G. Climate change amplifies the risk of potentially toxigenic cyanobacteria. *Glob. Change Biol.* **2023**, *29*, 5240–5249. [CrossRef]
61. Gevorgyan, G.; Rinke, K.; Schultze, M.; Mamyan, A.; Kuzmin, A.; Belykh, O.; Sorokovikova, E.; Hayrapetyan, A.; Hovsepyan, A.; Khachikyan, T.; et al. First report about toxic cyanobacterial bloom occurrence in Lake Sevan, Armenia. *Int. Rev. Hydrobiol.* **2020**, *105*, 131–142. [CrossRef]
62. Lee, S.; Park, H.-S.; Kim, M.-K.; Min, S.-K.; Hwang, H. Future increases in Eurasian mid-latitude winter temperature variability shaped by a weakened Atlantic Meridional Overturning Circulation. *Commun. Earth Environ.* **2025**, *6*, 22. [CrossRef]
63. Gevorgyan, A.; Melkonyan, H.; Abrahamyan, R.; Petrosyan, Z.; Shahnazaryan, A.; Astsatryan, H.; Sahakyan, V.; Shoukouryan, Y. A persistent surface inversion event in Armenia as simulated by WRF model. In *2015 Computer Science and Information Technology (CSIT)*; IEEE: New York, NY, USA, 2015. [CrossRef]

https://doi.org/10.3390/w18030352